ANCIENT METALS: MICROSTRUCTURE AND METALLURGY: VOLUME I. COPPER AND COPPER ALLOYS

ANCIENT METALS: MICROSTRUCTURE AND METALLURGY: VOLUME 1. COPPER AND COPPER ALLOYS.

David A. Scott

CSP
conservation science press

Printed by Createspace.Com and also by Lulu.com in different formats.
Set in 10pt Palatino Linotype.

10 9 8 7 6 5 4 3 2

Cover: Selected Microstructures taken from this Volume

Library of Congress Cataloging-in-Publication Data

Ancient Metals: Microstructure and Metallurgy
Vol 1. Copper and Copper Alloys.
Second Edition: 2012.

David A. Scott
p. cm.
Includes bibliographical references and index.
ISBN 978-0-9829338-0-0

This volume is dedicated to Lesley A. Moorcroft, Daniel S. Scott, Ellie R. Scott Kieran D. Scott, George A. Scott, Frances R. Scott, Valerie A. Scott and Diana Vera Scott.

.

CONTENTS

ANCIENT METALS: MICROSTRUCTURE AND METALLURGY: VOLUME 1, COPPER AND COPPER ALLOYS.

Author's Preface

This book is a completely updated and entirely rewritten version of the volume previously published by the J. Paul Getty Trust (Scott 1991) which is now out of print as a hard-text copy and available on-line from the Getty website as a *pdf*.

The present text takes a more coherent view of the different types of alloying systems important for ancient and historical metalwork and deals with many issues not addressed in the earlier volume, such as the development of metallurgy and how this relates to the use of metals, the ways in which they could be combined or alloyed, and something of their historical and societal context. However, this book is principally concerned with technical aspects of the alloys, and the application of metallography to their examination, it is not designed as a text whose aim is to account for the relationship between how societies developed in different parts of the world, and how that development influenced the use of metals and alloys.

This volume deals with the alloys of copper: the next volume in this series will be devoted to those of iron and steel. If time permits, a third volume will deal with the alloys of copper and gold in the South American context. One of the aims of these volumes will be to add significantly to the number of published microstructures of ancient metals, their corrosion and the morphology of their structures and patinations.

The world of metallography is the world of the aggregation of structure, a view frozen in time of the metallic components as they were cast, worked, heat-treated, patinated, quenched or coated from their original fabrication. Although prone to internal dissolution through general corrosion, galvanic attack, grain boundary precipitation and embrittlement, they preserve for us the record of fabrication and alloy manipulation which different cultures have employed at different periods of their existence. Corrosion and the ingress of corrosion products may be important factors which aid us in understanding the patination, corrosion resistance, or authenticity of corrosion processes, especially in ancient bronze and silver.

The science of metallography and ancient metals in particular is less of a general interest in the engineering world than it once was. This is because of the gradual decline in the importance of metals in the West in terms of manufacturing or research. The last metallurgist at UCLA, for example, retired in 2006 and currently there are no metallurgists at UCLA. Metallurgy Departments have increasingly been renamed as Materials Science Departments, as the importance of metals becomes less dominant and rather old-fashioned.

The history of the use of metallic materials is intimately connected with the properties of the materials available, and these properties are dependent on the microstructure and composition of the materials themselves. There are a myriad number of textbooks on metallurgy which can be consulted by the interested reader to augment the information given in the text here. The only drawback is that most modern textbooks have no interest in the kinds of alloy systems which are relevant to our discussion: copper-tin; copper-zinc; copper-arsenic and copper-silver alloys are all very important in the development of metallurgy, but barely merit a mention in 21st century engineering textbooks.

The order in which the alloy systems is dealt with has been deliberately chosen to begin discussion of phase diagrams with a simple binary eutectic system, so that readers are not immediately faced with the complex copper-tin phase diagram, which is more difficult to understand.

The system chosen is silver-copper, which is one of the most common and interesting alloy types. This volume then progresses through a discussion of copper, copper-arsenic alloys, tin bronzes, and copper-zinc alloys. Leaded alloys of the different copper systems are also examined. Case studies relating to the examination, in depth, of some artefacts are called out in the text and often include additional analytical data which formed part of the original investigation which is reproduced here from the original source or has been added to by the author for this publication.

Each volume in this series is printed on demand. After peer-review, it is very straightforward to simply amend the text with improvements, so that the subsequent emendations can be incorporated. Consequently, this text will be periodically updated. Suggestions for improvement to the text are therefore most welcome.

The important aspects of ancient metals dealt with here do not demand a rigorous mathematical treatment. The text can therefore be enjoyed by a wide variety of readers, from archaeometallurgists, interested layman, conservators, conservation scientists and archaeologists. The author would like to thank Jackie Burns,

Registrar's Department, The Getty Center, Brentwood for processing copyright requests for image reproduction from the collections of the J. Paul Getty Museum; Joseph Lehner, Graduate Student, UCLA Cotsen Institute of Archaeology for the translation of an article from German; Dr. Marion True, formerly Curator of Antiquities, Department of Antiquities Conservation, J. Paul Getty Museum, Malibu; Dr. Frank Preusser, formerly Head, Scientific Department, The Getty Conservation Institute, Marina del Rey; Dr Pieter Meyers, formerly Head, Conservation Department, Los Angeles County Museum of Art, Professor Gerhard Eggert, Academy of Fine Arts, Stuttgart, and the late Dr. Nigel J. Seeley, formerly Senior Lecturer, Department of Conservation and Materials Science, Institute of Archaeology, University College London.

Professor David A. Scott, Department of Art History, UCLA, Founding Director (2003-2011) IDP in Archaeological and Ethnographic Conservation, UCLA.

Acknowledgements and Technical Information.

Thanks are due to the following publishers or organizations for permission to reproduce a number of images in the text. To the J. Paul Getty Trust for permission to reproduce several images mentioned in the text. Especial thanks are due to MacMillan Company Inc., 175 5th Avenue, New York, 10010, for permission to reproduce several images from: Metallic Materials in Engineering, by Carl H. Samans (1963). To Prentice-Hall, Inc, Upper Saddle River, New Jersey, 07458, for permission to reproduce some images from: The Science of Engineering Materials (1969) by Charles O. Smith. To D. Van Nostrand & Co and successors for reproduction of Figure 18-13, Figure 16-12 and 14-12, from Butts, A., and Coxe, C. D. Silver: Economics, Metallurgy and Use (1967). To the Trustees of the British Museum for reproduction of Figures 2 and 4 from 2000 Years of Zinc and Brass (Ed. Paul P. Craddock) from the article Zinc in India by P. T. Craddock, I. C, Freestone, L. K. Gurjar, A. P. Middleton, and L. Willies 29-59 (1990). To the Getty Conservation Institute for permission to reproduce several images from the book Metallography and Microstructure of Ancient Metals, by David A. Scott, Published by the GCI in 1991 and which is currently out-of-print but available on-line. To the Journal of the Institute of Metals, for reproduction of nine plates from the Journal for 1904, which are out of copyright. All drawings are by the author unless otherwise captioned. All photomicrographs are by the author unless otherwise noted in the text. The author apologizes for the absence of a scale on the photomicrographs, as the software and microscope used at home did not allow for this to be placed automatically on each photomicrograph, as is standard for scanning electron photomicrography.

All percentages are weight percent unless otherwise specified as atomic percent. All temperatures are in degrees centigrade unless degrees absolute, or degrees Fahrenheit, are specified in the chart concerned. Etchants used for the samples are derived from the Appendix given at the end of the text. X-ray fluorescence analyses were carried out with a Kevex 0750A X-ray fluorescence spectrometer, at 50Kv 50mA for 240 seconds acquisition time employing a Ba/Sr secondary target. Electron microprobe analyses were carried out at the Getty Conservation Institute using a Joel Superprobe and a Cameca Electron Microprobe employing pure element standards with varying operational parameters, operated by Dr. Eric Doehne, formerly of the Getty Conservation Institute, Science Department. A Joel Superprobe was also used at the Department of Geology, UCLA, operated by Dr. Frank Kyte. Hand-held X-ray fluorescence data was acquired by the author using a Bruker Instrument operated at 40KV, 1.5 microAmps using a Ti/Al filter for 120 seconds acquisition time. ICP-MS analyses were carried out by West Coast Analytical, Los Angeles, between 1991-2003. Later analyses were carried out at UCLA at the Department of Earth Sciences. All photomicrographs were taken by the author with a Nikon D300 camera on a Nikon Epiphot inverted metallurgical microscope with apochromatic objectives.

Thanks are due to Molly O'Guiness Carlson, Private Conservator, Wiscasset, Maine, for reading the text and to making suggestions for improvement and to: Dr. Pieter Meyers, Senior Research Scientist, of the Conservation Department, Los Angeles County Museum of Art, Dr. rer. nat. Roland Schwab, Curt-Engelhorn-Zentrum Archäometrie, An-Institut der Universität Tübingen, C5, Zeughaus, D-68159 Mannheim, for peer-reviewing the text. Dr. Roland Schwab is particularly to be thanked for his review of the text and for suggestions for improvement. As the text is concerned principally with metallography, it is not meant to be a comprehensive account of the ancient metallurgy of the world, as this task is obviously beyond the author's capacity, but it is supposed to provide a broad sweep across this issue, without painting in too many of the details. The text therefore has weaknesses and omissions, but in concentrating on the properties of the alloys discussed it also fulfills part of an engineering or materials science approach to the topics at hand and brings together in one place several useful charts and diagrams which detail some of the mechanical properties of the alloys discussed.

Chapter One: Metallography and Metallurgy

Metallography and the morphology of metals have been an important aspect of our understanding of metallic phases, bonding and the techniques that are used to manipulate properties and attributes, hardness and colour. Many of these choices are culturally determined: they represent decisions made about the nature of materials and how they are to be used. Two of the metallic materials of antiquity have given their names to phases of societal development which we recognize as the Bronze Age and the Iron Age, a lens through which we can view the past and discern the sometimes illusory nature of the concept of continual progress, of the aggregation of technological development as one generation builds upon the achievements of the last (Childe 1944, 1950).

In the Old World we move from a premetallic phase, a Chalcolithic copper phase, an arsenical copper phase, a Bronze age, a pre-Roman Iron age, an Iron age, almost a Cast Iron stage in 19th century Europe, to the Age of Steel in which we are still enmeshed, although steel itself is giving way to new composites, polymers, and nano-engineered materials (Wertime, 1964;Wertime et al 1980; Hauptmann et al 1999; Roberts et al 2009; Thornton 2009). The concept of continual progression and societal development is firmly entrenched in this series of transitions from the hammering of small pieces of native copper to the production of carbon steels and beyond. It is a linear view that disguises a number of truths about metallurgical development in different continents and different societies, connecting style, systems of belief and technological choices (Lechtman 1977).

In the local development of different metallurgical traditions, independent inventions, skills introduced by diffusion, and local experimentation are the reality of the complex mix of events which make up the metallurgical past. There is not just one story, but many intertwined stories with different origins at different times. This, surely, should be an obvious facet of human history, but archaeologists in applying broad concepts to the development of metallurgy, have tried to simplify this complexity, arguing against the 'diffusionist' hypothesis of an earlier generation of archaeologists to expand on the theme of 'independent invention', which does not explain the complex tapestry of the metallurgical past.

Childe put forward the idea that metallurgy originated in the Near East and was adopted in Eurasia by diffusion. Renfrew (1967) argued for the independent invention of metallurgy in southern Iberia in the 4th millennium BC, a view which is now thought to lack evidence (Roberts 2009). Clearly, the *ex oriente lux* theory is an oversimplification of the complex societal events involved in the inception of metallurgy. The transubstantiation of minerals to copper metal and the early use of native copper has long been a source of fascination in attempts to unravel this mystery: Childe (1944) talked of "homotaxis," or modes of development, with this primary period of copper use corresponding to "mode zero" of our development over time, while Daniel (1965) talked of an "eochalcic episode." These neologisms, which now seem rather artificial, have faded into obscurity. More colloquially, and meaningfully, a whole age of humanity's past has been designated the Bronze Age, signifying the great importance of copper and copper alloys in a crucial stage of our development from hunting-and-gathering communities to settled villages and the rise of agriculture.

The use of native copper goes back in time at least ten thousand years and must have occurred in most geographic areas at a very early period in human prehistory. When Neolithic cultures first began using it, around the eighth millennium BC, they fashioned the metal into small ornaments and simple copper blades. The first undisputed evidence for human exploitation of native copper deposits comes from the aceramic Neolithic site of Cayonü Tepesi in southeastern Turkey, where beads of malachite and native copper were found that date back 8,750 to 9,250 years before the present day. In addition, worked copper dating from the eighth to the seventh millennium BC has been found in both Anatolia and Mesopotamia. Even at this early period, some of the copper objects had been heated to anneal them so they could be hammered into shape, ready ductility being one of the chief assets of native copper (Muhly 1986).

Evidence for the smelting of copper, in the form of primitive slags, has been found at such Anatolian sites as Catal Hüyük, dating to the sixth millennium BC, a date which probably needs further research to confirm the evidence. Scott (1912) maintained that extracted copper could first have been formed in a campfire, but the present state of knowledge suggests that the first steps to copper extraction were taken in ceramic vessels, perhaps special ceramics already functioning as crucibles (Craddock 1995). The deliberate or accidental smelting of copper ores with arsenic resulted in the first primitive versions of copper-arsenic alloys with golden colour and good hardening properties; some of these early alloys were perhaps derived from the smelting of mixed copper-arsenic minerals such as olivenite, $Cu_2(AsO_4)(OH)$, or clinoclase, $Cu_3(AsO_4)(OH)$ (Rapp 1986).

True alloys of copper with tin began to be made in the Old World from the fourth to the third millennium BC and were pervasive and universal during the Bronze Age. Since tin is found normally as cassiterite, SnO_2, however, the advent of tin bronze still holds many mysteries as to its sources and the development of knowledge needed to mine and extract the tin and use it to alloy with different metals. The use of native copper was common in the New World, especially in the period known as the Old Copper Culture of North America, which lasted from about 3000 to 1000 BC and produced ornaments and axes of native copper. Examination shows that New World metalworkers had discovered how to work and anneal copper. The Hopewell Culture of Mound City, Ohio, is one of the best known examples of these traditions. Because native copper was especially plentiful in the New World, it was still being used during the Christian era by the indigenous peoples of North America and Alaska from the time of the Spanish conquest of the Americas in the fifteenth century onward.

True bronze alloys with tin were made in the New World during the early centuries AD, much later than corresponding developments in the Old World. Tin bronze in the New World is especially associated with the Inca hegemony from the thirteenth to the fourteenth century AD, but some antecedents are found in regions where tin was plentiful, as in present-day Bolivia, Argentina, and Chile. This alloy remained unknown in North America.

There are complexities regarding the early use of metals in Western Europe, problems regarding the validity of the independent invention model, the transmission of metal objects, metallurgical practices and the role of metallic artefacts (Thornton 2009). New discoveries are continually shifting the picture which we have of the earliest use of copper. The earliest evidence for native copper and copper oxide use in Europe is now thought to date from the mid 6th millennium BC, from sites such as Divostin and Rudna Glava in Serbia (Borić 2009). These reflect broad patterns of increasing metal utilization stretching as far east as Pakistan (Roberts, Thornton and Piggot 2009). By the late 6th millennium BC, there is substantial evidence for native copper or copper oxide beads, hooks and needles in Europe (Thornton 2002). The earliest copper smelting is more difficult to determine due to lack of analyses and difficulties in distinguishing smelted copper from native copper objects (Wayman and Duke 1999). Recent research at Belovode, Serbia, has demonstrated the presence of copper smelting slag from the late 6th millennium BC (Radivojević 2007), and this site looks to be increasingly important. In the areas of central Europe and the central Mediterranean, there is evidence of copper smelting during the late 5th millennium BC at Brixlegg in Austria (Hoppner et al 2005).

Copper axes and ornaments are found throughout the plains north of the Alpine region as far as Scandinavia from the 4th millennium BC, and may represent the long distance trade of copper metal (Thornton 2009). By the 2nd millennium BC, metallurgy is common across much of Europe, even if on a small-scale, and mining sites such as Copa Hill in Wales, dated to 2100 BC, show that advances in this kind of technological process had reached to quite far-flung locations (Timberlake 2003). Roberts (2009) comes to the conclusion that there is no clear evidence to imply the independent invention of metallurgy in Western Europe. The radiocarbon dates, the technological requirements and the archaeometallurgical data do not provide a convincing challenge to the idea that skilled metalsmiths from the east introduced metallurgy to the region by exploiting its ore sources (Robert 2009).

It is obvious from many lines of research that smelting of the copper ores would have been initially small-scale, operating at relatively low temperatures, carried out under poorly reducing conditions on oxide or roasted sulphide ores in small stone or clay structures or crucibles without intentional flux additions, creating relatively little slag (Craddock 1999, Bourgarit 2007). Smelting of this kind usually creates prills of copper which have to be further refined, often by remelting in a crucible. Our ability to provenance this early copper is limited. For example, compositional data in regions lacking copper ore, such as eastern Britain and continental northwest Europe, have shown very different but coherent patterning, originally termed Bell Beaker metal (Butler and Van der Waals 1964), which may have originated from several distant geological sources, though it is at present impossible to define exactly where these are (Roberts 2009).

In broader terms there is 'independent invention' of metallurgy in the New World, which shows that metallurgy can be developed without contact with cultures which already have discovered metallurgical processes. The necessary materials required: ceramics, clay, charcoal, stone, wood and fire, are universally available. These materials needed to come first, especially ceramics and the ability to shape and fire large ceramic structures or make refractory crucibles in which metals can be manipulated or smelted.

In the New World, metallurgical development had not yet reached the stage of the extraction of iron by the 15th century AD, but given the parallels between human inventiveness in the Old and the New, there is no reason to think that iron metallurgy would not have come sometime after 1500 AD, when the continent was invaded by the Europeans and the native cultures destroyed.

Extractive metallurgy in the New World began in the Southern Andes, where the mineral resources are rich in copper, tin, lead and silver. Extractive metallurgy developed strongly in this region and 'diffused' up towards the North, with the dates for metallurgical activity in Costa Rica, being later than those in Colombia, which are later than those in Ecuador and Peru. The separation between different cultures in the New World, resulted in the smelting of sulphide ores being prevalent in Peru, but absent in Colombia (Lechtman 1984, 1988). The pre-eminent metallurgy of Colombia, Panama and Costa Rica is that of gold-copper alloys often with depletion gilding and a variety of different coloured gold alloys, especially during the period from the early centuries AD onwards. Silver was not extracted in this region, and occurs in the gold alloys as a result of its inclusion in the native gold, but was extracted in Peru and Chile, for example.

Copper may have been smelted from oxide ores, but not from sulphide ores in the Colombian area, while in Peru, complex metallurgical extraction processes had been developed for the cupellation and scorification of silver, the alloying of silver-copper alloys, the fabrication of silver-gold-copper alloys with extracted silver, and the extensive use of copper-arsenic alloys. One of the remarkable alloys made for decorative purposes is an extensive range of silver-copper alloys, which extends into Ecuador, but only crosses the Colombian border as far as the extreme south, in the Nariño region. These alloys are often surface depleted in copper, resulting in heavily debased silver-copper alloys which appear to be made of silver. This decorative use of metals is an important one: metals were not primarily instruments of war, in the same way that they were the in Old World. Bronze Age peoples made extensive of tin bronze for arrowheads, swords, daggers, and knives. Copper alloy maceheads are the most well-known South American war-like metallic implement, which we associate with the Inca Empire, but in earlier cultures in South America, the extensive use of metals for this purpose was practically unknown.

Many unique metallurgical achievements took place in South America. Perhaps the most impressive are the development of electrochemical replacement plating in Peru to deposit very thin coatings of gold or silver onto copper (Lechtman, Erlij and Barry 1982), and the other is the use of platinum plating or cladding in the La Tolita-Tumaco culture of Ecuador, which before the birth of Christ, had already made use of platinum, at that time unknown in Europe (Scott and Bray 1980, Scott 2011).

Each continent has its own especial creations or metallurgical developments depending on the culture and the geography, the ores available in the region concerned and how these ores are utilised or processed. Thus, metallurgical development in Africa encompasses 7,500 kilometres of distance and nearly five millennia (Killick 2009). Some practices extend through time remarkably, for example, isolated human finger bones were buried beneath prehistoric furnace bases in the Lowveld of eastern South Africa to ensure a successful smelt, and the same practice is recorded by early 20th century ethnographers in the same region. For various reasons, not much is currently known of early Egyptian smelting compared with sub-Saharan Africa, perhaps because excavations are more focused in these areas on iron metallurgy and more readily accessed. Copper objects first appeared in the Maadi culture near the Nile delta between 4000 and 3200 BC. Copper and gold artefacts first appeared in lower Nubia and are dated from 3600 - 3300 BC; the increased use of metals was a long slow process. As far as ferrous metals are concerned, their origin in Africa has been something of a mystery. A recent study of blooms from Rwanda and Burundi (Killick 2009) show furnace temperatures of at least 1250 °C being attained, although where all of the technology to enable this to occur came from is still unclear. But it is known that the southward expansion of iron-producing folk meant that iron made its way by the 2nd century BC to the areas of Angola, Tanzania, Zambia, Malawi and parts of Mozambique (Killick 2009).

Iron and unalloyed copper were the only two metals used in southern Africa before 1100 AD. Copper was used for much decorative metalwork, while iron was used for spears, axes, hoes and knives (Killick 2009). Surprisingly, there was no tin or bronze, no silver, no gold, and no lead in southern Africa before the eleventh century AD (Killick 2009). The lack of these developments may have a bearing on the discussion concerning the independent development of metallurgy in Africa, which could well prove to be an indigenous technology as was the entire metallurgical technology of South America.

The Mythology of Copper Alloys

Much of this volume concerns the use of copper alloys and copper. In about 75 AD, Pliny the Elder wrote:

> "...in Cyprus, where copper was first discovered, it is also obtained from another stone also, called chalcitis, copper ore."

Roman mythology links copper and Cyprus through a modified Egyptian ankh, which signifies copper and is also the zodiacal sign for the planet Venus. In Roman mythology, Venus was the successor to Aphrodite, the Greek goddess who maintained Cyprus as her principal place of worship. In modern times this connection between copper and Cyprus has been retained in the chemical abbreviation for copper, Cu.

The ancient use of copper is enshrined in Mesopotamian mythology. These tales told of many gods who lived deep in the bowels of the earth and were independent of the gods in heaven. Their leader was Enmesarra, or Nindara, a warrior god whose special mission was to combat demons, monsters, and plagues. Enmesarra, Lord of the Ghost World, was also the god of hidden metallic and mineral treasures. Covered in a skin of solid copper, he came from the mountains where this metal was manifest. Enmesarra was the night sun hidden during half his course, as was the ancient Egyptian god Ra in the form of Atum. Like the sun, Enmesarra waited to come out of the earth to reveal the splendor of metallic copper to the world of humans (Partington 1935).

At this early stage in human technology, copper was indeed a utilitarian as well as a decorative metal, easily extracted and worked into small ornaments, pectorals or *tupus* (South American cloak pins), or useful halberds (medieval weapons), palstaves (ax blades), or knives. Indeed, fire-setting in the mountains could be used to bring forth metal: gold, for example, could be melted from deposits by setting fire to the vegetation on a hillside, the heat of the conflagration producing rivulets of metal.

Diffusion and Value

If independent centres of metallurgical innovation did not just arise, it is difficult to assess how metal technology travelled, or was 'diffused'. The earliest presence of metal objects in Western Europe during the 4th millennium BC, argues Roberts (2009:273), did not provoke a significant material or technological transformation. Metals were just one of a group of prestige items at that time, which is hardly surprising as the quantity being produced must have been very limited and sporadic. Changes in society would come, in time, to be bound up with bronze and later iron, and then steel, but the beginnings were perhaps not so much a Chalcolithic revolution, as an addition of new and exotic materials to other decorative and prestige artefacts. In the broad context of metallurgical development over the past 10,000 years, there is still value to the idea of general progression in societal changes associated with native copper utilization onwards. A recent volume of the Journal of World Prehistory provides a series of reflections on the current state of the early developments in metallurgy across different cultures (Hanks et al 2009; White et al 2009; Killick 2009; Gullapalli 2009 and Roberts 2009). These issues form an important background to the text to explain how the alloying system being described developed, and also illuminate some of the cultural aspects of microstructural phenomena.

Chapter Two: Introduction to Metallography

The samples available for study in this metallographic compendium represent the most comprehensive such collection of mounted metallic samples in the world. From this study collection numbering over 1,000 samples, some representative types have been selected here for illustration and discussion. The book outlines the nature of metallography, the mounting of samples, some theoretical background to the alloy systems discussed and numerous examples of microstructures derived from a number of different studies carried out over the past thirty years. During the course of this study, a large number of metallographic samples have been prepared in the laboratory, of known composition or thermal history, which has substantially aided in the description of some of the alloy phase systems described in this volume.

Figure 2.1. Two Sphinxes, formerly with volute scrolls, from the collections of the J. Paul Getty Museum, Los Angeles. 86.SB.418. From Tuscany, Italy. Medusa head between their breasts. In the Renaissance these were associated with fire and death and often ornamented tombs and chimneys. By courtesy of the Getty Center Museum, Brentwood.

Figure 2.2. Heterogeneous bronze casting resulting from insufficient flow of bronze into the casting space, exemplified by extensive porosity in the two Sphinxes. Note the uneven distribution of the blue-grey eutectoid phase of the copper-tin system, large and variable interdendritic porosity, non-metallic inclusions, lead globules, as this is a leaded tin bronze. Coring is evident in this etched view of the section, which is coarsely dendritic. The microstructure affects the degree of perfection of the surfaces. They are not of the highest quality as a direct result of the microstructure of the casting, showing here large shrinkage porosity.

Metallography continues to be an essential component of many metallurgical studies. Since it is primarily a descriptive science, it is often used in detailed studies of individual artefacts and may be less common in the examination of ancient metallurgical sites and structures. As emphasis in metallurgy has shifted towards the exciting new developments in field research, the use of metallography has been seen less frequently in many journal articles. Polished sections are still very much in use however, not only for characterization of the metals concerned, but also for detailed analysis by electron microprobe analysis or ICP-MS.

Heavily corroded material is not only of interest to the conservation scientist, but may provide important clues to burial environment or preservation issues important for the field of conservation of metallic artefacts. For metallic artefacts in general, a good source for a classic compilation has been written by Cyril Stanley Smith is his "A Search for Structure" (1981). Many examples could be given of the application of metallography in recent times, some appropriate to this volume are the studies by Zhu and He (1993); Meeks (1993); Jett (1993); Pleiner (1998); Ryndina (2009).

Corrosion studies provide a wealth of information on polished sections of ancient metals and can help address authenticity issues and provenance, the use of sulphide or oxide ores in the smelting of copper and to answer quite sophisticated questions concerning the fabrication of ancient iron, steel and cast irons.

Metallographic samples from a site or culture frequently help to build a picture of the culture in terms of methods of alloying, surface finishing and alloying preferences, so it is very helpful if a range of material from a site is examined, rather than simply one sample.

It can be dangerous to draw conclusions based on one sample in some cases, and generally we try to work with a representative range of samples to gauge variations and different approaches which are sometimes taken to different types of artefacts. Worked sheet metal may be of different composition than decorative castings, silvered surfaces may represent a silver-copper alloy, or an application of silver to a copper alloy base.

Publication of metallographic information is most important, but rarely are the illustrations available in colour in most scientific or conservation journals. This is a pity, for some of the beauty of the metallographic image is lost in black-and-white reproduction.

To encourage the reader to produce even more pleasing photomicrographs, the photographs are reproduced here in colour. Increasingly, the author has become aware of the advantages of colour, or tint etching, as it is sometimes called, in the preparation of archaeological material. The reasons for this are, firstly, that grains can be delineated without deeply etching the sample, as in conventional etching, and that corrosion crusts are not necessarily disrupted by this technique. Secondly, the presence of casting segregation is very sensitive to colour etching, and since many alloys retain some aspect of this, it is clearly visible in the etched sections. The sample is not deeply etched: this is an advantage as the tint etch can easily be removed by repolishing on one micron diamond paste. Samples may etch rather deeply in alcoholic ferric chloride, our standard choice for copper and bronze alloys, and if repolishing is required here, there may be extensive dissolution which needs to be removed from the sample before it is ready to be etched once again.

Modern Metallography

The most recent and comprehensive volume on the subject is that published by George F. Vander Voort (1999). This textbook deals with macrostructure, specimen preparation for light microscopy, microstructure, light microscopy, hardness and quantitative microscopy. This text should be consulted, especially for matters concerning quantitative metallography which is covered very well indeed. There is also an extensive list of etchants for a large variety of different alloys, which is much more extensive than that provided in this volume, as here we are only dealing with alloys used in ancient times.

Quantitative Metallography

One of the first things that we notice about ancient worked metals is how much the size of the individual grains vary, often even within the same object. One of the functions of modern metallurgy is to make the size of the metallic grains consistent across the alloy, depending on the engineering application that it is intended to be used for. A method for measuring the size of grains has therefore to be adopted, and the most commonly used scale is that of the American Society for Testing of Materials (ASTM) grain size number. Of course, trying to decide on the size of a metallic grain which exists in three-dimensions, on the basis of a two-dimensional section, which is the usual approach, does have its problems, but as a first approximation it is a good way to compare the sizes of different grains.

In general, it is unusual for ancient metals to reach a grain size of ASTM 10. ASTM 1 is only found in rather soft alloys such as gold or silver. A typical range for ancient materials is between ASTM 6 to ASTM 3. The number of grain boundaries, especially across sheet metal artefacts seen in ancient examples may be inadequate for mechanical strength, as any fracture along a grain boundary results in a substantial weakening of the sheet, compared to an alloy with a much smaller grain size. For example, one sometimes sees gold-copper alloy sheet with only 3-4 grains across the thickness of the sheet, with an ASTM number of about 7-8. Compared with a similar sheet with 15-20 grains across the thickness, the larger grain size implies considerable weakness in the hammered sheet.

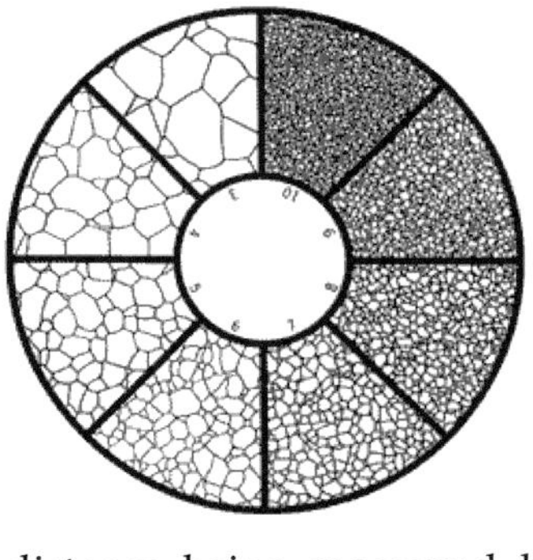

Figure 2.3 A grain size comparator eyepiece scale for a microscope at x10 objective lens magnification showing ASTM grain sizes between 3 and 10. A convenient way to estimate grain size.

Dendrite arm spacing is best obtained by measuring the number of arm intersections across a line traverse, the distance being measured by means of a stage micrometer or graticule, or on a photomicrograph of the area concerned.

Large, coarse dendrites generally mean that the cooling rate of the alloy in the mould was fairly slow, usually implying that heated or well-insulated moulds were employed. Finer dendrite arm spacings usually imply faster cooling rates. Grain sizes on worked objects can be measured using a grain size comparator eyepiece such as the graticule shown in

Figure 2.3. By direct visual comparison at an objective lens magnification of x10, giving a magnification of x100 overall, the ASTM number can be determined. Great accuracy in this type of measurement is seldom required; what is useful is the approximation to an ASTM number that can readily be compared by another investigator. The ASTM has prepared two typical series for ferrous and non-ferrous materials which are reproduced here in Figures 2.5 and 2.6. Strictly speaking, ferrous alloys are estimated on a logarithmic scale based on the formula:

$y = 2^{(n-1)}$ where n is the ASTM number, and y represents the number of grains per square inch. The most useful numbers in this series are from ASTM 1 to ASTM 8, with 8 being the smallest defined grain size. As well as ASTM numbers, it is possible to refer to a scale which gives the average grain diameters, usually from 0.010 to 0.200 mm. With twinned nonferrous alloys, the structure of worked grains can sometimes be confusing because the grains appear fragmentary.

The concept of grains, grain structure and the nature of dendrites, the snow-flake like crystals which characterize many ancient castings will be discussed further in Chapters 5 and 6. Since metals are crystalline solids, grains are what usually make up worked and annealed alloys.

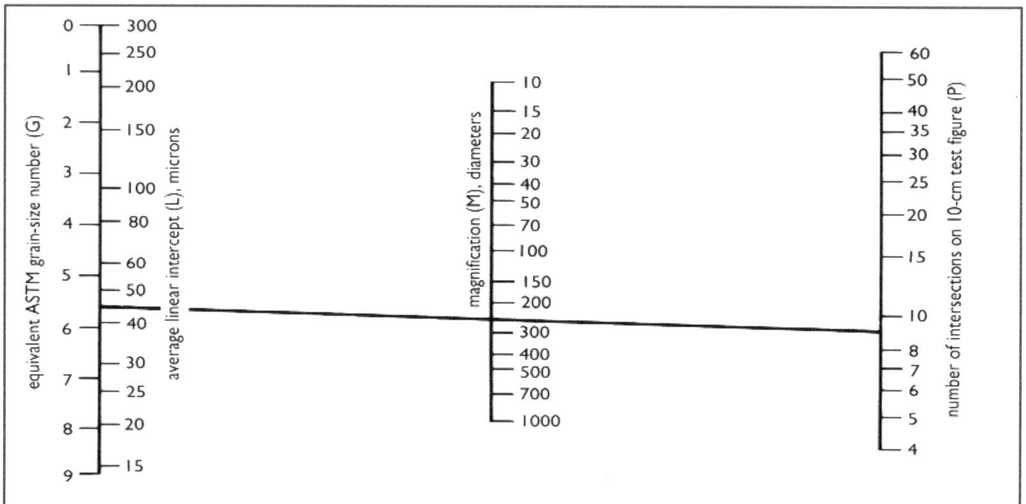

Figure 2.4 A nomograph for grain size estimation based on Hilliard's method which is convenient to use. After Hilliard (1964).

The intercept method can also be used to determine the average grain size of the material. The equation which gives the relationship between the factors required is: N(D) = length of intercept (mm)/ magnification x no. of grain boundaries.

Hilliard's circular intercept method can also be used and is referred to in the nomograph of Figure 2.4. In this technique, a photomicrograph of known magnification is selected and a circle, drawn on a sheet of plastic film of diameter 10 cm is placed over the photomicrograph. The circle should intersect more than six grain boundaries for the method to give a reliable grain size estimation. The circle is used until at least 35 grain boundary intersections have been counted. One of the advantages of the circular intercept method is that it is easier to apply to deformed structures. The ASTM grain sizes for the steels are illustrated in Figure 2.5. The idealized grains as hexagonal shapes are shown here from ASTM 1 to 8.

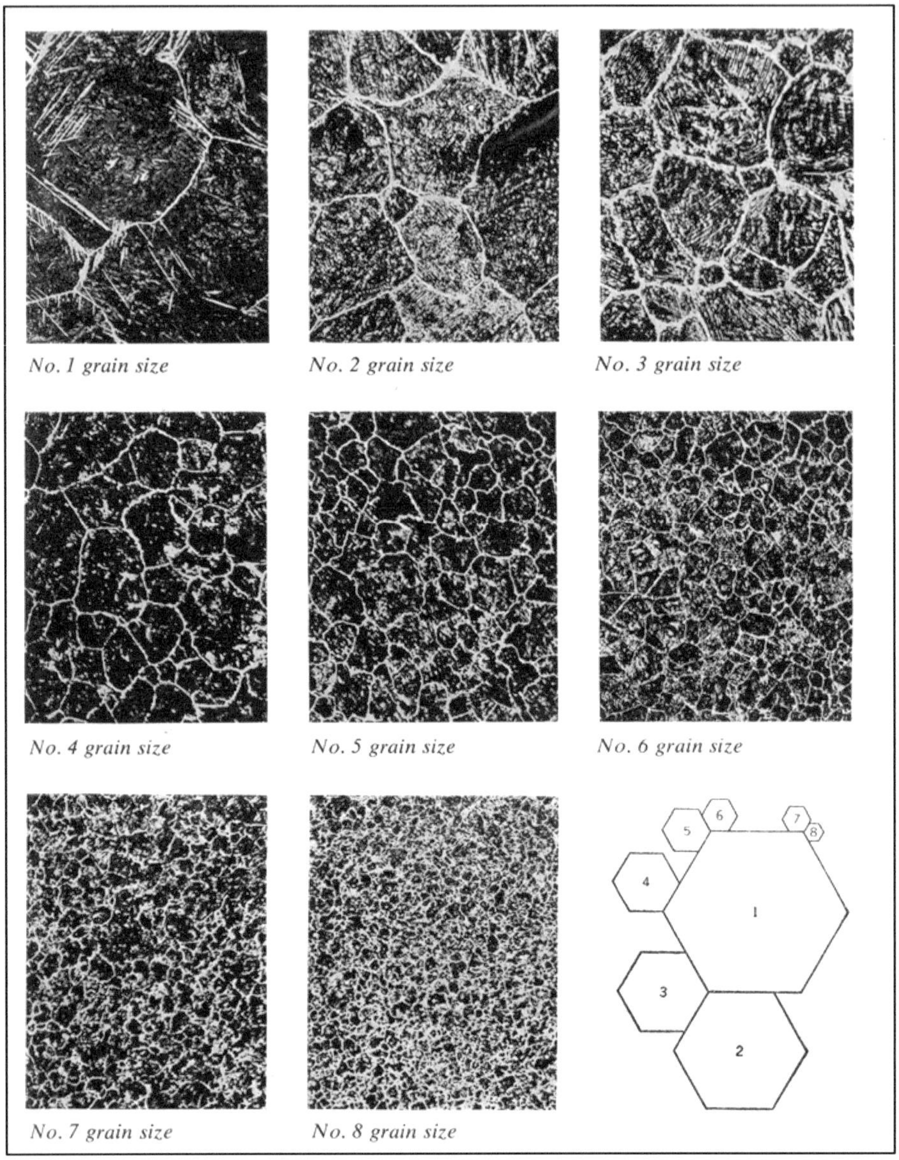

Figure 2.5 Typical standards for estimating the austenitic grain size of steel from the ASTM. These are photomicrographs of samples carburized at 930 oC for 8 hours and slowly cooled. Etched in nital, x100. Original photomicrographs from the Timken Roller Bearing Co, after Samans (1963: Figure 1.19).

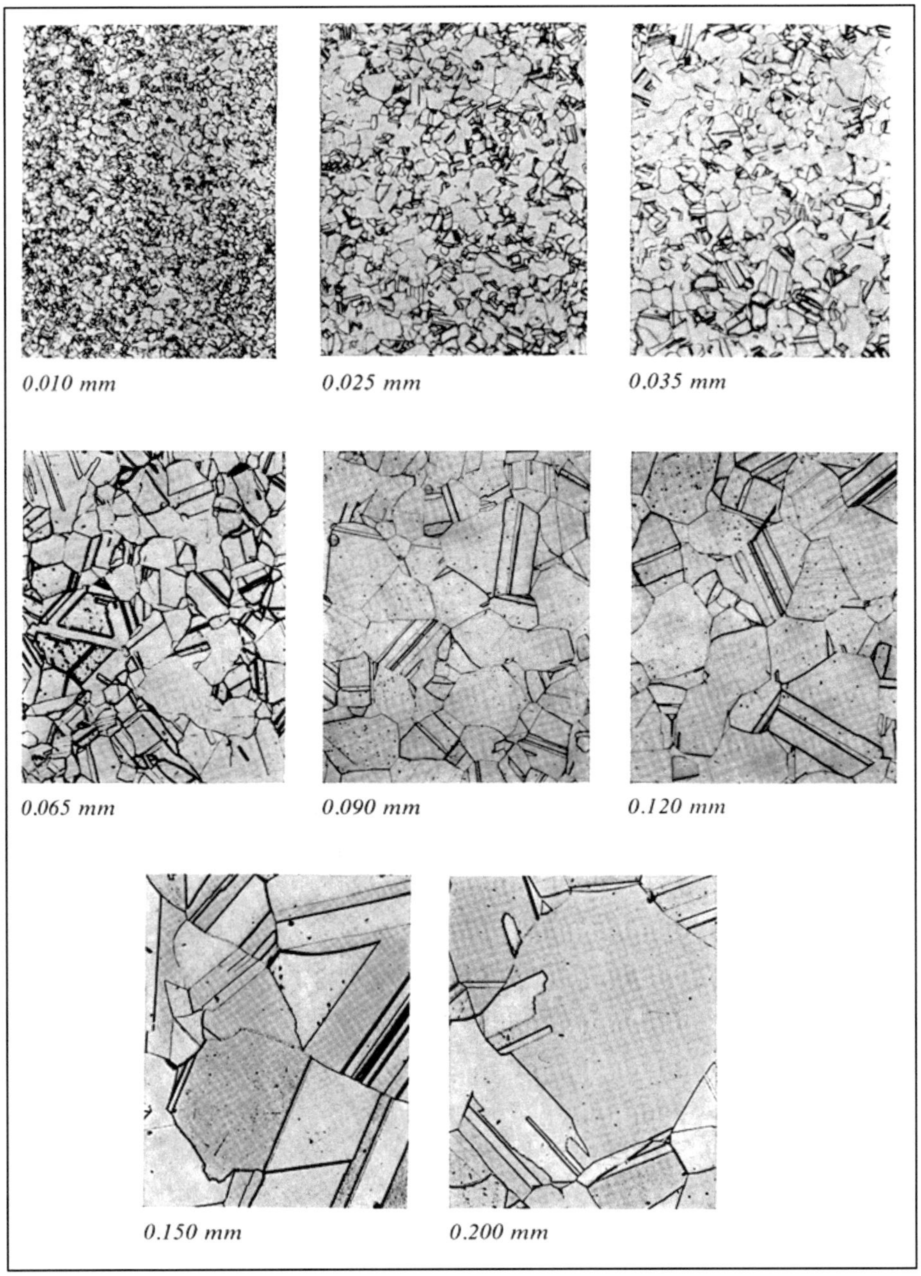

Figure 2.6 Typical standards for estimating grain size of annealed non-ferrous materials such as brass, bronze and arsenical copper, x70. Actual diameter is noted on the photomicrograph. (After Samans 1963 Figure 1.20)

Chapter Three: Applications of Metallography

Metallography is the application of microstructural studies to sections or polished surfaces of ancient metallic materials for the purpose of gaining information concerning composition, microstructural components, extent of corrosion or method of fabrication. As applied to archaeological studies, metallography came into use in the first few decades of the 20th century and has been of great importance to archaeologists, corrosion scientists, conservators and metallurgists ever since. Comparatively few studies were published between the 1st and 2nd World Wars, a dearth of young researchers meant that very few advances occurred in conservation science during this period. Shortly before the 1st World War, the foundations of metallography were initiated by the researchers of scientists such as Henry Clifton Sorby (1826-1908), particularly in the study of the phases in iron and steel alloys.

The principal difficulty in creating these advances was the limitation on the kind of microscopy available at that time. In metallurgical microscopy, the light has to pass through the objective lens and be reflected back from the polished specimen surface to the eye. This is called brightfield illumination and is essential for metallographic studies. During the late 1950's and early 1960's the situation began to improve and has continued to go from strength to strength since that period.

Some of the authors in English who began to have an impact in the 1960's were: Herbert H. Coghlan (1967; 1971; 1975; 1978; 1979; 1980), Ronald F. Tylecote (1962; 1968; 1973; 1990) and Cyril Stanley Smith (1965; 1971; 1981; 1992.) Important writers in the metallographic field of ancient metals emerged in many countries from the 1950-1960 period onwards, such as Jerzy Piaskowski (1958, 1993, 2002); Carlo Panseri and Massimo Leoni (1957, 1959, 1965); Albert France-Lanord (1949, 1952, 1964); and Radomir Pleiner (1963, 1969,1979) .

Figure 3.1 Etruscan bronze kouros, about 490 B.C. 8 7/8 in. Photograph by courtesy of the J. Paul Getty Museum No. 85.AB.104. The folding of the drapery is unusual: most Greek kouroi were nude with their hands falling to their sides. The youth wears a tebenna, a forerunner of the Roman toga, that is wrapped around his right hip with one end of the long garment pulled over his left shoulder from behind, and the other end thrown over his left forearm. After casting, the sculptor incised the individual strands of hair. This kouros was examined metallographically because art historical opinion at one time regarded the kouros as dubious. This will be discussed further in Chapter 9.

There are many recent publications concerning metallography in the wider context rather than purely concerned with ancient metals, of which the most comprehensive is that produced by ASM International, such as the ASM Handbook, Volume 9, Metallography and Microstructure (Vander Voort 2004). The interested reader is referred to this multi-volume work for a host of material relevant to the study of metals, their metallographic preparation, microscopy, and theoretical metallographic background information. Volume Nine alone is 1184 pages in length, so coverage is comprehensive. The metallographic sections used to be published only in black-and-white in earlier editions, but the 2004 edition now includes an excellent account of colour etching by Vander Voort (2004: 512) . Many of these colour etchants, also used extensively in these volumes, were originally devised by Klemm (1952, 1968),

Vilella (1959), and Beraha (1965, 1970). Detailed recipes for some of these are given in Appendix 1 to this volume. Advances have also been made in terms of polishing compounds and making available a number of different polishing materials for the modern metallographer. The reader can find a great deal of this information from online catalogues from such companies as Leco Corporation and Buehler Ltd. Care has to be taken in applying standard industrial metallographic practice to ancient metals, as often several minutes of grinding of samples is recommended for industrial alloys, which would be impractical for ancient specimens, whose size frequently precludes such thorough preparation techniques; there would simply be no sample left to study!

Metallography can help to inform us about the past use of metallic materials as well as aspects of their extraction, as the methodology is also applied to waste products, furnace linings and metallurgical slag, the latter often including tiny prills of metallic materials along with the silicate minerals which make up the slag.

Authenticity studies frequently make use of metallographic study and the examples shown in Figures 3.2 were shown to be old while those in Figure 3.3 were determined to be of recent manufacture as a result, even though the composition of both was acceptable, the patina was not in the case of the mirrors in Figure 3.3, and this must be determined through a metallographic examination.

Metallography can often be used to check on the results of analytical determinations which may be affected by inclusions or heterogeneity of the alloy. The only way to gauge these effects on the analysis is by metallographic examination.

Figure 3.2. Decorative elements on the top of a large Korean bronze, of 10th -13th century AD. Metallographic examination reveals that the bell could well date from this period as the corrosion crust and cuprite penetrate into the alloy.

The primary distinction to be made in a great number of initial studies is whether the metal or alloy has been cast to shape, or whether it has been worked and annealed to shape.

Figure 3.3. Two Chinese bronze mirrors with a dark patina, but clearly not of the special black-patinated Chinese bronze mirrors which are very hard to replicate. These mirrors are modern reproductions. Made in an alloy which is essentially of the correct composition, 70% copper, 25% tin, 5% lead, surface examination, metallographic and patina studies have to be undertaken in order to come a conclusion as to whether the mirrors are authentic or not.

Often, artefacts are selectively worked by hammering to harden them, and this kind of evidence may show cultural or time-period effects on production. The kind of copper which was used in early societies may be studied by metallography to show how it was shaped and worked, sometimes providing evidence for the use of native copper, as a result of folding of creased ridges of metal back on itself, and sometimes showing clear evidence of the use of smelted copper. This transition marks an important stage in early metallurgical advances and metallography is one of the essential tools for these kinds of investigation. Metal objects may be plated or coated with other metals, which limits the extent to which non-destructive analysis can be used to provide a definitive answer to the question of their identity. In such cases, metallography has to be used to gain further information and can determine, in conjunction with analyses of the polished cross-section, the nature of the surface coating.

One of the most powerful applications of metallography is in its application to ancient ferrous materials. This is because in the west, early iron was made by the bloomery process in which the iron was never really molten, while in Africa, Iron Age material may predate the introduction of bronze, and the importance of iron and steel at an early date, and its variations, can best be explored by metallography. Metallography can determine if the material is a steel or a wrought iron, whether it has been made of a number of pieces joined together, whether the cutting edges have been quenched to harden them or whether the artefact is made of a low-carbon steel. In China, early ferrous materials were often made in cast iron and softer tools and weapons could be fabricated by decarburizing white cast iron. Information relating to these developments and utilization of materials must be obtained by metallography.

Precious metal alloys such as silver-copper and gold-copper alloys are often finished by surface enrichment, or depletion gilding or some other form of surface treatment or coating. For the investigation of these types of alloys, metallography is invaluable.

Numerous problems relating to the authentic nature of ancient metallic artefacts, particularly silver and bronze artefacts can be addressed by metallography. Metallographic studies are an integral part of a comprehensive investigation of the elemental composition, lead isotopic data, surface binocular examination, and microstructure of the artefacts in question. In cases of authenticity, it is often possible to adduce enough information from the metallographic section to avoid the need for further studies, particularly in the case of ancient bronzes, where the extent of patina penetration and the type of patina may be crucial and sufficient evidence to condemn the artefact as a modern reproduction.

Chapter Four: Samples and Metallography

Usually a small sample has to be taken from the object under study in order to carry out the metallographic examination, although at the other end of spectrum of sophistication, it is now possible to probe metallic artefacts with neutrons to non-destructively determine phase composition. The artifact is often sampled by cutting out a small section with a jeweller's saw and the resulting sample is then embedded in a suitable mounting media, such as Buehler © epoxide resin in a one inch or inch and a quarter diameter casting mould. The Buehler © epoxide resin is used by weight, five parts of resin to one part of hardener, and takes eight hours to set.

Samples are therefore best mounted one day and then finished one day later. The orientation of the sample in the mould should be carefully noted. The mounted sample is then ground and polished on wet silicon carbide papers, usually from 240 - 600 grit in size, before polishing on synthetic napped cloth wheels using diamond powder suspension in water as the final polishing agents. These are usually six micron diamond and one micron diamond. If required, a finer finish can be obtained using colloidal silica or 0.25 micron diamond in water suspension, but for most purposes, polishing down to one micron is sufficient. As colloidal silica has a tendency to coagulate, the supernatant liquid should always be used for final polishing: do not shake the bottle in this case before use. Colloidal silica is useful for removal of scratches from the one micron diamond stage of polishing if they cannot be otherwise eradicated. The best diamond polishing compounds to use are those that are available in aqueous emulsion spray bottles. These are not suitable for some furniture cross-sectional studies in which grounds may be affected by water, but for most metallurgical application they are the preferred polishing medium. Each bottle of polishing compound is differently coloured and generally six micron and one micron paste are sufficient for most purposes.

Figure 4.1. The author grinding samples on wet silicon carbide papers. It is most important to remove the scratches from each previous stage. Grit sizes are 240, 320, 400 and 600. Finish lightly on 600 grit to avoid deeper surface scratches before moving to 6 micron diamond. Photograph by Lesley A. Moorcroft.

Figure 4.2. The author polishing metallographic samples employing six micron diamond and one micron diamond aqueous emulsion sprays on napped synthetic cloth. This is generally sufficient to polish archaeological specimens. Colloidal silica is sometimes useful for really fine finishing after one micron diamond paste. Samples should be washed in ethanol and then dried. Store in a desiccator if not immediately examined. Photograph by Lesley A. Moorcroft.

Examination of the polished section can utilize the light microscope, preferably a metallograph, and samples can be placed in the electron microprobe (EPMA) for quantitative analysis, in the scanning electron microscopy (SEM), for semi-quantitative analyses or imaging, environmental scanning electron microscopy (ESEM), or analyzed with x-ray fluorescence (XRF) techniques, all of which are invaluable in a comprehensive study of metallographic sections. If careful work is done it is possible to produce fully quantitative analysis with an EDX system as long as the operator is trained in the aspects of detection limits and high standard deviations of point analysis. An experienced operator should be able to obtain quantitative results

It is possible with large objects, or with the edges of coins, to polish selected areas of objects to examine the metallographic structure *in situ*, in cases where it may be impossible to cut or remove a section from the object.

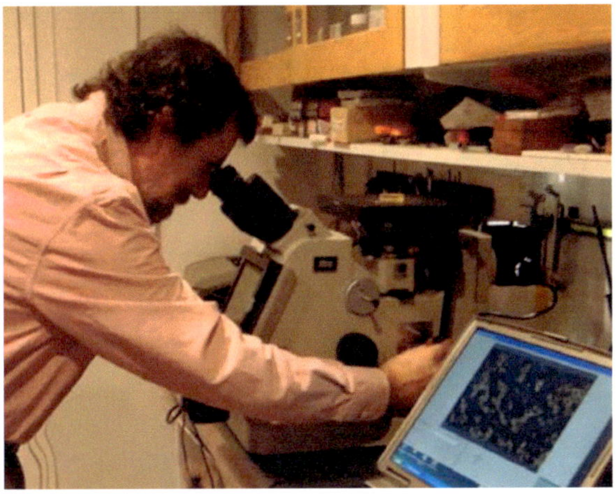

Figure 4.3. A Nikon Epiphot metallograph purchased on e-bay for $3,800 allows the polished sample to be viewed directly on the inverted stage. Digital recording of the image is being made with a Nikon D300 camera back using a Dell laptop D520 computer. Total cost of this equipment $5,500 in 2008. Images are adequate but will be surpassed in future recordings as sensor size increases with advances in digital technology. Photograph by Lesley A. Moorcroft.

Microstructural features may be selectively etched or pseudomorphically preserved on metallic surfaces, which allows this technique to reveal micromorphology directly. A list of suitable etchants is given in Appendix 1. A useful on-line source of scores of different etchants is that provided by the Etchant's Database (2010). One should remember that a polished section is a two-dimensional view of a three-dimensional structure. In the pattern-welding of steels, supplementary techniques such as X-radiography to reveal the internal pattern, should be employed and careful attention to the orientation of the mounted sample given.

When mounted samples are studied, the surface of the metallic specimen should be examined first in the freshly polished state, followed by etching with a suitable etching reagent. Very important samples where the interface between the corrosion crust and the metal may be reserved for examination in the polished condition only, as etching may attack the corrosion crust while it reveals metallic structure, but the importance of the corrosion interface may be paramount, especially in authenticity studies, frequently required in the case of ancient bronze and silver artefacts.

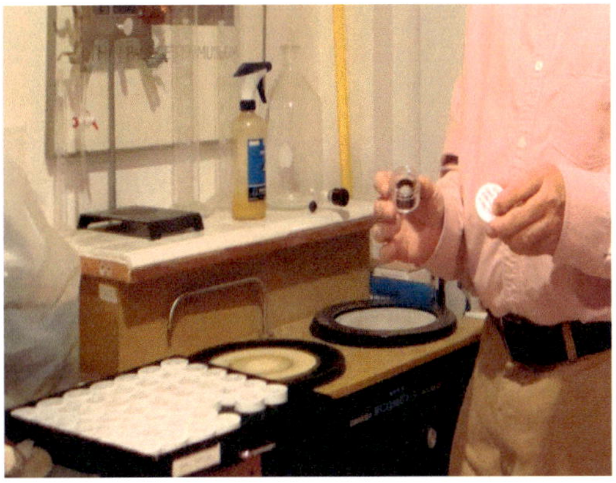

Figure 4.4. Mounted samples are stored in plastic jars with screw cap lids. Each white jar lid is labelled to correspond to the specimen label. These squat cosmetic jars are ideal for metallographic specimen storage of diameter 1.25" and 1.0." Superior to laboratory tins in a number of respects, they are corrosion resistant and allow samples to travel safety contained without the danger of lids becoming detached or galvanic corrosion between sample and metallic surface. A desiccant can be added to each jar for travelling if required. Photograph by Lesley A. Moorcroft.

There are often severe restrictions on the quantity of metal which can be removed from an artifact for metallographic examination, especially non-ferrous metalwork. On the other hand, every a very small sample, smaller than a pin-head, can be mounted and polished for examination, although great care has to be exercised at all stages of preparation if samples are that small. It is much easier to work with larger samples although, by archaeological standards, samples the size of a pea are already unusually large, unless whole artifacts or substantial fragments are available for sectioning.
There are a number of criteria that should be considered before any sampling is undertaken:

1. The microstructure of the samples should not be altered in the process of removal.
2. The sample should be representative of the object as a whole or of a selected feature or area of the object.
3. The orientation of the sample in its dimensional relationship to the object concerned should be marked on a photograph or a drawing of the object.
4. If it is not obvious from where the sample was taken, the position on the object concerned should be marked on a photograph or a drawing of the object.
5. The sample should be assigned a laboratory number together with sufficient documentation to enable its identity to be preserved.
6. The object should be photographed or drawn before the sample is taken. This is especially important if the dimensions of the object are fundamentally altered by the material removed.

Metallographic studies should be integrated as far as possible with archaeological data, provenance, chemical composition, and technology of manufacture to extract the maximum benefit from the removal of a small sample from the object. There are still far too many metallic artefacts lacking in detailed provenance information because they were obtained from dealers, museum collections which lack documentation, or uncontrolled excavations.
The range of features that may be made visible in prepared samples is variable, depending on the type of specimen examined and how it is prepared; detail not apparent using one etchant may become visible only after another reagent has been employed. The following are some of the major microstructure features which can be examined:

1. The range and the type of grains present. Their size can be compared with an eye-piece marked with grain sizes for comparison or with American Society for Testing Materials (ASTM) standard grain size numbers.

2. The presence of different phases which can be observed either in the corrosion crust or the metal.
Gross heterogeneity or differences between various areas of the sample should be noted: especially in the case of iron and steel artefacts.
3. The grain shapes or surface deformation features or heat-treated zones at cutting edges and worked surfaces.
4. The distribution of inclusions, weld lines, slag particles, or porosity and if quantitative metallographic tools are available, volume fractions.
5. The presence of any surface coating or gilding. Sometimes, careful examination at high magnification is necessary to establish the presence of surface coatings, lead gilding, amalgam gilding, tinning, etc.
6. The distribution of any corrosion products present and the existence within corrosion layers of pseudomorphic remnants of grain structure or other microstructural features, remnant metallic grains, and layering or unusual features.
7. Indications of grain-boundary thickening or precipitation of another phase either within the grains themselves or at the grain boundaries.
8. The presence of twin lines within the grains and whether these are straight or curved. In face-centered cubic metals, such as copper, silver and gold alloys, these are common.
9. The presence of stain lines (or slip lines as they may also be called) within the grains: common in copper and bronze alloys.
10. Whether dendrites (a structure which is common in cast alloys and which may look like an intersecting snowflake pattern) shows indications of coring and the approximate spacing, in microns, of the dendritic arms, if these are clearly visible.
11. The size and shape of any non-metallic inclusions present in the metal. Most inclusions cannot recrystallize and become elongated or broken up along the direction of working of the metal.
12. The presence of intercrystalline or transcrystalline cracking in the polished section.

Further details concerning many aspects of metallographic preparation can be found in the volume edited by Vander Voort (2004), referred to in Chapter 3 and in Scott (1991).

Chapter Five: Low-Cost Aids for Metallography

The preparation of cross-sections is an essential part of many different kinds of technical investigation of art objects in order to reveal detail of microstructure, pigment layers, corrosion or grain structure. The material which is used for a cross-section from an art object must be carefully examined before a decision is taken to embed it in resin. There are a number of non-destructive techniques of analysis which can be used on small samples removed from artefacts, before they are embedded in resin for grinding and polishing.

For example, unmounted samples may be positioned in an x-ray diffraction instrument to obtain corrosion product identification or pigment characterization. The same sample can then be placed on a diamond cell ATR instrument and the infra-red spectrum obtained. The sample can then be examined in the SEM or ESEM, which together with EDAX, provides a useful complement of analytical information which may be necessary to understand more about the sample which has been removed. It is increasingly important to bear these factors in mind before proceeding to encapsulate the sample in a resin which can never be satisfactorily removed.

Once a decision has been made to proceed with the embedding, there are a number of different choices which the conservator can make. One of these is to decide on whether the sample is to be treated as a 'painting' cross-section, or a 'metallic' cross section. The reason for this differentiation is that we usually mount tiny samples from paintings, frescoes, furniture, wall-paintings, etc., in small rectangular blocks of resin, while samples of slag, metal, crucibles, and larger painting samples are mounted in cylindrical plastic blocks. A recent review which illustrates some of the methods of mounting standard specimens has been published by Khandekar (2003). The technique of preparing mounted metallographic samples goes back to the 19th century and the origins of metallography (Smith 1960). Painting cross-sections did not become standard until popularized by Plesters (1956).

With the standard small rectangular block employed for painting samples, the block is half-filled with resin and allowed to set. The sample is then laid on this prepared resin bed and sandwiched between a new layer of resin poured over it to fill the mould. This means that the orientation of the sample is normally fixed as a cross-section of the layers of paint when the block is removed from the mould and ground and polished.

When the sample is embedded in a cylindrical block of resin, it is easier to position the sample in the mould to obtain the kind of section required, especially with samples of varied shape and size. Here we describe the use of home-made equipment in order to facilitate the preparation of samples in cylindrical moulds especially, although aspects of the equipment may also be useful for painting cross-sections as well.

The use of home-made equipment has a long pedigree in the history of sample taking and sample preparation of art objects. From the use of ice-cube trays for casting small resin embedded samples, used at the Department of Conservation and Materials Science, Institute of Archaeology, London, in the 1960's and 1970's, to a free-standing coring device for paint sectional studies invented by Gettens in 1932. There is much scope for further innovation and adaptation by the conservator of techniques of sample taking and sample preparation, especially given the increasing reluctance of curators to allow any sampling whatever, particularly from metallic artefacts already within museum collections, and as a result, the status of the sample has become considerably enhanced. It is important that the sample is retained with full documentation in the original object's archive or that photographic documentation of the results of the metallographic study be included with the object files for future reference.

The Equipment

Specimen mounting jig.

There are a number of ways in which small samples can be held in place for mounting in resins. One method is to hold the sample in a small plastic clip, which is embedded in the resin with the sample, and which then holds it in position, preventing the sample from moving when the resin is poured over it.

The disadvantage of this technique is that it creates a differential background for microphotography which is sometimes obtrusive. The small clip may also crush some samples. If they are very corroded, even slight action of the plastic clip is already enough to break them.

Painting cross-sections are frequently very tiny, less than $1mm^3$, or even $0.5mm^3$, and for these types of samples it is important to ensure that the mounting resin which is used is not too hard. The epoxy mounting resins, such as Buehler Epoxide©, are harder than most acrylics or polyesters, and are more suitable for mounting metallic samples rather than painting cross-sections. For the latter, Bioplast© is a good alternative. It sets quite clear, and is much softer than the epoxies. This is important, since otherwise too much time is expended in grinding down the epoxy resin in the rectangular block mount in

order to expose the sample for further polishing. For the grinding or polishing of paint cross-sections, it may be necessary to avoid the use of water altogether, if the components of the ground are water-soluble, and micromesh cloths with white spirits can be used instead. This is rarely a problem with metallic samples in cylindrical epoxy mounts, and these can be polished in a variety of oil or water-based polishing compounds. We prefer to use synthetic diamond suspensions in water-based carrier for this kind of polishing.
The specimen mounting device, built entirely from Meccano©, shown in Figures 5.1 and 5.2, has proved to be of considerable practical benefit to the author in over thirty years of use.

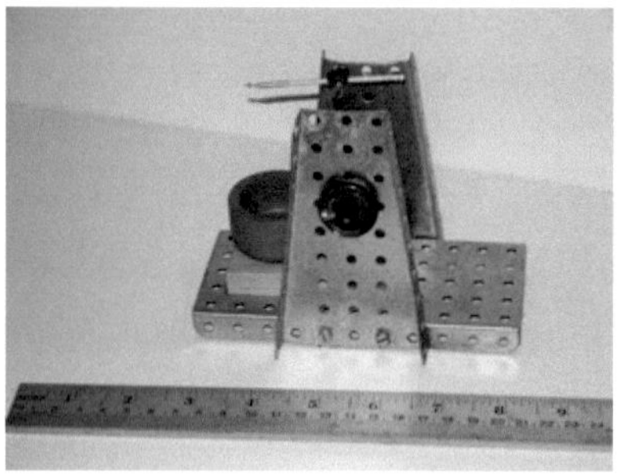

Figure 5.1. Specimen mounting jig with small sample in raised position attached to a cocktail stick with adhesive. Photograph by the author.

The device enables small samples to be attached to a cocktail stick, which is held by two small set screws to the holder, mounted on a rod which can be rotated either to the horizontal position for viewing using a stereo-microscope, and without any further disturbance to the sample, it can then be rotated on the stick until it rests in the desired position in a one inch or 1.25 inch silicon rubber or polyethylene mould. The epoxy resin or other suitable mounting media can then be poured into the mould without disturbing the orientation of the sample. .
The usual *modus operandi* is to adhere the small sample to the cocktail stick using a small blob of HMG© cellulose nitrate adhesive, which has a suitable tack to enable the sample to stay in position on the end of the stick. Occasionally it is helpful to just remove the tip of the stick or to position the sample to one side of the edge so that it is held securely prior to mounting.
Once adhered, the sample and stick can be placed in the jig and the screws which hold the stick in place tightened. The sample can be further examined or checked at this stage under the binocular microscope by rotating the holder to bring the sample into view, or it can be placed at any angle desired in the mould, as for example, in the preparation of a taper section. This kind of careful positioning of a mounted sample may be a critical issue in obtaining the kind of section through the artefact sample required to obtain the information sought.

Figure 5.2. Specimen mounting jig with sample being lowered into silicon rubber mounting cup. Silicon rubber cups require the use of a mold release agent for prolonged employment. Photograph by the author

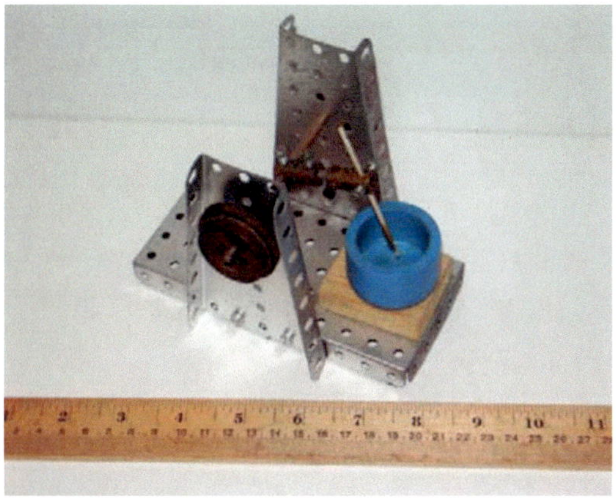

Because epoxy resin mounts require a considerable effort to grind away even a millimetre of set resin, it is good practice to ensure that the mounted sample on the stick just touches against the surface of the silicon rubber cup, or is only the smallest fraction of a millimetre above it. This reduces tedious grinding later on. The HMG adhesive is unaffected by the epoxy mounting resin often used for metallographic preparation, namely Buehler© epoxide resin which sets in eight hours and gives good edge retention.

If the sample is required to be set in the mounting resin at a particular angle, then the device allows for every possible angle to be accommodated. The set-screw can be released and the sample rotated around its axis if needed before mounting at an angle, allowing the maximum degree of flexibility. The label can be provided of a sufficient size at the back so that it can be added to the mount after pouring the resin, and without allowing the label to be held away from the resin by the cocktail stick. The cocktail stick [or toothpick] is simply broken off from the back of the mounted sample before grinding and polishing in the usual way. There is no commercial equivalent on the market and the device has therefore to be hand-made; the cost is minimal; the author has been using this particular apparatus for twenty-five years and it has proved to be very useful.

Specimen Dryer

There are many industrial or laboratory dryers on the market which suffer from a number of deficiencies in design for drying of small mounted samples. Firstly, they are often too powerful, so that the mounted sample, on being placed in front of the dryer, simply falls over from the force of the air. Round metallographic samples or square mounted painting cross-sectional samples should be dried with the sample facing the dryer so that remaining washing fluid or etchant is forced away from the surface of the sample in all directions and not dried across its exposed face. Some samples are very prone to staining or patchy drying, leaving drying rings, if the washing fluid, such as ethanol, is not uniformly evaporated away from the surface of the mounted specimen, or cannot fall away from it through the dual action of gravity and air drying.

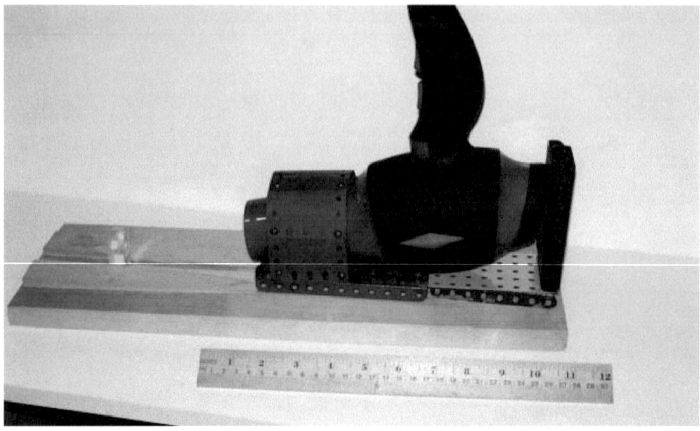

Figure 5.3. Convenient sample dryer. When the wooden part becomes stained, it can simply be replaced. Commercial dryers use stainless steel racks, if they have a drying surface at all, and the steel becomes corroded through droplet spray from etched sample surfaces. Photograph by Lesley Moorcroft.

Many commercial dryers do not provide any means of securing the sample to prevent movement so that it has to be held with a pair of tongs in the blast of air from the dryer. Some dryers, designed for metallography, are provided with a small drying rack at the front of the instrument, which is often inadequate, as it fails to prevent the specimen from falling over, and is made from stainless steel or aluminium alloy. Since the samples for metallography are often etched in very corrosive solutions before washing and drying on the specimen dryer rack, traces of the etchant rapidly corrode the rack with unsightly results.

The specimens are often best dried in a current of warm air, but many commercial dryers often come with only two options: heated or unheated. The heated option may be so intense that there is a danger of burning one's fingers if the sample is held a few inches from the dryer! The unheated option may fail to prevent drying rings or may be too slow.

All of these deficiencies for the conservation laboratory are overcome by using the device shown in Figure 5.3. It is a conventional hair-dryer, held in position by Meccano© parts bolted together and placed on a wooden board with slightly angled wooden wedges, which has the following advantages over commercially available instruments:

(1) Round samples can be positioned on the wooden wedges and do not fall over when the dryer is turned on. The slight gap underneath the center allows excess washing fluid to easily fall away from the sample.

(2) The wedges are five inches in length, which allows the sample to be positioned close to, or further away from the dryer depending on need.

(3) The base underneath the sample and the wedges are made of wood, cannot corrode, and if they become degraded, can easily be replaced as the dryer simply detaches from the board on which it rests.
(4) Samples can safely be left on the drying board and do not need to be held in place with tongs.
(5) The warm air from the hair dryer is considerably less intense than the hot air from laboratory dryers, which means that samples are never too hot to touch with the fingers and warm air can be used without the danger of burning the hands.
(6) Small painting cross-sections can be rested on the wooden wedges at a convenient distance from the dryer for gentle drying and will not fall over.
(7) The instrument costs next-to-nothing to fabricate, especially compared with the high cost of metallographic equipment.

Specimen levelling press.

For conventional microscopy, rather than for inverted stage microscopy using a metallograph, samples often have to be mounted on a glass slide with a small ball of Plasticine©, and levelled in a suitable press to produce a surface parallel with the plane of the objective lenses. This is accomplished using a leveling press of which there are many types on the market. The home-made device shown in Figure 5.4, allows the sample to be leveled in the same way as conventional presses, but is also provided with a flanged wheel option at the press head so that the metallic end itself does not impinge on the sample. Even using a piece of lens tissue between the levelling press and the sample surface can be problematic for gold and silver samples.

Figure 5.4. Specimen levelling press. Not required if a metallograph is available for use. Inverted microscopy for metallographic examination is generally to be preferred.

The home-made press is made entirely from Meccano© parts and costs about $20.00 to make compared with $250.00 for a commercial unit. Used equipment can also be purchased on e-bay or other second-hand source.

Chapter Six: The Nature of Metals: Equilibrium Diagrams: Featuring Silver-Copper Alloys.

Atoms can be stacked in different ways to create a variety of simple lattice types.

Metals are an aggregation of atoms that, apart from mercury, are solid at room temperature. These atoms are held together by "metallic bonds" that result from sharing available electrons. A negative electron bond pervades the structure, and heat and electricity can be conducted through the metal by the free movement of electrons. There are three common types of lattice structure that most metals belong to: face-centered cubic, body centered cubic and close-packed hexagonal. In the face centered cubic lattice (FCC), layers of atoms can be built up, modeled in spheres, so that the third row of spheres does not occupy the same position as the spheres in the first row, the structure then repeats every third layer (ABCABCABC....). In the close-packed hexagonal structure (CPH), a model can be made up of spheres stacked in layers as well, but in the hexagonal structure, the spheres repeat the pattern every second layer (ABABABAB....). In the body centered cubic metals (BCC), the atoms are arranged at the corners of a cube with one atom in the centre of the cube. This arrangement is strong, but allows substitution or other atoms to enter the cube if they are small, such as carbon in steels.

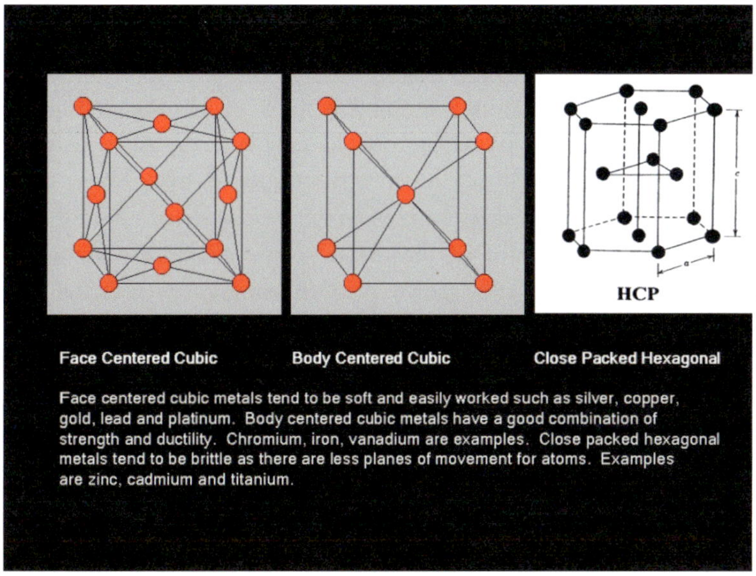

Figure 6.1 Some common lattice types relevant to ancient and historic metals.

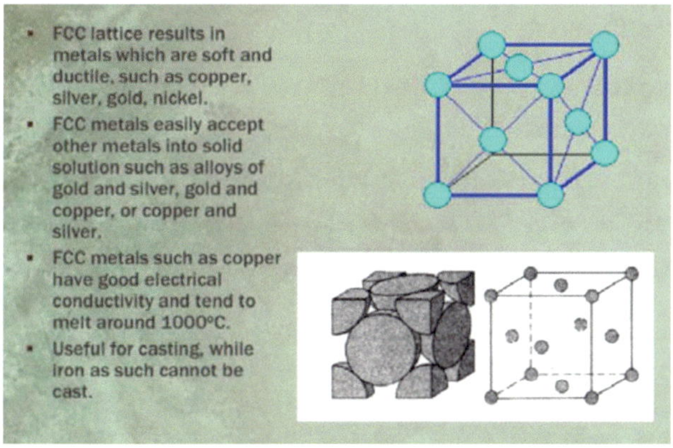

Figure 6.2 The face-centered cubic metals are largely those dealt with in this volume. When alloyed together, the kinds of phase diagrams produced by FCC metals with each other, tend to be simple solid solutions or eutectics.

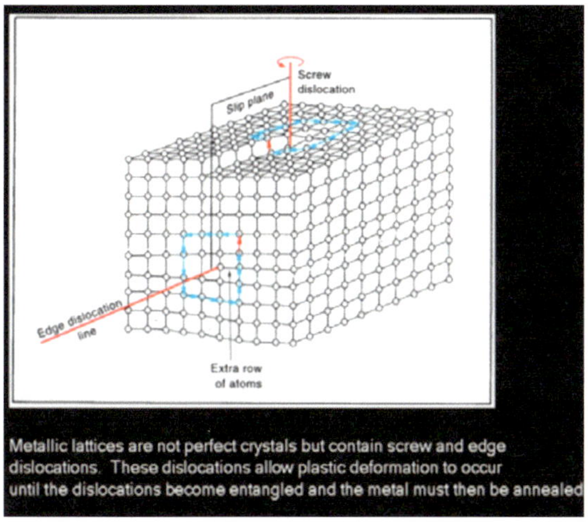

Figure 6.3 Dislocations are important in metals as they allow plastic deformation. Here are shown both edge and screw dislocations.

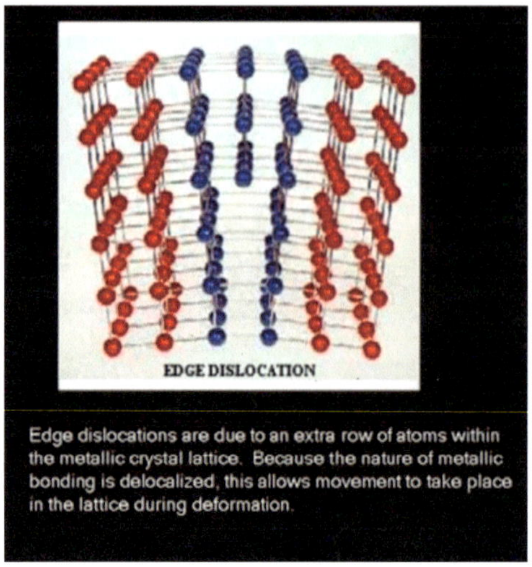

Figure 6.4 Another view of an edge dislocation. Some crystal defects may be only 'point' defects in which one atom is substituted by another. In line defects, such as an edge dislocation, a row of atoms form an incomplete plane extending through the crystal lattice. Another common type is the screw dislocation shown in Figure 6.3, where the displacement is parallel to the dislocation. In this dislocation, the crystal no longer consists of parallel planes of atoms, but instead has a single plane in the form of a helical ramp.

Phase Diagrams

One of the most useful theoretical aids to the study of ancient metals is the equilibrium diagram, also called the phase diagram. Since many alloys are mixtures of two or three different components, the phase diagram is used to plot temperature against composition and to map out the areas of the different phases which occur at different compositions or temperatures in the system concerned.

A phase is a component of an alloy system whose composition can vary within certain limits, so phases can be of rather variable elemental composition, for example the alpha phase in the bronze system could have 2% of tin or 10% of tin dissolved. It could not have 20% of tin, because that is outside of the stability limits of the alpha phase.

Alloys do not have a particular melting point. They soften and pass through a pasty stage between temperature zones shown on the phase diagram as solidus and liquidus curves. The solidus is the line in the phase diagram that separates the pasty stage of the alloy, usually a mixture of solid and liquid, from the completely solid alloy below the temperature of the solidus line. The liquidus is the line on a phase diagram that shows the temperature at which solidification begins on cooling from the melt.

The boundary region between these lines can be narrow or broad. In some alloys of importance in antiquity, such as copper-tin, gold-copper-silver, or copper-arsenic, there is considerable separation between liquidus and solidus curves, which exacerbates the difficulties of attaining equilibrium cooling conditions from the melt, and enhances the segregation effects that can be observed in these alloys. This is because the alloy has a longer period of time in a semi-solid region of partial solidification in which time for segregation to become fixed in the solid is accentuated compared with alloys with a narrow separation between liquidus and solidus. Some metals are completely soluble in each other, as shown in Figure 6.5, the phase diagram for the gold-silver system. At any composition of the alloy as it cools down, it reaches the liquidus line and begins to solidify. Most alloys pass through a pasty stage with some solid and some liquid, until the solidus line is reached, at which point the entire alloy is a solid mass of crystals. This is a completely homogeneous solid solution, shown on the diagram as the alpha phase, and the grains are usually close to their equilibrium predicted state. The gold-silver system in which complete solid solubility occurs between two components is actually uncommon: most metallic elements are not completely soluble in each other, and at different compositions, different phases form and these can be seen in the polished and etched microstructures, and are modelled by the equilibrium constraints affecting their composition and temperature stability.

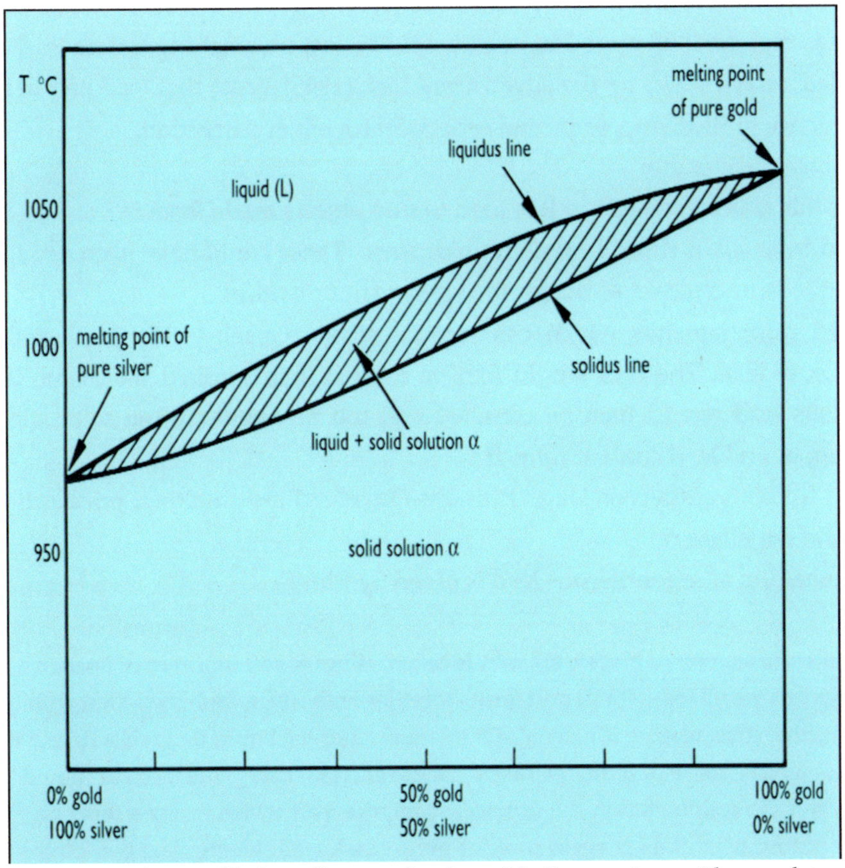

Figure 6.5. The equilibrium diagram for the gold-silver system with complete solid solution across all compositions.

Silver and its alloys

As an example of the phase equilibria in an alloy system important in antiquity, we have chosen the silver-copper system for study here. Extracted silver metal has been produced for millennia. By the time Pliny was writing in the 1st century AD, silver had been used in the Old Word for thousands of years. He writes about one of the Roman mines:

"......one of these mines, now called after Baebelo, furnished Hannibal with 300 pounds weight of silver a day......there is also a humour, in round drops, that is always liquid, and is called quicksilver. It acts as a poison on everything and breaks vessels by penetrating them with malignant corruption....." [Pliny Natural History Vol IX, Book XXXIII, 99-101].

These Baebelo mines mentioned above by Pliny are probably Rio Tinto in Spain. The Phoenicians, Carthaginians and Romans worked the jarosite deposits there whose silver content could be as high as 1000 ppm, but most would be lost on smelting unless lead or lead ore was added, which soaks up the silver. Craddock (1995) states that lead had to be brought some distance to Rio Tinto, imported especially for silver extraction.

The sources for ancient silver are:

(1) Native silver: this was very rare and it is hard to find objects made from it.

(2) Silver smelted from silver chloride and sulphide ores. These could have been used as very easy to smelt to extract silver, although the ores are not plentiful.

(3) Silver smelted from jarosites which can contain silver. Usually, this implies the deliberate addition of lead. The ores would first be roasted then smelted with lead, the newly argentiferous lead would then be cupelled and the raw silver, often with some bismuth, gold, copper and lead content refined.

(4) Silver smelted from argentiferous lead. Processes involved are smelting, purification or scorification, and cupellation.

A version of the smelting of argentiferous lead is given by Pliny:

"...there are two different sources of black lead, as it is called, either found in a vein of its own and produces no other substance mixed with it, or it forms together with silver, and is smelted with the two veins mixed together [this mixture is galena]. Of this substance the liquid that melts first in the furnace is called stagnum [or stannum, an alloy of silver and lead], the second liquid is argentiferous lead, and the residue left in the furnaces is impure lead which forms a third part of the vein originally put in; when this is again fused it gives black lead, having lost two-ninths in bulk." [Pliny, Natural History, Volume IX, Book XXXIV, 159.]

The word *stagnum* is a general term for undefined white base metal. Here it is argentiferous lead, typically with some silver up to about 1%. The black lead referred to here, is the lead, regained with losses after the smelting of the litharge residues, left after cupellation. The cupellation stage, the final one in the process, involves heating the lead in a strong air blast to create litharge, with silver floating on top, as the lead oxide reacts with the clay of crucibles to make lead silicates, which are dense and viscous.

The earliest silver artefacts were made during the 4th millennium BC and are found in the areas of Persia, Anatolia, Mesopotamia and Egypt. The native metal is too rare to account for much of this early use, but was employed when available. Meyers (2003) states that the earliest silver objects were made from silver smelted from these silver ores directly. Silver and gold both became trade commodities by the middle of the third millennium. It is generally believed that silver metallurgy derived from lead smelting technology, being extracted from cerussite, $PbCO_3$ and galena, PbS, using the cupellation process. The possible use of other ores such as cerargyrite, AgCl, and argentite ,Ag_2S, as well as aurean silver, electrum, and pyrites is also probable.

While an extensive literature of historical accounts of silver production exists and ancient sources do state that silver was extracted from lead ores, no specific ore is identified. The reworking of mine sites and old slags confuses attempts at historical interpretation. Meyers (2003) notes that the gold/silver ratio can help indicate the metal's origins, and the absence of lead is evidence that the silver was not produced by cupellation. Meyers (2003) cites detailed analyses of Early Bronze Age silver objects determined to be made of silver extracted from cerussite and other oxidized lead ores, as well as Chalcolithic and Early Bronze Age objects made from silver ores. The evidence indicates that the earliest silver objects were smelted from silver ores, sometimes using nonargentiferous lead to purify the silver by cupellation.

These objects are characterized by low gold content and low iridium content, determined by neutron activation analysis (NAA), which has been used by other workers for silver analyses apart from Meyers. An overview of NAA concerned with the numerous investigations for the analyses of silver is given by Kuleff and Djingova (1990). NAA is probably not the best method for the analysis of silver because the concentrations of lead and bismuth cannot be measured and also because of the high background of radioactivity created in the silver as a result (Schwab 2010). When initially abundant silver ores were exhausted, argentiferous cerussite was the next choice, since it was already being used to produce lead for cupellation.

This material is characterized by a much higher gold content--0.25-1.0%, high lead levels, high bismuth levels, and higher iridium content. Subsequently, galena was used as the main source for silver. The overall picture is that as some resources were gradually depleted, others were substituted.

The chlorides and sulphides of silver are easily reduced back to metal with a mix of charcoal and a low temperature furnace or crucible. However, these ores are rare, and most silver was obtained by cupellation (Craddock 1995: 230). The elements which accompany the silver as a result of cupellation, seen by the author in numerous XRF analyses, are typically 1-3% gold, 0.5-2% lead, 0.2% bismuth and a little iron. These elements can be determined by non-destructive XRF analysis, a useful tool for silver analysis as a result, especially in preliminary screening before more detailed studies.

Pure silver is very soft, too soft to be practicable for many purposes, and from an early stage, metalsmiths learnt to alloy some copper with their extracted silver to produce a harder product. 3-4% of copper is a common addition in the Old World, and this continued to the production of "Sterling" silver which usually has 8% copper content. In Prehispanic South America, a whole range of silver-copper alloys were produced, and were used for a variety of artefact types. By adding more copper, silver alloys could be debased, which could later be surface-enriched in silver by pickling in plant juices or mineral salts, or used as solders with lower melting points than the two constituent components, often for joining silver or gold components together.

Occasionally, artefacts made of solid, cast silver are found, such as the silver statuette of a naked female dating to the 26th Dynasty in the collections of the Metropolitan Museum (Becker and Schorsch 1994).

Figure 6.6. The silver-copper equilibrium diagram

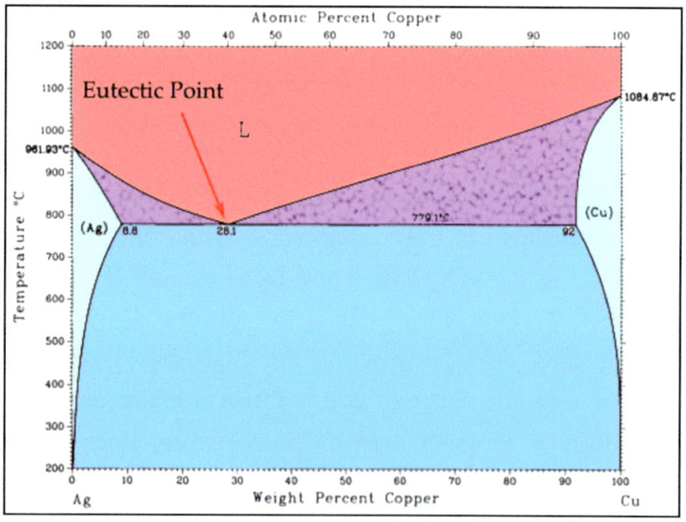

Pure silver for such a use is rare in solid castings, as it is very expensive. One way to economise is to use silver-copper alloys, or coat the underlying metal with silver. One of the most common types of equilibrium diagrams is that which includes a eutectic, a mixture of two phases which have their minimum melting temperature at the eutectic point. The silver-copper system is shown in Figure 6.6. The pink area shows the region where the mixture is liquid, and the mottled blue regions, the areas where the alloy is a mixture of liquid and solid phases. The light blue region is a mixture of two solid phases, the alpha phase, which is silver-rich and the beta phase, which is copper-rich. The two light blue areas designated as (Ag) and (Cu) refer to the α phase and the β phase respectively. At the eutectic point, the molten alloy passes directly from liquid to solid, at 71.9% silver, 28.1% copper.

Notice that the melting point of the eutectic is much lower than the melting point of either of the pure constituents, at 779 °C. This is of great practical importance, since silver artefacts can then be soldered together with silver-rich alloys without the danger of the silver artefact itself becoming molten. Various proportions of copper and silver for this kind of operation are recommended by classical recipes, such as those described by the Renaissance craftsman, Benvenuto Cellini (1967:93):

"...the solder I used was *ottavo*, that is a solder composed of one-eighth part of an ounce of copper to one of silver.......I bound the two legs to the body. Then I laid it on the wall over a good fire, and applied *quinto* to it, i.e., solder composed of one fifth of an ounce of copper to one of silver. As I worked with eleven-and-a-half silver to half copper I had nothing to fear as far as the latter was concerned and I would have everyone aware that if he wish to make his job succeed he must not employ inferior silver..."

The silver statue Cellini was working on was therefore cast in an alloy of 99.6% silver, 0.35% copper. This is practically pure silver with a melting point of 961 °C. The *ottavo* solder has a composition of 91.1% silver, 9.9% copper, with a liquidus of 880 °C, while the quinto solder, used for heavier attachments, has a composition of 86.1% silver, 13.9% copper, and has a liquidus at 845 °C. All of these alloys will precipitate silver-rich phases first during solidification, and the phase diagram shows that the *quinto* alloy melts at a temperature over 100 °C less than that of the silver Cellini used for the statue itself.
The microstructure of these alloys, all greater than 71.9% of silver consist of grains of α with an infill of the eutectic, while alloys with less than 71.9% silver will consist of grains of β with an infill of the eutectic. The eutectic is usually precipitated as a fine mixture of one phase in another, so the appearance of the eutectic ($\alpha + \beta$) is a fingerprint-like matrix. Phase diagrams, such as the silver-copper diagram shown in Figure 6.6, often do not label the phase regions, which with more complex alloy systems can become very annoying. The proper labelling of the silver-copper system is shown in Figure 6.7, below. Notice that this diagram is drawn as the mirror image of the diagram above. Since these eutectic diagrams are roughly symmetrical in shape, there is a tendency for no conventional representation of them.

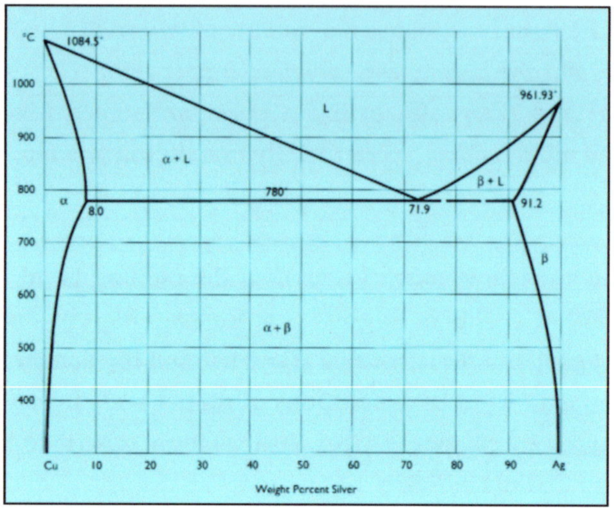

Figure 6.7. Another version of the silver-copper phase diagram. Here the phases are properly labelled and represented as a mirror image of Figure 6.6. After Butts and Coxe (1967).

It does not matter which way round they are shown, except that the designation of the alpha and beta phase are also reversed. In more complex diagrams this situation is anomalous: we always draw and represent the copper-tin system with copper at the left hand side of the diagram and tin at the right. As a map of a complex phase space, it becomes incomprehensible if viewed as a mirror-image, but there is no fixed convention for the silver-copper alloy diagram, or the lead-tin diagram which is a similar eutectic type, and they are illustrated in either orientation in many different textbooks.

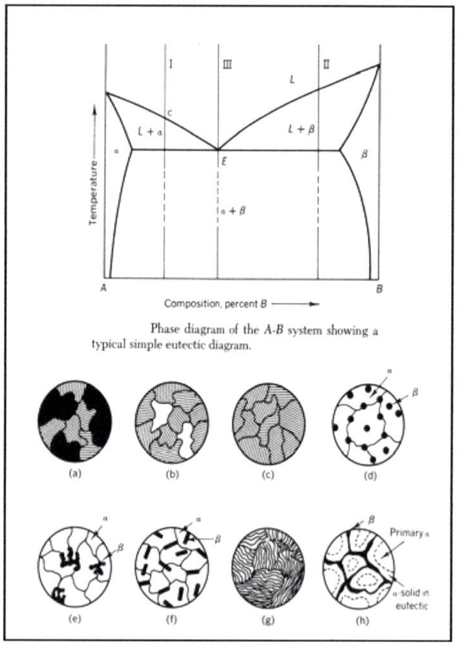

Figure 6.8 Some different types of microstructure in the eutectic system. After Smith (1969, Figure 7-13).

We can examine a few features of these eutectic diagrams as they relates to the morphology of cast alloys, at equilibrium, as shown in Figure 6.8.

Alloy I is called a *hypoeutectic* alloy; one having less of component B than the eutectic. If cooled under equilibrium conditions, at point c on the diagram, solid begins to precipitate. This solid is a primary solid solution of B in A. As cooling proceeds, the composition of the solid moves along the solidus line and the liquid cools until it reaches the point E at which it breaks down into the two different phases of alpha + beta as a fine eutectic mixture. This is shown in drawing (a). Similarly, if an alloy of composition II cools down, then grains of beta phase will form and the remaining infill will consist of the eutectic, as shown in drawing (b). This alloy can be referred to as *hypereutectic*, as it contains an excess of component B.

Alloy III, corresponds exactly to the eutectic point and will pass directly from liquid to solid, forming a fine intermixture of the eutectic, as shown in drawing (c). If the eutectic composition is close to the composition of one of the phases making up the eutectic, then a variety of different microstructures are possible. These structures are the nodular, often spheroidal, structures, shown in drawing (d), the irregular precipitates of drawing (e), acicular precipitates of drawing (f), coarse lamellae of drawing (g) and the divorced eutectic of drawing (h). In the divorced eutectic, there is no distinct boundary between the primary solid solution and the portion of the same solid solution in the eutectic. There is a tendency for the two phases to separate into large regions so that there is considerable divergence from a normal eutectic structure.

The silver-copper alloys are often heavily segregated, discussed below, in which snow-flake like crystals called dendrites begin to precipitate out between the liquidus and solidus regions, and these may remain fixed in that state and composition when the alloy cools down. The consequence of this on the morphology of an alloy with about 40% silver, 60% copper is shown in Figure 6.9. In drawing (a) the dendritic structure dominates, while the equilibrium structure for the same composition is shown in drawing (b).

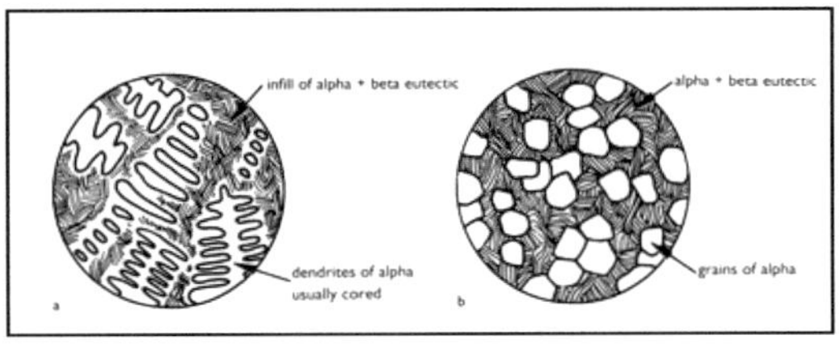

Figure 6.9. The dendritic and equilibrium microstructures for an alloy with about 40% silver, 60% copper. After Scott (1991). Equilibrium is less commonly found in these alloys.

With alloys containing about 5% of silver 95% copper, the phase diagram shows that the alpha grains of copper-rich phase should form with an infill of the beta phase between the grains. This is illustrated in Figure 6.10(b). In drawing 6.10(a) the same composition in the non-equilibrium state would be represented by copper-rich dendrites with some beta films or an infill of a little of the alpha + beta eutectic. The eutectic itself, representing the alloy with 79% silver 21% copper tends to be formed in the same way regardless of whether the conditions are fully equilibrated or not, and is shown in drawing 6.10 (c).

Figure 6.10. Three different microstructures of binary eutectic type. After Scott (1991).

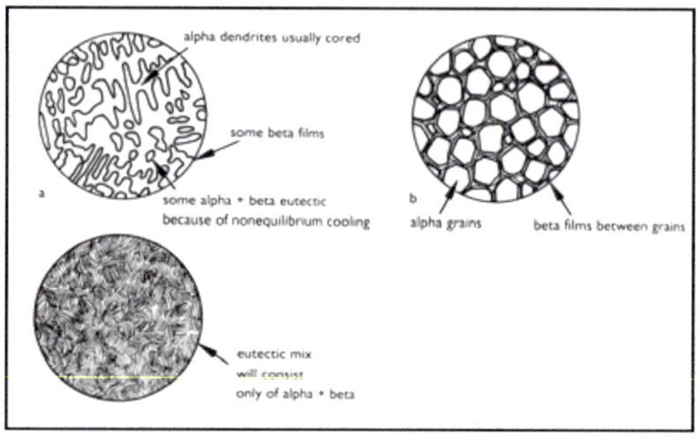

If the alloy composition is about 20% silver 80% copper, then the usual cast morphology of the alloys is that shown in Figure 6.11 on the left. The dendrites, which may be either copper-rich or silver-rich depending on which side of the eutectic point the composition falls, form with an infill of the eutectic and on hammering and working, the dendrites become flattened and elongated along the length of the artefact as dendritic stringers, with an infill of the eutectic between them as shown in Figure 6.11 on the right.

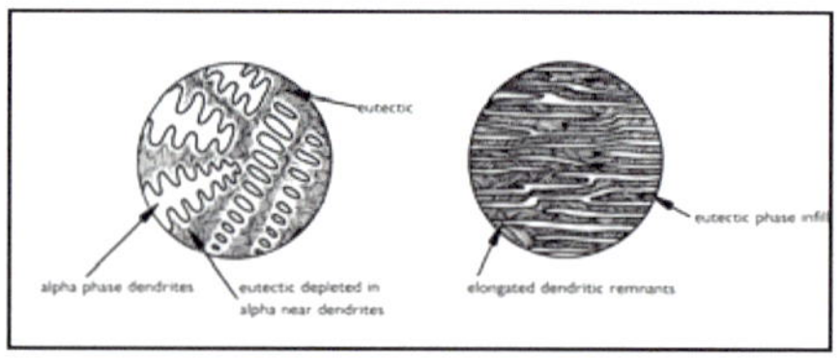

Figure 6.11. Some typical cast and worked morphologies for an alloy with about 20% silver 80% copper. After Scott (1991).

Some Properties of the silver-copper system

Microhardness

Figure 6.12 Conversion chart for Vickers and Brinell hardness values. From http://www.gordonengland.co.uk (2010).

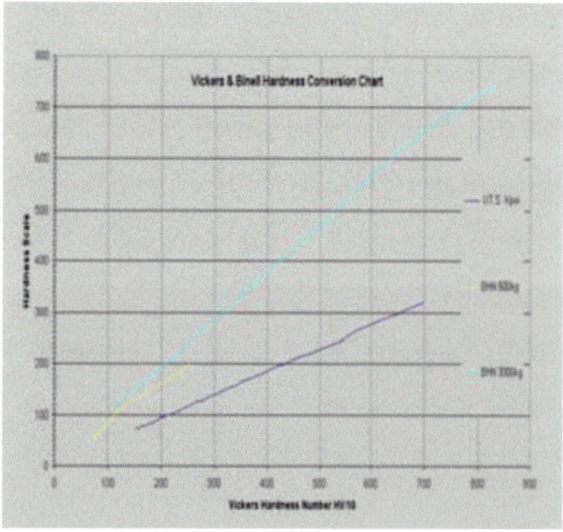

The hardness of metals cannot be determined with any accuracy using the Moh's hardness scale. Instead a variety of other scales have been devised. Three common measures are the Brinell hardness, HB, the Vickers Diamond Pyramid Hardness, HV, and various scales of the Rockwell hardness, HR, the latter being more common in industrial practice rather than laboratory studies. While the hardness, obtained on a microscale using tiny indenters and microscopes to measure the size of the indentation, do not give large differences between the Brinell and Vickers numbers, there are systematic variations in the two numbers which are addressed in the comparison charts shown in Figures 6.12, 6.13 and 6.14.

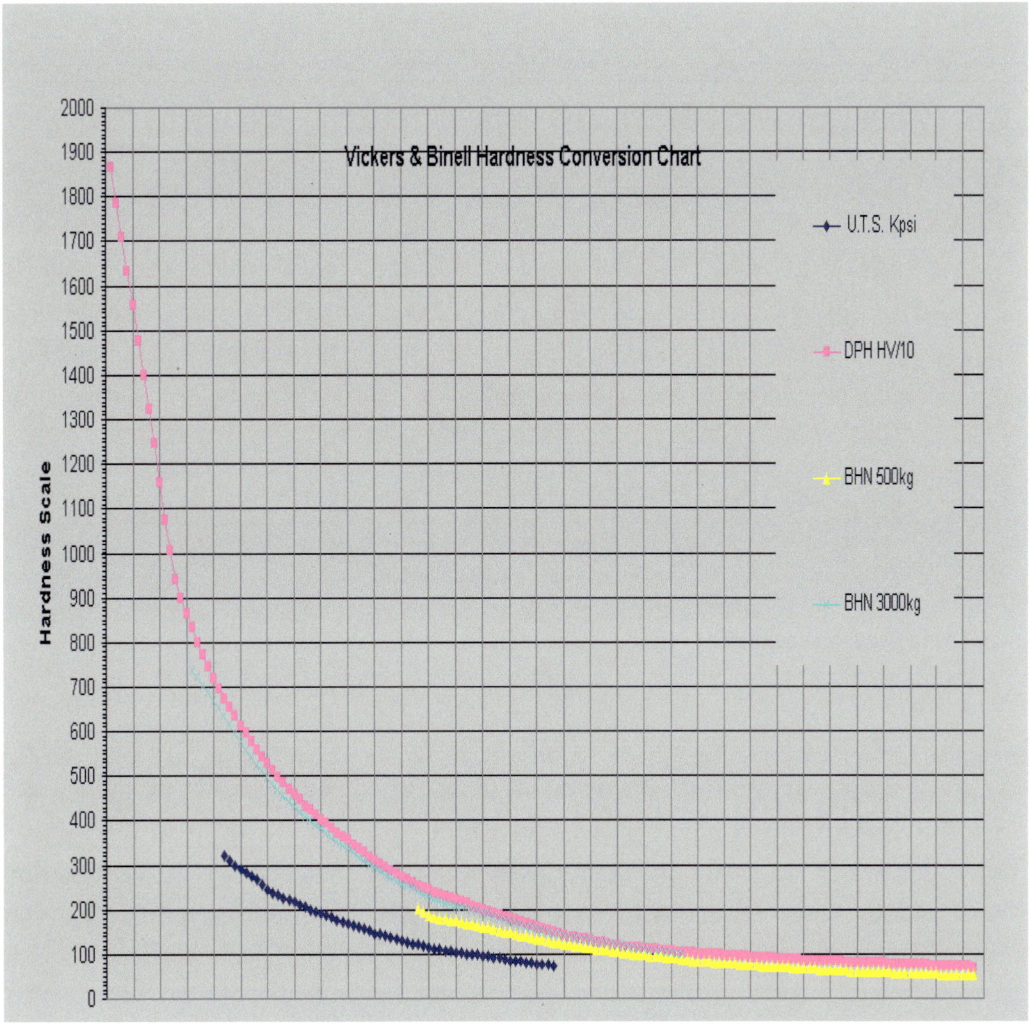

Figure 6.13 Vickers and Brinell Hardness Conversion Chart. From http://gordonengland.co.uk. (2010).

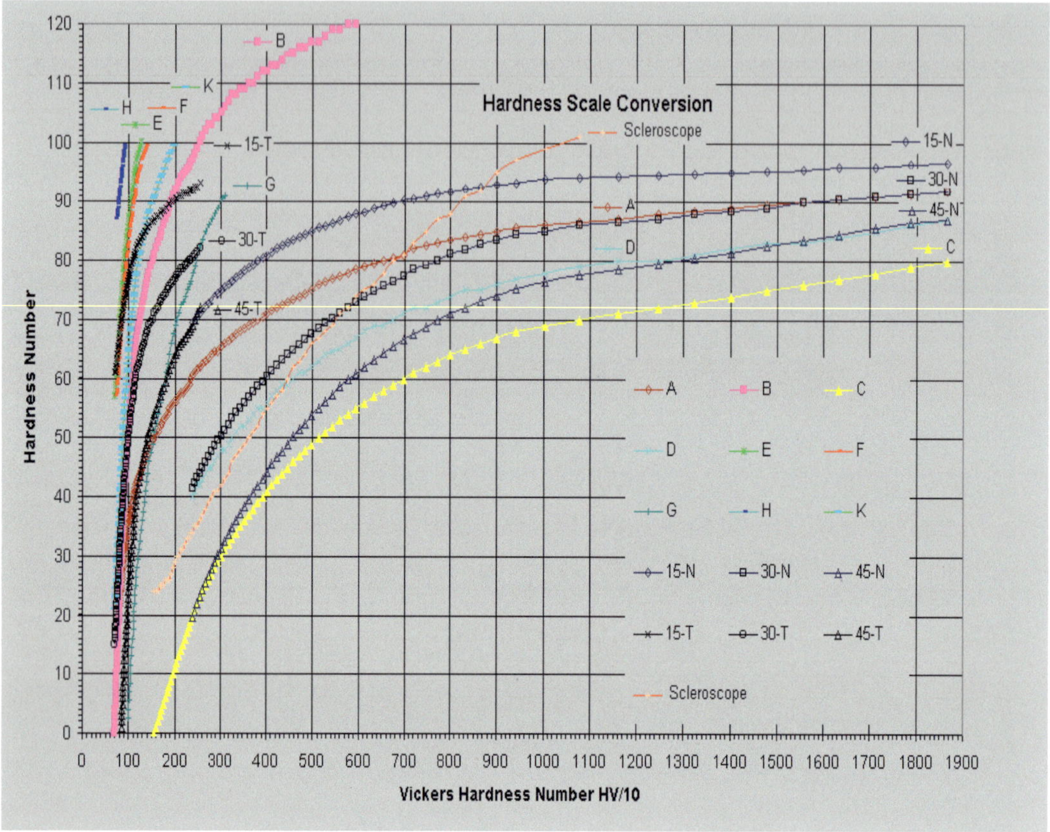

Figure 6.14 Comparison of the Vickers Hardness with 10 gram loading compared with Hardness values in a number of different scales. For example, a Vickers Hardness of 110Hv gives a BN of about 70. From http://www.gordonengland.co.uk. (2010)

These charts will be useful in further comparison of data related to microhardness discussed later in the volume or given as data for individual alloying systems. Hardness is usually inversely proportional to brittleness.

As alloys are cold-worked, dislocations pile up and the ability of the alloy to be cold-worked further drops. The hardness increases until the alloy may become too brittle to actually be used in practice without cracking. Annealing of the alloy will reform the grain structure and allow the dislocation density to be relieved.

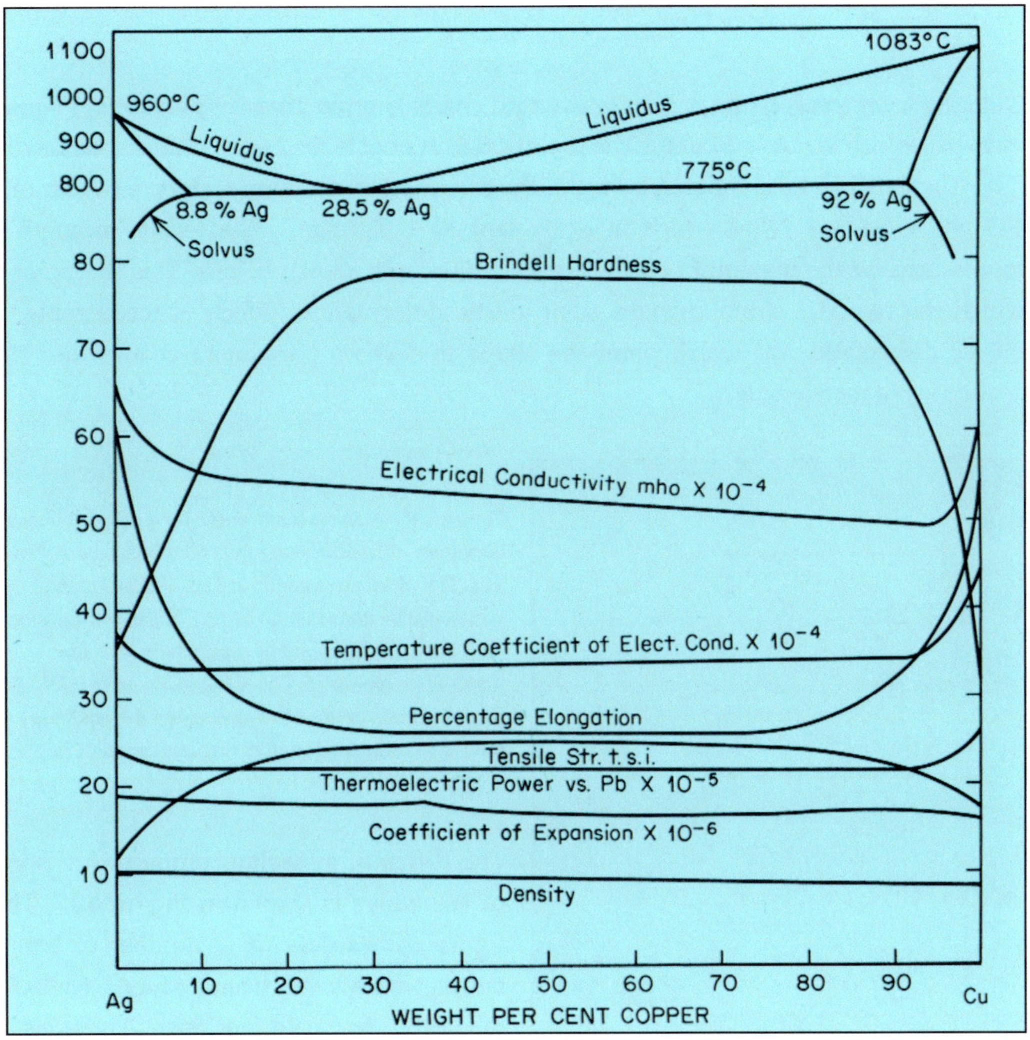

Figure 6.15. Variations of properties with composition in the silver-copper system. After Rhines (1956).

Figure 6.15 details a number of physical properties of the silver-copper alloys, of which the most relevant here are the Brinell hardness, and the percentage elongation. The hardness of the alloys in the unworked state reach values of 80 Brinell, particularly in the region of the phase diagram between 28% and 87% silver. At the regions towards the copper rich or silver rich end of the diagram, the hardness values fall away rapidly to that of pure silver at 32 Brinell and pure copper at 37 Brinell. The percentage elongation falls as the hardness of the alloy increases, from around 60% at pure copper to about 28% in the duplex area. If silver-copper alloys are hammered and annealed, followed by further hammering, then the microhardness will be much greater than that shown on this diagram. Values of 180 Hv or greater can be achieved depending on the amount of cold-working.

Old data from stress-strain or tensile strength charts is often given in pounds per square inch, psi, which is not an SI unit. The preferred unit now is the Pascal (Pa) or the Newton (N). The data can be compared by using the following conversion factors, 1 N/sq.mm. = 145 psi, 1 Pa = 1 N/m^2 which is equivalent t0 1 $Kg/m.s^2$. The tensile strength is represented by the maximum on the stress-strain curve shown below. It is the point at which the material strain changes from elastic deformation, which is recoverable, to plastic deformation, at which point the shape undergoes permanent change and this change is not recoverable.

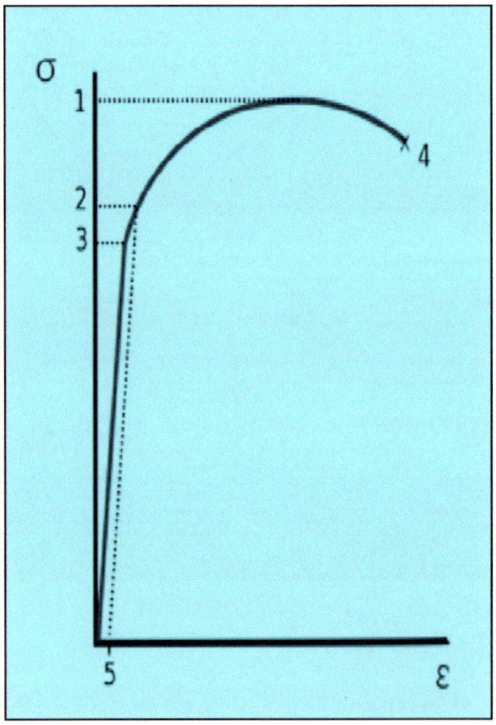

Figure 6.16. A stress strain curve for a soft metal such as copper. The different points on this chart are shown as 1. The ultimate strength. 2. The yield strength at which plastic deformation begins. 3. The proportional limit stress. 4. The point of rupture at which the material breaks and 5. The offset strain often 0.2%. For further details of this kind of diagram the reader is referred to standard engineering texts such as Samans (1963).

The different annealing properties of some of the alloys is shown in Figure 6.17. This figure shows data for a number of binary alloys which have been reduced to 49.7% thickness by cold-working. There is a different response to annealing depending on the composition.

Alloys with 95% silver, 5% copper only reach a tensile strength of $50x10^3$ psi (pounds per square inch), while those of 72% silver 28% copper reach $74x10^3$ psi.

The increase in tensile strength of some of the alloys when the annealing temperature exceeds 593 °C (1100 °F) is of interest. This is due to age-hardening which occurs when these alloys are air-cooled from above 593 °C. Wires of the silver-copper eutectic alloy when drawn to a 99.5% reduction in area, can reach a tensile strength of 130 x 10^3 psi (Butts and Coxe 1967: 254).

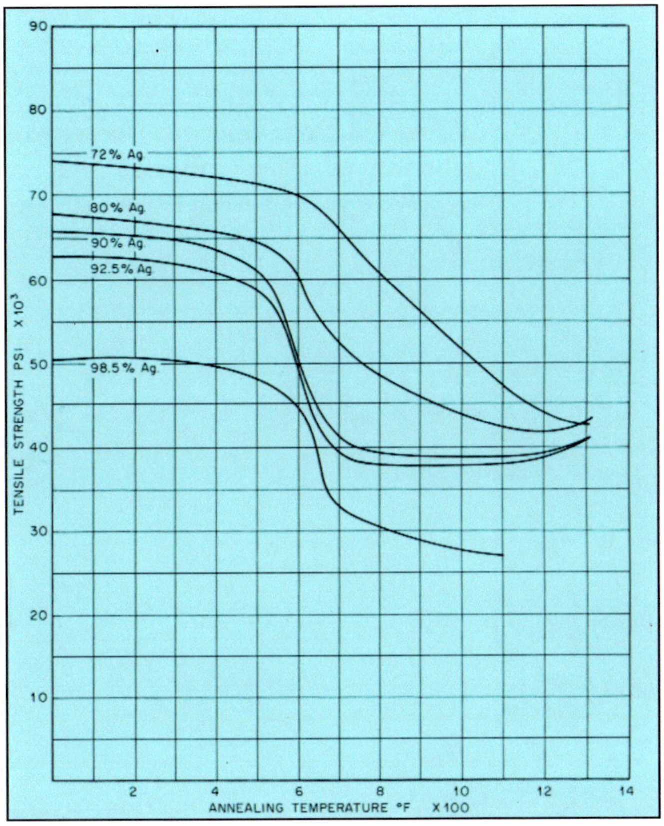

Figure 6.17. Silver-copper alloys cold-worked to 49.7% of original thickness annealed at different temperatures show a variety of responses depending on annealing temperature. After Butts and Coxe (1967, Figure 18-13).

Gordon and Knopf (2007) carried out some investigations of the mechanical properties of silver-copper alloys in the cast condition to replicate studies of ancient alloys from the site of Machu Picchu, excavated in 1912 by Hiram Bingham, in which several tupu pins were found to be made of silver-copper alloys with about 68.6% copper, 29.4% silver, while a head band was made in 74.7% copper, 24.5% silver. Three laboratory alloys, S (25.5% Ag 74.7% Cu 0.5% Sn); M (29.2% Ag 68.3% Cu 2.2%Sn) and E (72% Ag 28% Cu) were examined. Microhardness measurements of alloy E under different cooling rates showed those with large eutectic lamellae to have a hardness of Hv 66. With a spacing four times smaller, the hardness went up to Hv 89 and with barely resolvable lamellae, the Hv was 130. Tensile tests were carried out on these alloys, all breaking with cup-and-cone fracture, typical for FCC metals in ductile condition. A stress-strain curve for alloys S and E compared with pure copper is shown in Figure 6.18. They break with 30% strain and lack the extended deformation of pure copper. The tensile strength of alloy E was 391 MPa; M was 296 MPa and S, 308MPa. Pure copper gave 222MPa (Gordon and Knopf 2007: 43). The silver-copper alloys are strengthened by the silver-copper eutectic or beta phase precipitation.

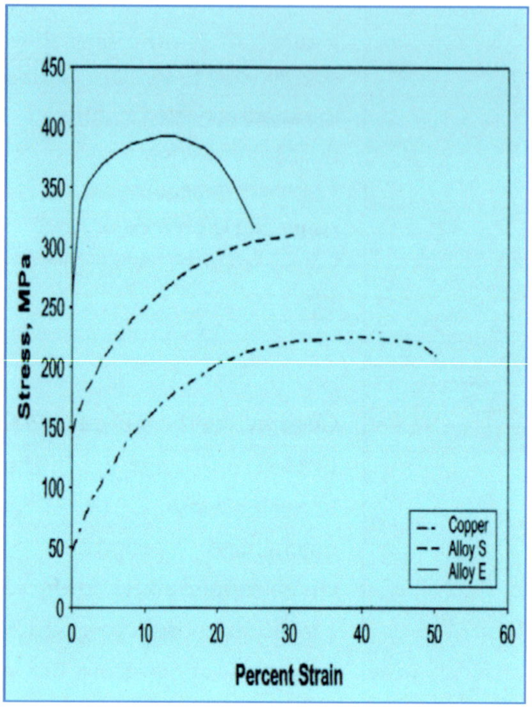

Figure 6.18 Stress-strain curve for samples of silver-copper alloys compared with pure copper. After Gordon and Knopf (2007:Figure 8)

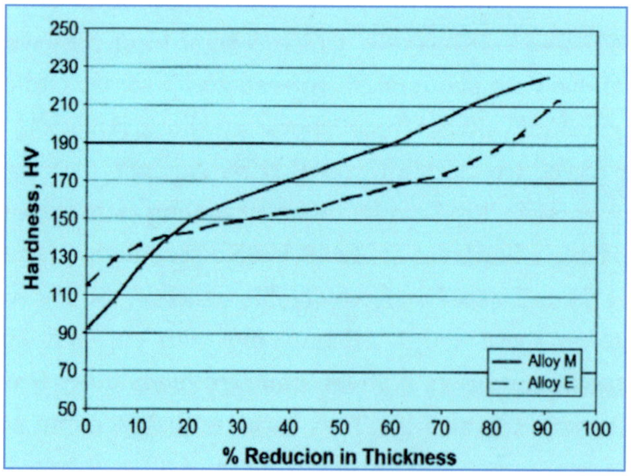

Figure 6.19 The microhardness values of silver-copper alloys M and E compared with extent of reduction in thickness. After Gordon and Knopf (2007: Figure 9). Hardnesses in the range of 200 Hv can be obtained with 80% reduction in thickness.

Alloys M and S had a lower initial hardness than B or E, but harden more rapidly than B or E. Alloys M and E, shown in Figure 6.19, attain hardness values of about 200 Hv after 75% reduction in thickness. The experiments showed that hammered shawl pin stems with 80% silver could have been shaped without annealing. Mathewson (1915) concluded that holes in the shawl pins were cast, suggesting no need for further working. Heavily elongated artefacts could have been annealed between 500-600 °C at least three times to shape the artefact. This work shows that the silver-copper alloys achieved a significant increase in hardness values if cold-worked. Pickling of the surface in acidic plant juices will easily dissolve some of the copper from the surface, resulting in silver-enriched surface layers, effectively a form of *depletion silvering*.

Deviations from Ideality

The phase diagram is an equilibrium system. It can only predict the phases which will form at equilibrium. If there is some reason why this ideal state cannot be reached, then the actual microstructure which is observed in ancient artefacts will be potentially different from that predicted by the phase diagram. The phase diagram is however, the essential skeleton on which the details of the appearance of the system hang, even if the ancient alloys in question are far removed from an equilibrated state. One of the major reasons why phase diagram predictions may be invalidated by ancient silver-copper alloys is because of *segregation*. Not only silver-copper alloys: many alloy systems are heavily segregated, such as copper-arsenic, copper-tin, gold-copper and lead-tin.

Segregation

There are several forms of segregation. The most critical for copper alloys is *dendritic segregation*. As the liquid cools through the region where the first solid begins to form, small snowflake-like particles of solid begin to solidify. These are the dendrites. Another common form of segregation produces *columnar* grains, long radiating crystals which are usually seen in *chill* castings, that is castings which have been made into an unheated mould. The columnar grains of chill castings are not common in ancient cast materials, because of cooling slowly in ceramic moulds or the presence of substantial alloying elements such as arsenic or tin, which result in the solidification of higher-temperature phases first of all, and tin-rich or arsenic-rich liquid remains which then fills in the spaces between the dendritic crystals. If alloys with low amounts of these alloying elements are annealed, then equi-axed grains can be produced as the dendritic segregation is removed as shown in Figures 6.20 and 6.21.

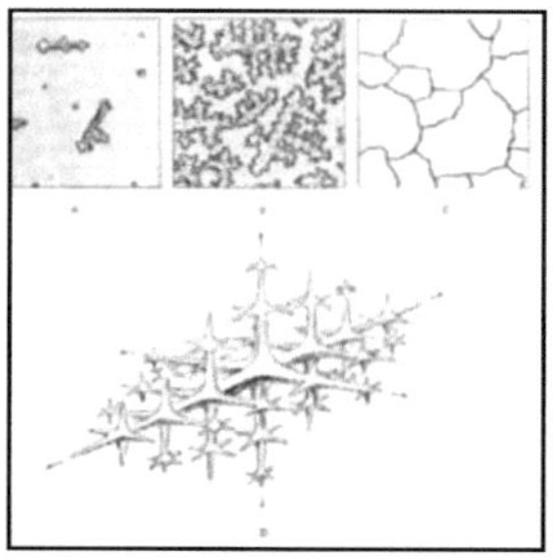

Figure 6.20: The shape of dendritic growth in copper alloys. In most ancient alloys we get dendritic growth in the cast state. This results in very uneven distribution of alloying elements, far out of equilibrium as assumed by the phase diagrams. If these dendrites are annealed, we would get equi-axed grains as seen in the top right hand drawing. Coring in the grains is a remnant of this dendritic segregation, which is a common feature in ancient copper alloys.

Dendrites may continue to grow until most of the liquid has been consumed. Typical dendrites are shown in the drawings of Figure 6.20. If this cast structure is deformed by hammering, then the dendrites may become distorted, as in Figure 6.21b. Full annealing of an alloy with a small amount of another component may homogenize the resulting dendritic structure and produce a collection of equiaxed, hexagonal grains, shown in Figure 6.21c. This state represents the fully-homogenized equilibrium state, assuming that there is no second-phase which has to form as a result of the solid solubility limits for the system concerned. When these *equiaxed* grains are cold-worked, then the grains become flattened as illustrated in Figure 6.21d. Annealing will reform the original grains, but in metals belonging to the face-centered cubic system (*FCC*), such as gold, silver and copper, the crystals reform by a *twinning* operation, and the resulting crystals are called *twinned grains*, or twinned crystals. These are shown in Figure 6.21e.

Figure 6.21. The sketch shows the dendrites as initially formed (a). Distorted dendrites as a result of cold-working (b). The completely homogenized equiaxed grains (c). Grains deformed by hammering (d). Grains which are twinned by recrystallization by annealing of the cold-worked state (e). Grains which are subsequently cold-worked and hammered, showing deformed twin lines and strain lines in the grains. After Scott (1991).

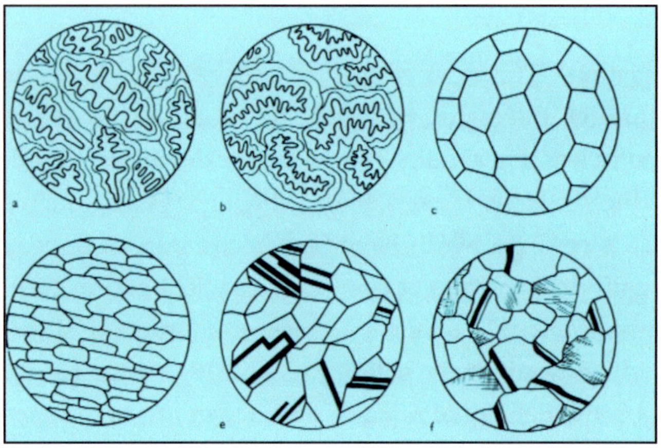

If these twinned grains are then cold-worked, the grains become flattened, the twins become distorted or bent and *strain lines* may appear in the structure. These strain lines are an indication that heavy or localized cold-working of the structure has taken place and a typical result is shown in Figure 6.21f. If this metal is then fully annealed, the twin crystals will re-form and the *strain lines* will disappear, resulting in a structure similar to Figure 6.21e will be generated. If the process of annealing is carried out very carefully, then *grain refinement* may occur and the grain size of the successively worked and annealed metal will become smaller. If over-annealed, *grain growth* can occur and the grains will increase in size, weakening the entire structure of the artefact. In FCC metals, it is not possible to get back again to the structure shown in Figure 6.21c, unless the alloy is essentially re-melted. Strain lines are visible on etching as a fine collection of parallel lines or criss-crossing lines through the grains and are sometimes outlined by corrosion, because the energy stored in the crystal results in preferential corrosion along these lines of weakness. Annealing twins do not occur in other crystal lattice types, such as the body-centered cubic metals (*BCC*), the most notable example being that of iron and its alloys.

Etching can reveal the twinned grains as well as strain lines in the crystals. A number of different etchants for use with copper and silver alloys is given in Appendix I in this book. Most of these etchants reveal the shape and detail of the grains by selectively etching the polished surface. Some etchants deposit a coloured film on the surface of the polished sample and reveal detail by selective deposition as well as by surface etching.

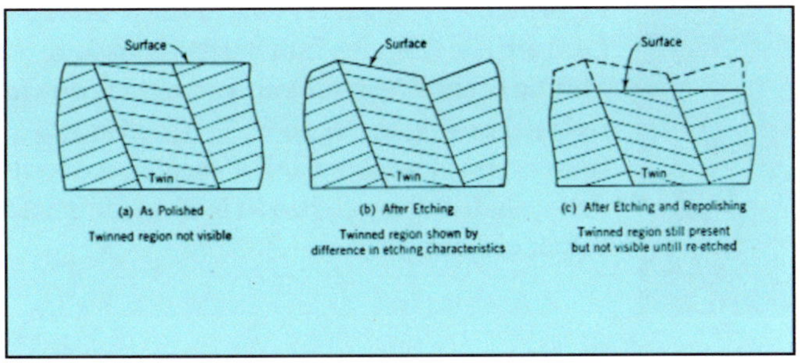

Figure 6.22. Twinned crystals are not evident in the as-polished state, but etching preferentially attacks different crystal faces, making the twinned crystals visible. After Samans (1963).

Some twinned crystal microstructures are shown, after etching, in Figures 6.23 and 6.24. Figure 6.23 shows the well-formed and twinned recrystallized grains in a Luristan axe with about 6% tin, revealing a background to the twinned grains of some segregation from the initial cast ingot from which the axe was shaped. The degree of working and annealing was not extensive, so the grains are typically large and the definition clear. One of the unusual phenomena associated with copper is that when the metal is deposited from solution, the grains are twinned. Why copper deposition from solution should be twinned is not explained in standard metallurgical textbooks. We assume that this represents the minimum energy requirement for the precipitate to form. There are four occurrences of twinned grains in FCC metals:

(1) Twinned grains formed as a result of working and annealing to shape. As the hammered crystal stores energy, the temperature at which it will anneal, or reform into new grains, becomes lower. With many cycles of working and annealing there is a tendency for the grain size of twinned copper alloys to become smaller.

(2) Twinned grains formed as a result of hot-working. In this high-temperature process, the grains are annealed and deformed and annealed at the same time, resulting in twinned grains. Some alloys cannot be worked hot in this way and are then described as "hot short"

(3) Twinned grains formed during natural corrosion processes in which copper may be redeposited on the surface or in the interstices of an object as a result of variations in the

oxidation-reduction potential of the environment, when copper deposited from these solutions will be twinned.

(4) Twinned grains formed as a result of deposition in the formation of an electrotype. A process often used, especially in the past, to produce a very exact copy of an original form. The grains may also have a *columnar* aspect. *Columnar* grains often form at the sides of moulds in casting an alloy, but are also seen, together with twinned grains in some copper objects, created by electrolytic deposition of copper from solution.

Figure 6.23 Twinned crystals in a low-tin bronze Luristan axe from the Iran Bastian Museum, Middle Bronze Age. Colour variation is due to residual casting segregation. Twinned crystals from cycles of working and annealing are well delineated. Colour etched in lead thiosulphate colour etchant, x280.

Figure 6.24. An electrotype copy of a decorative head, added to a Roman mirror case in the collection of the J. Paul Getty Museum, by restorers working at the British Museum in the 1920's. (Podany and Scott 1997). The confection was not suspected until a metallographic sample of the head was taken, showing that it was made in pure copper, and with an electrotyped grain structure as shown here, etched in alcoholic ferric chloride , x 140. Note that the electrodeposited grains have a columnar aspect.

As the dendrites grow, the incremental growth of the solid may be of slightly different composition than the first solid which formed. This compositional variation is called *coring*. Coring may persist even through recrystallized grains formed by the annealing of such alloys later on. Segregation created by dendritic growth in many silver-copper alloys may be so severe and persistent that they dominate the microstructure of the alloys at the expense of grain development.

Inverse segregation can afflict many copper alloys. In this form of segregation, one of the components, usually the lower melting point phase is forced preferentially to the exterior of the casting. In bronze alloys, copper-tin binary mixtures, this can affect the distribution of the tin, resulting in cast surfaces which look as if they were tinned but which have formed, or could have formed, by inverse segregation of tin rather than a deliberately tinned surface.

Normal segregation results in the concentration of one constituent towards the centre of the casting and is less commonly encountered in archaeological material. The most common segregation is that of dendritic growth in copper alloys, the persistence of coring, and the variation in composition introduced by this form of segregation.

An example of a 60% silver 40% copper alloy in the cast state is shown in Figure 6.25. The first dendrites to form are copper-rich and these essentially do not change in composition from the first moment of solidification. The infill is the fine eutectic mix of alpha + beta. When this alloy is cold-worked by hammering and then annealed, the dendrites become flattened along the length of the section or sheet, but they do not disappear: the segregation persists through these stages of fabrication.

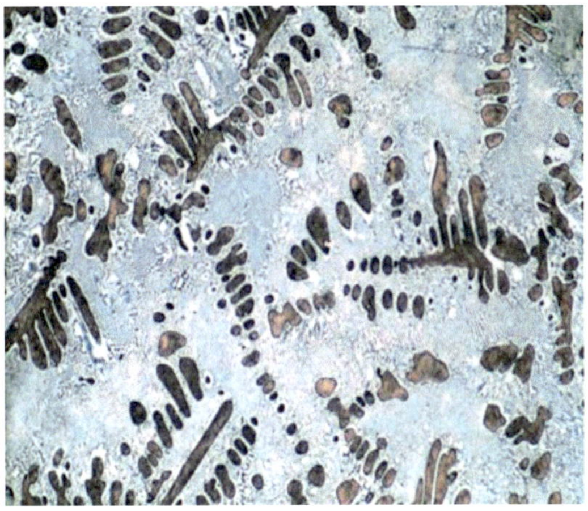

Figure 6.25. Photomicrograph of a cast laboratory ingot close to the eutectic composition. Slow movement towards equilibrium in this binary system results in a copper-rich dendritic precipitation with an infill of the eutectic mixture of the alpha and beta phase as a fine interlaced mixture. Etched in potassium dichromate, x 260.

A good example of the application of metallography to fabrication techniques is given by Pistofidis et al (2006), who examined three silver coins of the 3rd century BC, minted in Korkyra, Dyrrachion and by the Illyrian king Monounios. The coins have different chemical compositions and different microstructures. The Korkyra coin was probably hot-worked or cold-worked and annealed, then after cooling, cold-worked again. The coin of King Monounios is composed of equiaxed grains with numerous slip lines. Probably, this coin was cast and then cold-struck. Finally, the coin from Dyrrachion is characterized by small-sized grains and strain lines, and as a result, this coin was probably cold-struck from a well-worked silver sheet blank.

Examples of heavily segregated silver-copper alloys are common in ancient South America, because the silver alloys used there were often heavily debased and then surface enriched to remove copper or copper oxides, resulting in a silvery-coloured surface even on alloys containing as little as 10% of silver. In the alloy microsection shown below in Figure 6.26, which is a fragment of silver sheet from the site of La Compañia, Los Rios Province, Ecuador, from a 10th century chimney urn burial, the interdendritic silver-rich eutectic remnants are strung out and flattened like a fabric of netting along the length of the section, while the dendritic copper-rich areas have been almost totally corroded and replaced with cuprite. Corrosion in these debased silver-copper alloys is frequently severe, due to the electrochemical difference in potential between silver and copper. As they are in intimate contact and segregated, the tendency for the copper-rich phase to be corroded is high.

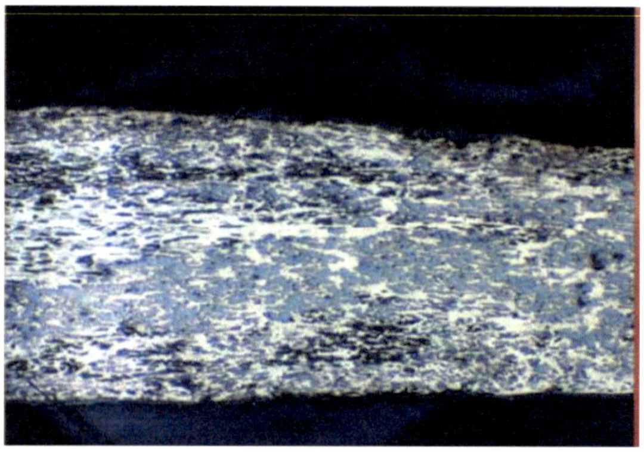

Figure 6.26. Silver-copper alloy ornament from the 10th century AD. La Compañia, Los Rios Province. A heavily debased silver-copper alloy with ribbons of the interdendritic silver-rich phase between the alpha dendrites. Despite working and annealing the dendritic segregation still persists. Unetched X54.

Another example from the La Compañia site is shown in Figure 6.27, part of a cast sessile blob used to make the stud component for the nose-ornament. The silver-rich component is the interdendritic fill around the copper-rich dendrites. The silver content here is about 30% silver 70% copper.

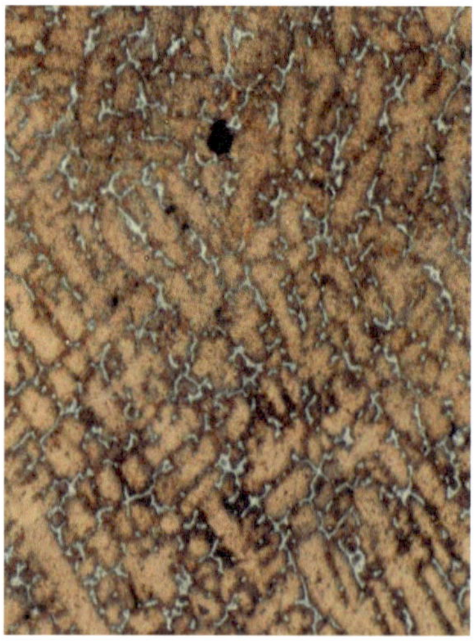

Figure 6.27. Part of the microstructure of a cast sessile blob used to make the stud of a silver-copper face ornament. La Compañia. Unetched, x94.

At lower silver concentrations, the proportion of the silver-rich phase becomes proportionally less as can be seen for the alloy of 10% silver, 90 % copper in Figure 6.28.

If we compare the morphology of the sessile drop shown in Figure 6.27, with that of Figure 6.26, the diminution in the proportion of the silver-rich phase is evident.

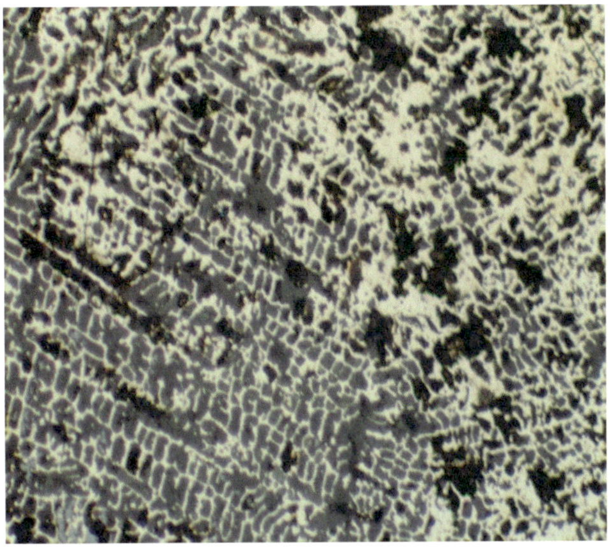

Figure 6.28. Cast alloy of 10Ag90Cu as cast. Etched in alcoholic ferric chloride, x95.

A laboratory prepared alloy of 20% silver, 80% copper is shown in Figure 6.29. Etching reveals some coring to the dendritic structure with a regular array of interdendritic silver phase.

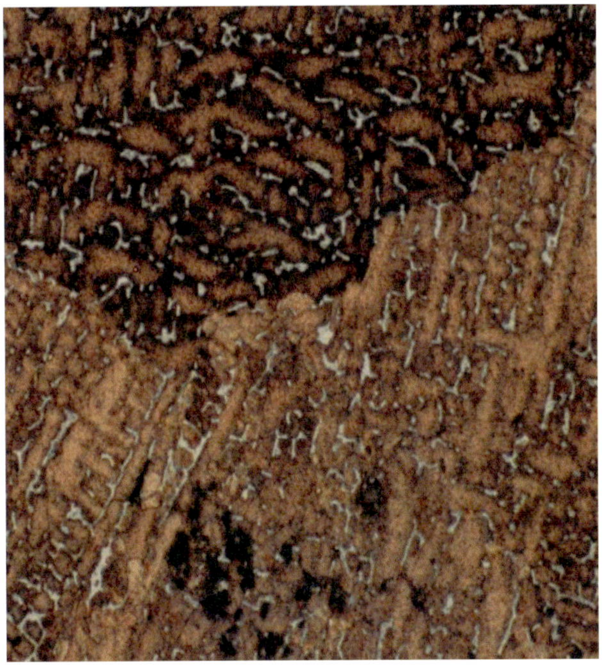

Figure 6.29. 20% silver 80% copper alloy as cast. Etched in alcoholic ferric chloride, x 150.

Figure 6.30. Cold-worked and annealed to about 50% reduction in thickness, the whiter stringers passing along the length are the flattened dendritic remnants of the cast structure. Etched in alcoholic ferric chloride, x120.

On cold-working and annealing of this 20% silver, 80% copper alloy, the whole structure becomes elongated and compressed along the direction of working, as shown in Figure 6.30.

Coating with silver

A considerable number of different technologies have been used in different cultures for coating of silver on to different substrates. Anheuser and Philip (2002) examined the silver plating on 3rd century AD Roman coinage based on a study of a hoard found at Rogiet, Monmouthshire. This bronze coinage had a very thin silver wash. The plating was rarely thicker than 1-2μm and often too thin to be visible in cross section. There was no eutectic layer or diffusion zone at the interface between plating and small amounts of silver phase within the copper core.

From a comparison of these results with replication specimens, it was concluded that the plating was applied in the form of a silvering paste, possibly based on silver chloride. Hot-dipping into molten silver chloride proved to be impractical in the replication experiments and was ruled out. No mercury was detected on any of the coins. A contrast is provided by Ma and Scott (2004), who examined gilding and silvering techniques in China. Amalgam gilded and amalgam silvered leaded bronze horse harnesses dated to the Western Han dynasty (221 BC-8 AD) were studied using the metallurgical microscope, x-ray fluorescence, x-ray diffraction, and environmental scanning electron microscopy (ESEM). Metallography showed that a sublayer of gilding was prepared on the leaded bronze surface before the silver layer was applied. Metallic compounds of $Ag_{0.36}Au_{0.52}Hg_{0.09}Cu_{0.2}$, (Ag,Au)Hg, $(Ag,Au)_3Hg_2$, and Au_3Hg were detected by XRD, showing that amalgam silvering and amalgam gilding were used in the Western Han dynasty.

Scott (1986) examined a number of gold and silver coated copper artefacts from Ecuador. Some small items such as needles and fish-hooks revealed that "fusion-gilding" or "wash-gilding" was employed in the manufacture. The coatings found included silver alloy coatings over copper and gold alloys over copper. The coatings are often themselves superficially enriched at the surface; they are thick, and completely different from the surfaces found on depletion-gilded objects which are far more common in ancient Colombia. The most extraordinary discovery is that reported by Lechtman (1979) for the Vicus and Moche metalwork from northern Peru. Silver was deposited by electrochemical replacement plating, resulting in thin films 0.5-2 microns thick being plated onto copper sheet.

Similar methods can be used in the laboratory by dissolving silver grains in an aqueous solution of sodium chloride, potassium nitrate and potassium aluminium sulphate. This early use of electrochemical replacement plating of precious metals, without the use of strong acids or batteries, remains one of the most underrated achievements of the ancient Peruvian cultures.

Almost since the invention of coinage, imitation silver coins have been manufactured by attaching a silver layer, with the same composition as the contemporary official coinage, onto the surface of a base metal core. In the case of copper cores, this process could be carried out, in Roman coinage, by coating the surface of the core with a layer of solder consisting of a silver-copper alloy, or by attaching a layer of silver foil to the surface, either by soldering or by heating to form a layer of eutectic at the interface of the silver and copper (Zwicker et al 1993). The latter technique is, in effect, a self-soldering process which is known today as Sheffield plating. In the case of iron cores, however, the self-soldering process could not be used, and silver plating had to be carried out either by using hard (i.e., silver-copper) solder alone or by soldering silver foil to the surface of the iron.

In the Roman Republican and early Imperial periods, a common technique of manufacturing silver plated forgeries was by the application of silver foil; after the debasement of the official silver coinage, which began at the end of the 1st century AD, plated coins with a high purity silver layer on the surface became rare. Instead, forgers used hard solder coatings for the blanks.

The introduction of Sheffield plate in the 18th century AD revolutionized the silver plating industry for it offered an effective, relatively cheap, supremely versatile and durable plating technique which could convincingly imitate solid silver. (Turner 1993), but after the introduction of electroplating, the Sheffield plate went into decline. The metallographic research reported here shows that both the electroplating of silver, and the coating of copper substrates with silver, were not new ideas: both had been utilized thousands of years before.

The silvering of lead coins used by fakers in the 3rd century BC was studied by Giovannelli et al (2006). Material characterization revealed a bilayer coating with a copper inner layer and a silver top layer. Scanning electron microscopy (SEM) morphology and x-ray diffraction (XRD) textures of the copper and silver layers indicated that thermal or mechanical procedures were not used to produce the coins, but that they exhibit typical features of electrochemical displacement processes. In the electrochemical replacement or displacement process, some of the substrate ions are dissolved in the media and are replaced by the deposited silver ions. Usually, such processes demand very clean substrates, free of oxide films or tarnish. The authors replicated a coating process with materials that are documented for the historical period of interest and that lead to the formation of a bilayer coating exhibiting essentially the same morphological and structural properties as the archaeological object.

The direct displacement coating of lead with silver has been shown not to be feasible, while the duplex coating could be achieved by sequential use of a cupric acetate solution, displacing lead, and of a silver solution obtained by dissolving silver chloride in dilute ammonia. The double displacement process yields a cauliflower-like agglomerate structure, showing close similarities with the morphology of residual white-metal coating of the coin. Copper and silver displacement layers exhibit a typical texture that cannot be reconciled with that obtained by thermal or mechanical plating techniques and that matches the texture found in the counterfeit coin. Karmazin et al (1996), carried out metallography and electron probe microanalysis of silver temple rings found in the Slavic cemetery at Musov, Czech Republic. Four different materials could be identified: (1) a copper core covered by a silver foil; (2) solid silver; (3) a hypoeutectic silver-copper alloy with a thin silver surface layer; and (4) a hypoeutectic or eutectic silver-copper alloy with a thick silver surface layer.

The wires were made by hammering, rolling between two plates, and drawing. To produce the silver layers in materials 2 and 3, the alloys were repeatedly heated and cleaned of the copper oxide with a chemical agent--an intentional precipitated corrosion of copper in the surface layers. The silver layer on material 4 was made by dipping a silver-copper wire into pure molten silver. The different materials can probably be attributed to craftsmen working in different areas, though the rings were found in the

same cemetery. Ingo et al (2006), investigated Greek and Roman silver and gold plated coins.
The metallurgical techniques used by Romans and Greeks for coating the copper core of coins with a thin or thick layer of gold or silver are studied by means of the combined use of scanning electron microscopy combined with energy dispersive spectrometry and metallography. The results indicated that several methods were used by Greek and Roman craftsmen, including the mechanical application of thin malleable gold or silver foils to be welded via thermal treatment.
Silver reuse is common, and apart from the determination of lead isotopic composition, which is expensive, it is sometimes possible to probe the origin of the silver with XRF. For example, the composition of some 6th century Kentish silver brooches was investigated by Leigh et al (1984). Twenty-five sixth century silver square-headed brooches were studied by XRF, from the cemeteries of Howletts in Kent and Chessel Down on the Isle of Wight. The results support the theory that the metal source might have been scrap Roman silver and brass, as well as melted down Germanic metalwork. Comparison between paired brooches suggested the casting of some pairs from a single batch of metal.

Coatings over Silver

One of the few common coatings over silver is, of course, gold. The malleability of gold permits it to be easily hammered into very thin sheets, and when pure, into extremely thin leaf (Oddy 1991). The beating of gold into foil was developed in the Bronze Age, the foil then being used to gild silver and copper by wrapping it around the base metal and folding over the edges. This technique was in use by about the beginning of the 3rd millennium BC. From about the middle of the 1st millennium BC, attachment of the foil was improved by inserting the edges into grooves cut into the surface of the base metal and then hammering them closed. Use of this technique continued for about five centuries.
Techniques for purifying gold were developed around 2000 BC in Mesopotamia and somewhat later in Egypt. The purer gold permitted preparation of thinner gold leaf, which was attached to surfaces probably by an organic adhesive, or by burnishing. This process is especially effective on a hot metal surface, and is called *hot cladding* or *diffusion bonding*. Leaf may also be attached by punching it into the substrate. From about the beginning of the 1st millennium to the 2nd or 3rd century AD, a period of experimentation took place in the Old World in which variations on essentially the same theme were employed. From about the 3rd century AD, fire-gilding, using an amalgam of mercury and gold, came into widespread use, although some examples are known

from the 1st century BC (Schwab 2010), and remained the method of choice in Europe and the East throughout the Middle Ages until it was finally replaced by the introduction of electroplating in the mid-19th century.

The gilding of Roman silver plate was investigated by Oddy (1988), who examined artefacts up to the Romano-Byzantine period. Foil gilding, leaf gilding, diffusion bonding, cold mercury gilding and fire gilding were all employed at different times, and will be discussed further in Volume II.

A study of the gilding of silver in Byzantium was carried out by Dandridge (2000). Byzantine goldsmiths used partial gilding extensively as a decorative embellishment on silver, but there have been few scientific analyses or technical descriptions of the methods utilized to adhere gold to the silver, and those that exist are focused exclusively on the Early Byzantine period defined as beginning with the founding of Constantinople in 324 AD and terminating with the end of iconoclasm in 843 AD. Two recent acquisitions by the Metropolitan Museum of Art provided an opportunity not only to review the pertinent literature, but to supplement the existing studies and extend the time frame covered in the literature into the later part of the Middle Byzantine era, 843-1261 AD.

The study of three samples from chalices of the Attarouthi Treasure (late 6th-early 7th century AD) and a processional cross of the first half of the 11th century AD showed different microstructures and different distributions of gold, silver, and mercury as shown by microprobe (EPMA) analysis. The gilding technique started with a mercury amalgamation of the silver surface or application of a gold-mercury amalgam to the silver surface, heating, and then a final application of pure gold leaf over the mercury gilding. A richer gold tone was therefore achieved.

Gilded silver from ancient Nubia, Sudan, in the collection of the Museum of Fine Arts, Boston, was examined by Gansicke et al (2000) to identify gilding materials and techniques. The material is primarily from the Twenty-Fifth Dynasty and Napatan Period (ca. 750-300 BC) to Meroitic times (300 BC-350 AD), with a few earlier pieces. The objects originate from the Harvard University Museum of Fine Arts expeditions in the Sudan in the early 20th century and form an important body of provenanced ancient gilded silver. Diffusion bonding was apparent on most pieces with varying composition and thickness of gilding layers and silver substrates. Gilding of silver in the ancient Sudan and perhaps the entire Nile Valley appears to be much more widespread and elaborate than previously assumed. Further discussion of this topic will be subsumed under gold, and discussed in Volume III.

Chapter Seven: Precipitation Phenomena & Silver Alloys

In many alloys, precipitation of different phases may occur due to a variety of factors. Two of the most common are termed *continuous precipitation* and *discontinuous precipitation*. In continuous precipitation, the growth of another phase occurs within the grains themselves. In discontinuous precipitation, a metastable solid solution is formed, in silver-copper alloys usually as a result of quenching or fast cooling of the alloy following working and annealing to shape, and this precipitate of copper is dormant within the grains of the silver-rich solid solution. Alloys in which precipitation can occur as a result of aging or heating are called, *age-hardening* alloys, or *precipitation-hardening* alloys. The silver-copper system is an example of an age-hardening system, which can be used industrially to harden certain alloy compositions.

Over long periods of time at room temperature, ancient alloys of silver and copper can undergo age-hardening (Schweitzer and Meyers 1978a). Some of the copper within the grains migrates to the grain boundaries and these become altered, ultimately producing a jig-saw appearance to the grain boundaries of the silver grains. A diagram of the morphology of the precipitates which should be formed is shown in Figure 7.1. (After Schweitzer and Meyers 1978b). Notice that here the silver-rich grains are labelled as the alpha phase, the opposite convention of the usual form of showing the phase diagram given in Figure 6.6.

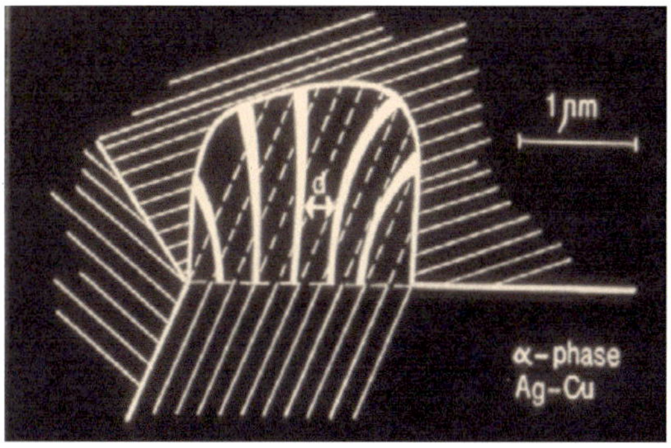

Figure 7.1. Precipitation in the silver-copper alloys. The copper precipitate is shown here as white within the grain boundary region where this change slowly occurs. After Schweitzer and Meyers (1978b).

The spacing, d, between the copper precipitates which grow from the grains should be indicative of the age of the growth in terms of the time which it takes them to form and develop. In practice, obtaining data of this kind from ancient artefacts is very difficult to do. A good example of the potential for useful data, from unpublished work carried out by Hazel Croome at the Metropolitan Museum of Art some twenty years ago is shown in Figure 7.2, at low magnification in which the space bar represents 37 microns.

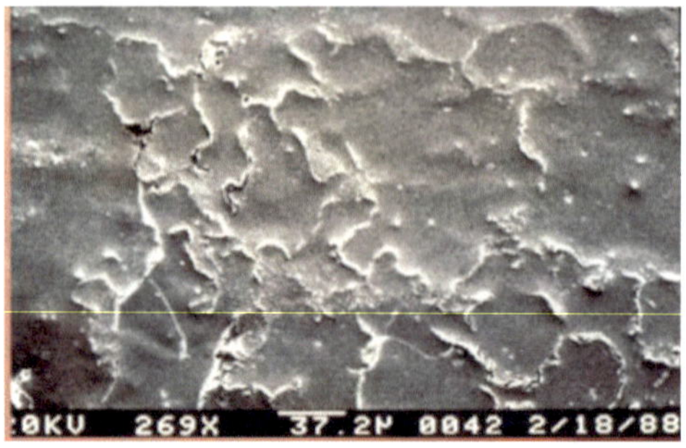

Figure 7.2. Altered grains in silver due to discontinuous precipitation of copper. Photomicrograph by Hazel Croome. Scale bar is 37.2 microns.

At higher magnification, where the bar represents 4.5 microns, the two-phase nature of the precipitate can be seen in Figure 7.3.

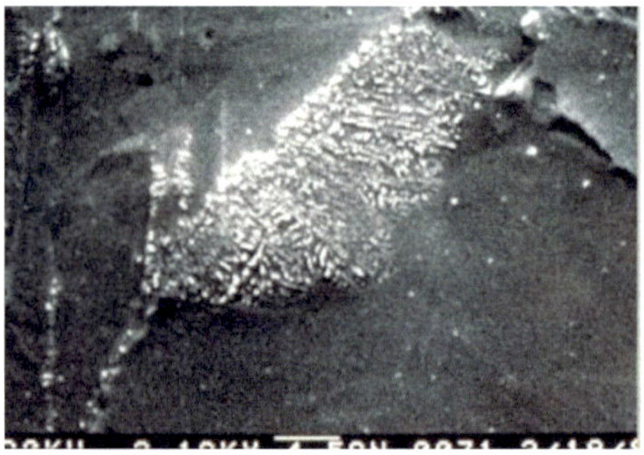

Figure 7.3. An enlarged view of the precipitation. Photomicrograph courtesy of Hazel Croome. Scale bar is 4.5 microns.

Finally the view below, in Figure 7.4, where the bar presents one micron, at 20,000x magnification, the individual lamellae of the copper precipitates can be seen and measured.

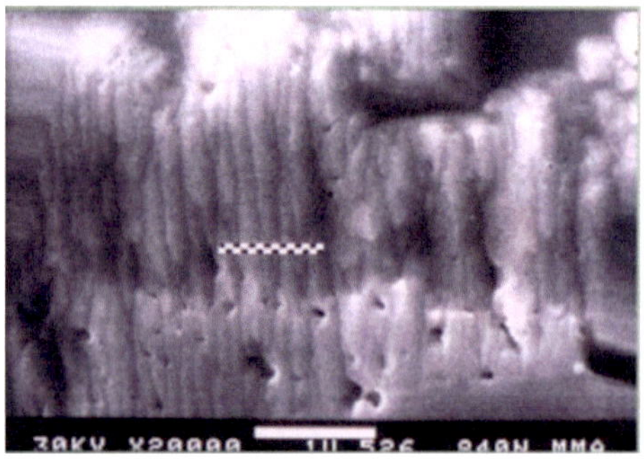

Figure 7.4. SEM view of the spacing between the silver and copper lamellae in the zone of discontinuous precipitation at the grain boundaries. Photomicrograph courtesy of Hazel Croome.

The age embrittlement of ancient silver could be due to discontinuous precipitation of copper from the supersaturated silver-rich silver-copper solid solution during the long sojourn at ambient temperatures. Microstructures of a number of ancient silver samples (500 BC to 600 AD) have been examined with the optical microscope and by microhardness measurements.

The microstructures are compared with the precipitation behavior of modern, artificially aged silver-copper alloys. Based on kinetic data the growth-rate of the precipitating cell at 25° C is estimated to be approximately 10 μm/year. The temperature-dependence of the interlameller distance of the segregated copper in the precipitated cell could be used to distinguish between ancient objects (aged at ambient temperatures) and artificially aged modern objects (aged at elevated temperatures). High quality silver artifacts which had belonged to ancient nomads (Sarmatians) that became embrittled as a result of burial conditions were studied.

The processes taking place during annealing in the course of conservation of the artifacts were also studied by Schweizer and Meyers (1978b). Metallography was the main research method of the study. Intercrystalline corrosion was shown to have resulted in brittleness. During annealing of ancient examples, partial recovery of ductility occurred due to coagulation of corrosion products and recrystallization.

A recent re-evaluation of the microstructural reasons for the very brittle nature of ancient silver has been undertaken by Wanhill (2003). The severity of the embrittlement is enhanced by large grain sizes and there is a synergistic effect between corrosion-induced and microstructurally-induced embrittlement. These microstructural-induced effects do not affect silver alloys with substantial copper contents. Pernicka et al (2007) investigated several different silver alloys dating from Roman and early Mediaeval sites, from Iron Age Italy, the Kurgan culture from Russia, and mediaeval Cambodia. Discontinuous precipitation was only observed once, within a sample of a Roman helmet. The SEM instrumentation used in this study was not new, so it was hard to check what kind of precipitations there were in the alloys. The only kind found were continuous lead precipitations (Schwab 2010).

As most ancient silver objects from the Old World have a composition of about 96% silver, 3% copper, 1% gold, with lesser amounts of lead, bismuth and iron, the grains are primarily of silver with metastable contents of copper: when the copper concentration is over about 8%, the alloy system becomes two-phased and the same kind of grain problems do not occur, instead general or galvanic corrosion is enhanced as the copper-rich phase is anodic to the silver-rich cathodic phase.

Difficulties are created in quantitative examination of ancient silver artefacts where the ability to take samples, and to make measurements of the discontinuous precipitation are infrequent. The measuring of the lamellae distances is therefore impossible in many cases. The jig-saw effects on the grain boundaries, when pronounced, is often taken as a proof of the age of the silver object under study, and this is frequently a reliable indicator. A good example is the proto-Elamite silver bucket shown in Figure 7.5. The bucket is 5,000 years old, has a silver surface which is practically uncorroded, is of unattractive shape, and has deeply incised symbolic writing which could not be read. The art historians therefore initially pronounced the bucket to be a forgery. But is it?

Figure 7.5. Proto-Elamite silver bucket. Probably found in a dry Iranian tomb. Photograph by courtesy of Pieter Meyers, Los Angeles County Museum of Art.

XRF analysis showed that the silver contained a small amount of lead, gold and copper, the first two elements intimately associated with the extraction of silver from lead ores by the process of cupellation. This is potentially useful in distinguishing old silver from new, but the crucial evidence is provided by a metallographic section taken from the edge of the bucket shown in Figure 7.6 below. The grain boundaries can clearly be observed as jagged from the discontinuous precipitation. There can be no doubt that this silver is ancient.

Figure 7.6. Microstructure of a proto-Elamite silver bucket dated to about 3500 BC, showing cracking through some of the grains, with occasional twinned crystals and discontinuous precipitation at grain boundaries which results in the meandering boundaries seen here. Etched in acidified potassium dichromate, x 270.

Art historical opinions were later to accord with the metallographic interpretation, as a visiting German scholar at LACMA could read one of the symbols as the name of a proto-Elamite king. Art historical opinion therefore went into reverse and declared the bucket to be authentic, although some archaeologists still have doubts about the veracity of the inscription and feel that the inscription could have been added later, but rarely is this the case for such instances as examined here. The bucket must have been sitting in a dry Iranian tomb for 5,000 years, accounting for the almost total lack of external corrosion products on the silver surface. The typical composition of these Iranian and Sasanian silver alloys is 95% silver, 4% copper, 0.3% gold, 0.2% lead with small traces of various other elements (Meyers 1978a). Another example is that of a gilded silver ear-ring from the site of Tell Farah in Jordan dated to the Middle Bronze Age, the first half of the second millennium BC. One of a pair, the artefact examined here is drawn in Figure 7.7.

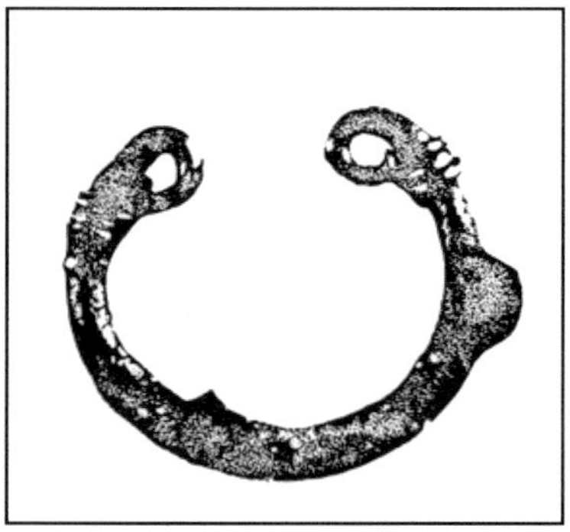

Figure 7.7 Gilded silver earring from Tell Farah, Jordan. 22mm across. Middle Bronze Age.

These earrings were also decorated with granulated gold spheres attached to the gilded surface. The terminals of the earrings end in loops, which are secured with twisted gold wires. Before conservation, both earrings were covered in grey corrosion of a waxy consistency. X-ray powder diffraction (XRD) showed that the main constituent was silver chloride. Conservation was carried out both chemically and mechanically using 15% ammonium thiosulphate and hand tools. The corrosion products had erupted through the gilding in several places forcing the gilding upwards. Characteristic features of foil gilding were observed, in particular the overlapping and corrugating of the thick gold foil. Metallography revealed a corroded silver substrate covered with a thick layer of gold. An interface several microns thick in some areas suggests that a diffusion bond was made between the applied foil gilding and the silver core. The silver of this Jordanian earring contains about 3% of copper.

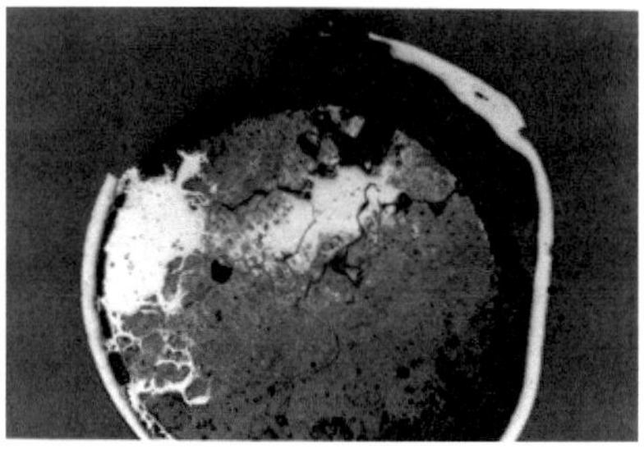

Figure 7.8. Part of the gold foil covering with overlapped join in the outer gold foil. The foil has been diffusion-bonded to the silver beneath. Note that the silver of the earring is heavily corroded with few sound metallic grains remaining. Unetched x21.

The silver base metal had heavily corroded and in many areas no metallic silver remained at all, but enough survived to produce the section shown in Figure 7.9. Undulating grain boundaries and the very large size of the grains themselves can be seen here. There can be no doubt that the silver is ancient.

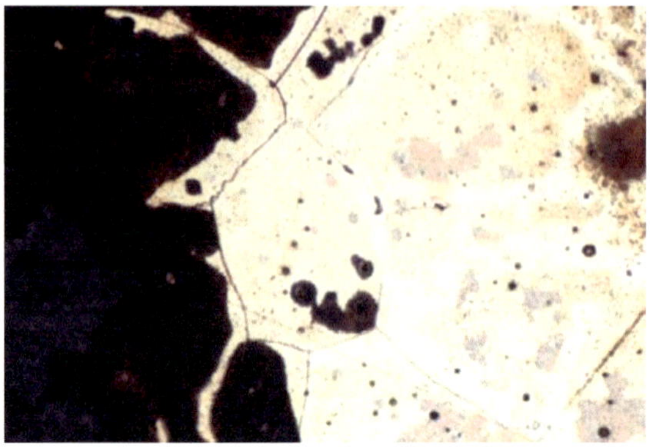

Figure 7.9. Discontinuous precipitation of copper can be seen at the grain boundaries. Etched in acidified potassium dichromate, x500.

Sterling silver is an alloy of 92% silver, 8% copper, and was much used in the 19th and early 20th century period for silver trays, cutlery and ornaments. Alloys which are more debased than this are increasingly subject to general corrosion attack and tarnishing of the silver surfaces. The term "Sterling" has a historical origin. In the 12th century AD, five towns in eastern Germany joined together to form the Hanseatic League, an entity which engaged in substantial commerce with England. In payment for English cattle and grain, the League used its own currency, the Easterling, adopted under the reign of Henry II as the standard English coinage and its name contracted to 'Sterling'. Many alloys are made with 92.7% silver rather than 92.5% silver to compensate for slight loss of silver during processing. The degree of copper solubility in the solid solution of the silver-rich grains depends on the heat treatment used.

The overall physical properties of the alloys can be affected, not only by heating the silver to different temperatures, but also by using different cooling rates. Silver alloys are usually used in a soft state for ease of working, but can be made in various degrees of hardness by working the alloy without annealing. Many sterling silver artefacts have a considerable degree of homogeneity, resulting in an even, large-grained microstructure which is soft and ductile. Heat treatment can be used to increase hardness, and this is a deliberate example of *precipitation hardening*. This involves heating and cooling the silver to cause the copper to precipitate out of the solid solution, producing a fine binary structure. Precipitation hardening of silver alloys involves: (1) heating the alloy to between 745 – 760 °C; (2) holding at this temperature for 15 minutes, and (3) rapidly quenching in cold water. The alloy is now in a softened condition and can be re-hardened by heating to 316 °C for 30-50 minutes followed by air-cooling. The resulting hardness is equivalent to the value obtained by cold-working to a 50% reduction in area.

Annealing of silver alloys

It is always preferable to anneal silver alloys in a non-oxidizing environment to prevent excessive oxidation of the copper content. When silver alloys are annealed in the open air, copper oxides tend to form. This is because oxygen is rather soluble in hot silver. The upper layer which forms is tenorite, CuO, which is black overlying a layer of cuprite, Cu_2O, which is red, and which may penetrate into the surface. The black oxide can be removed by pickling in acidic plant juices or mineral acids. Repeating these actions will tend to naturally result in surface enrichment of silver at the surface of the artefact, and in many ancient silver alloys, *surface enrichment* in silver may easily have occurred as a result of this surface oxidation of copper.

The best annealing temperature for normal softening of sterling silver is between 593-649 ^{o}C. Temperatures above 649 ^{o}C tend to dissolve the copper-rich phase, and unless the cooling rate is controlled, maximum softness is not achieved. At temperatures above 704 ^{o}C, the artefact, if worked, develops an "orange peel" texture on the surface due to internal oxidation problems. The small lumpy skin created by "orange peel" effectively ruins the smooth surface of the silver artefact, for which the only remedy is to grind away the internalized oxidized zone, wasting silver in the process. These "orange peel" effects can be hard to prevent in heavily worked sheet as we will see later in this section. In the casting of these alloys, they must be heated to at least 899 ^{o}C to be completely molten and complete solidification does not take place until 779 ^{o}C.

Another example of a typical silver-copper alloy from the 10th century AD site of La Compañia, Los Rios Province, Ecuador, previously illustrated in Figures 6.26 and 6.27, is shown in Figure 7.10. The site is a very rich chimney urn burial from the Milagro-Quevedo period which is called the "Cacique Guayas" site. It is an Integration period burial, dating from around 850-1532 AD excavated by Meggers but never properly published (Meggers et al n.d.). At the site a group of thin copper sheets called "feathers" were found, thought to have been used as headdress ornaments (Hosler et al 1990: 22).

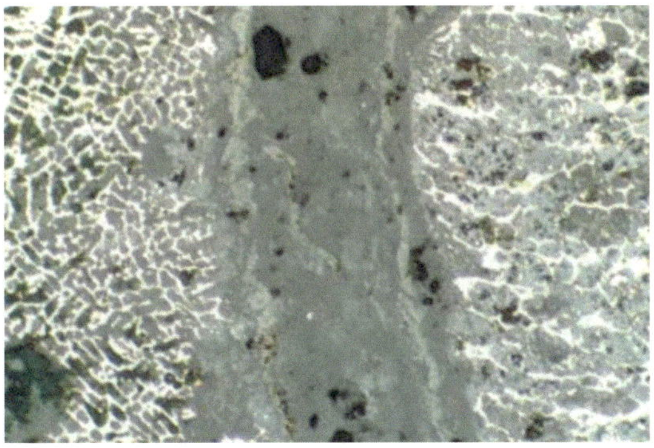

Figure7.10. Photomicrograph of a section through a face stud from the site of La Compañia, Los Rios Province, Ecuador, dated to about the 10th century AD. On the left hand side is the cast structure of the spherical ball of

silver-copper alloy used for the head, and to the right is the worked and annealed structure of the shaft of the face stud, which has been joined by depressing the worked shaft into the heated ball of the head. In the photomicrograph, corrosion has occurred extensively at the join area. Unetched, x 220.

The first set of metallurgical analyses from this site was carried out by McCall and Buchheit (1971).

The uncorroded whiter regions of the structure correspond to the interdendritic remnants of the cast silver-rich phase of the eutectic. When examined at high magnification, it is possible to see that some of these areas do encompass the eutectic mix of the two phases.

But the phase which has been corroded away is the copper-rich dendrites, whose composition essentially remained unchanged from the first moments of their solidification from the melt, and also become flattened and elongated along the length of the section from hammering and annealing to shape. These alloys are often then surface enriched, creating silver-coloured surfaces on alloys which may contain only 20% of silver.

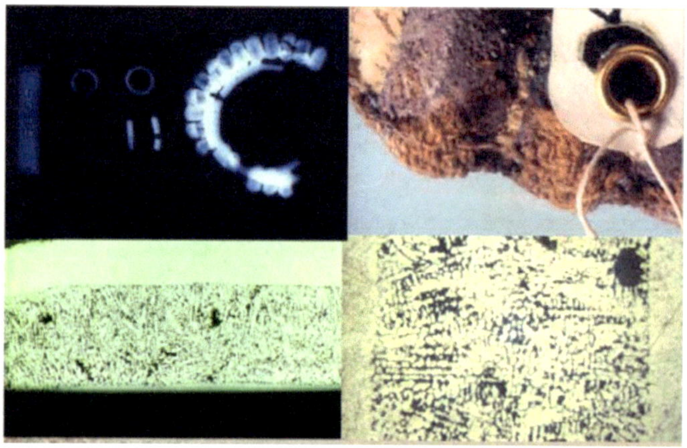

Figure 7.11 Chimu silver beads from Peru in the collections of the Cambridge Museum of Archaeology and Ethnology. These skillfully made beads are made in silver, cupelled from lead ores, and joined with a fine silver-copper alloy solder join close to the eutectic composition, showing sophisticated appreciation of melting points of different alloys by the Moche.

Case Study Number 1: The Philosopher Plate and the Fisherman Plate in the collections of the J. Paul Getty Museum. These plates were disputed as being of Renaissance date and a modern forgery respectively by art historian, Anthony Cutler (1989), and were proposed to be of Byzantine date by others (Scott (1989). Metallographic examination was carried out to resolve the issues regarding date, authenticity, and manufacture, which clearly indicated that the plates could not be from the Renaissance period. The plates themselves have few, if any parallels and remain enigmatic and very unusual silver artefacts. Both of these plates are currently on display at the Getty Villa Museum.

Figure 7.12 Philosopher plate. Late Antique, Eastern Mediterranean, 500-600 AD. Silver Height: 45cm, Width: 28cm, Weight: 2.5475 kg. 83.AM.342. Photograph by courtesy of the J. Paul Getty Museum, Villa Collection, Malibu, California..

Two seated philosophers, Ptolemy and Hermes, engage in a spirited discussion on this fragmentary plate. A woman stands behind each man, gesturing and partaking in the exchange. The woman on the left is identified as Skepsis. Above the two seated men, an unidentified enthroned man is partially preserved. The scene on this plate has been interpreted as an allegory of the debate between Myth and Science: Ptolemy, the founder of the Alexandrian school of scientific thought, debating Hermes Trismegistos, a deity supporting the side of myth (Getty Handbook 2002).

The two plates were acquired together and are said to have been found together in the Mediterranean, off the coast of Gaza.

Figure 7.13 The Philosopher plate from the reverse, showing that the plate was not shaped by repoussé working as the back is completely flat! This is a great surprise. The plate is not made in two-part construction either. Photograph by the author, 1991. The fact that the plate is not made in two-part construction, and that it remains flat on the reverse shows that the plate must have been carved to shape it. A remarkable discovery.

The problem with two-part construction, sometimes seen in Sasanian forgeries, is that the worked front plate of silver is soldered to a plain back plate, but these are in general not ancient examples, therefore one of the important aspects of this plate is that it is made in one piece. The conclusion is therefore that the plate must either have been cast to shape, or that it has been worked principally or exclusively from the front, by carving the silver to shape.

Figure 7.14. The philosopher plate before restoration when it was infilled with coloured epoxy resin. Photograph by Anna Bennett.

The fill was carried out for conservation purposes to ensure that the ragged broken edges were not subject to further loss, and the central hole which has been eaten through the plate is visually quite distracting. A decision was therefore taken to infill these areas, to produce an aesthetic appearance of the plates which would not be too distracting. These fills have, in general held up well, although they appear slightly uneven and look as if they are made in coloured epoxy resin which are still rather visible on display.

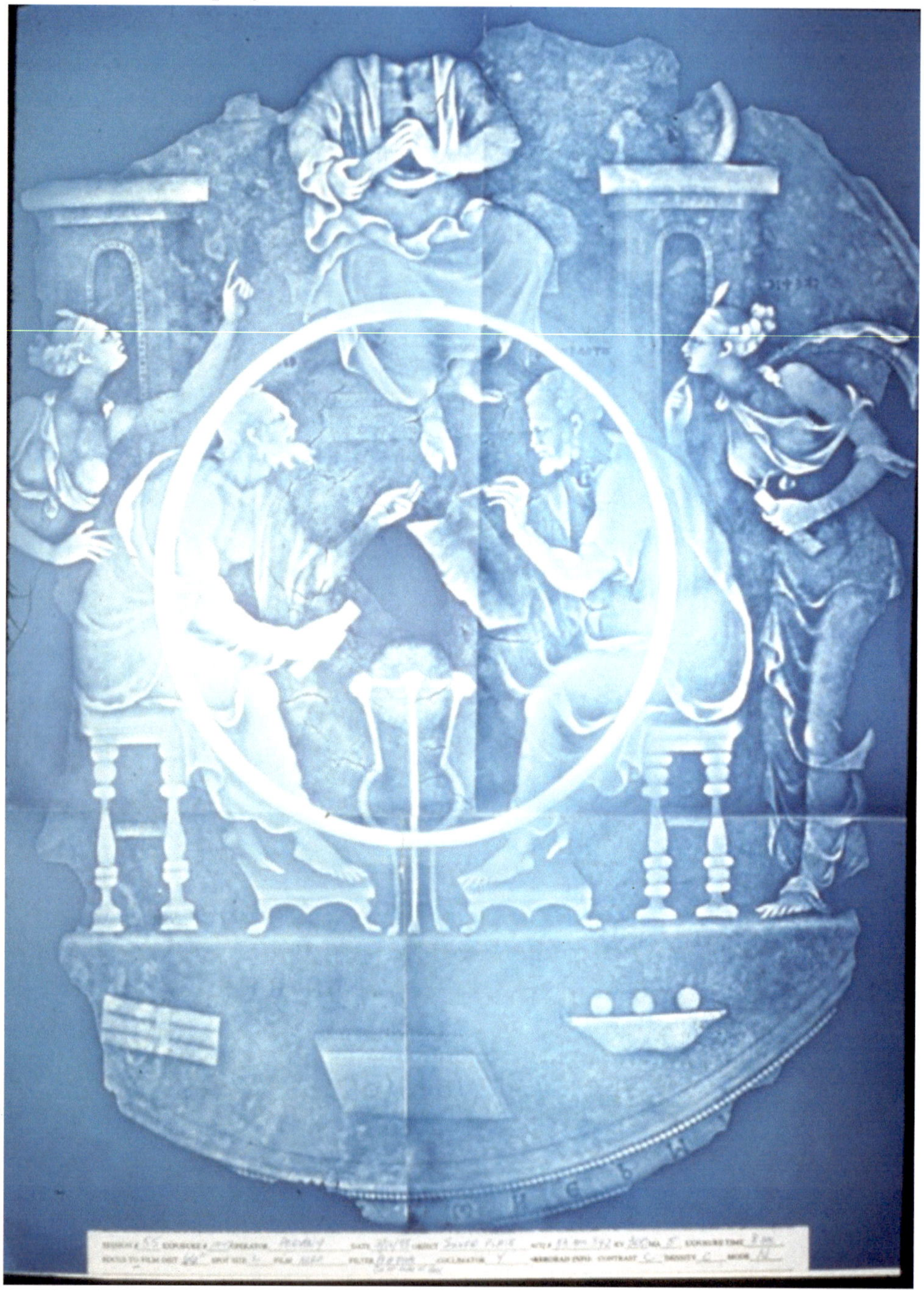

Figure 7.15. Part of a zero-radiograph of the Philosopher plate which resulted in the discovery of a faint floral decoration around the rim of the plate. Photograph by courtesy of the J. Paul Getty Museum, Department of Antiquities Conservation.

Figure 7.16 Partial view of the Philosopher plate. Photograph by the author 1992. The detail here has been produced by carving to shape. Metallographic sample taken from the edge of the proper left hand side in the scalloped area of the folds of the leg. Photomicrograph by the author, 1992.

Figure 7.17 Heavily etched microstructure of the Getty Philosopher plate showing discontinuous precipitation at the grain boundaries. Etched in acidified potassium dichromate, x 190.

Figure 7.18. Etched microstructure of the Philosopher plate under partially oblique illumination to accentuate grain detail. Etched in acidified potassium dichromate, x 160.

The surface appears to be heavily worked as decorative figures are present in relief, however the microstructure shows that these figures have been carved to shape as there is no deformation of the worked structure of the sheet itself. Carving is an unusual form of shaping silver which was postulated by Maryon (1948) to have been used in shaping the Mildenhall silver plates in the British Museum.

Both the Philosopher and Fisherman plate are remarkable because they have surface detail on the front, which would normally be expected to be carried out by repoussé working, but the reverse of the plates are flat. The microstructure of the Philosopher plate is shown in Figures 7.16, 7.17 and 7.18. These show that the plate silver consists of small recrystallized grains, with ASTM numbers between 7 and 8, and some orientation to the structure along the length of the plate section as a result of working and annealing. Some small, nonmetallic inclusions are present that are probably cuprous oxide, and these two are flattened and elongated along the length of the section.

The jagged grain boundaries and the extent of corrosion (discussed in more detail in Scott 1989: 38), would be very unusual for a Renaissance silver piece and suggest an earlier date of manufacture. In fact, we know from comparable examples of ancient silver that have been studied that grain displacement and discontinuous precipitation occur slowly in silver alloys. The question still remains as to how the plate was made. Herbert Maryon, writing in 1948, reported on the technical study of a group of 4th century AD Roman silver objects found at Mildenhall, England and known as the Mildenhall treasure. One of these silver plates, the Neptune plate, was carefully studied in an attempt to discover how it had been produced. It too has figures in relief design on its front which was not made by repoussé working from the back. Maryon noted that the Neptune plate had subtle hollows under each raised figure, and he concluded that the designs on the plate had been fabricated by working solely from the front. Maryon (1948) wrote:

"...if these works were neither cast nor produced by repoussé work, were they carved? To carve the designs upon works such as these it would have been necessary to support the work firmly-probably upon a pitch block, possibly upon an anvil. The removal of part at least of the background with chisels would be first undertaken. The carving of the figures and other ornament would follow. Some small amount of chasing and scraping would accompany the carving. Upon removing the dish from its pitch bedding, or lifting it from the anvil, one would find very little evidence upon the underside of the work done from the front..."

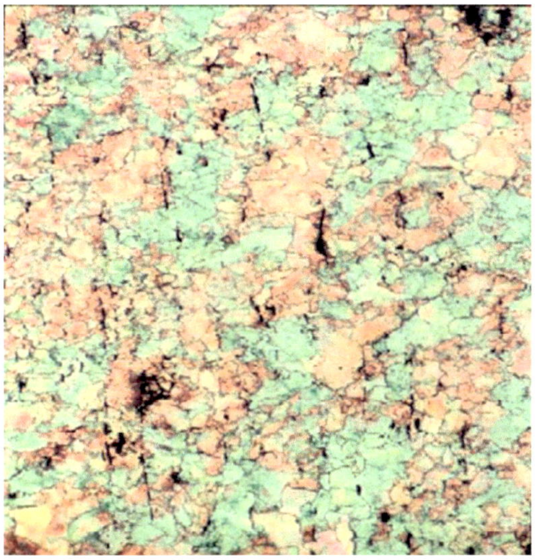

Figure 7.19. Photomicrograph of part of the Philosopher Plate, in the Getty Museum, showing worked and annealed silver-rich grains with meandering grain boundaries as a result of natural aging, together with darker bands of copper-rich phase strung out in the direction of working of the silver plate, which has therefore not been cast to shape. Etched in acidified potassium dichromate, x 240.

In his technical study of Sasanian silver, Meyers (1981) notes that carving was employed, especially for the removal of metal around decorative elements, leaving the decoration in low relief. The "Sion" treasure in Dumbarton Oaks, the Metropolitan Museum and the State Hermitage Museum in Leningrad, were examined by Meyers and also were determined as carved to shape. Study of the Hormizd II plate by Gibbons et al (1979) revealed some interesting aspects of the working and manufacture of Sasanian silver objects in the collections of the Cleveland Museum of Art. Most of the high relief detail was made by the addition of separate pieces of worked silver, but those in low relief were carved to shape from the plate itself.

By the time of the Renaissance, silverworking techniques were quite well established; the methods of repoussé working, soldering, chasing, punching, engraving, and casting were well known. Extensive carving of the surface to shape, is not a known technique from this period.

The evidence discussed here shows that the Philosopher plate was made from a cast ingot of silver, the ingot being extensively worked and annealed to shape, followed by surface carving to create or accentuate the design. The microstructure of the silver in the areas examined suggests a typical striated and heavily worked silver alloy with elongated copper oxide inclusions.

Non-metallic inclusions cannot recrystallize during cycles of working and annealing and so act as additional messengers of change from the cast to the worked state. As working proceeds, inclusions tend to become flattened and elongated along the length of the section, and they retain this characteristic even if the silver grains themselves lose all evidence of segregation from the cast ingot. Here we have both forms of evidence to show extensive working. A small sample could be removed from the plate for study by XRF, EPMA and ICP-MS analysis.

The XRF data is shown in Table 7.1.

X-RAY FLUORESCENCE ANALYSIS BY ENERGY DISPERSIVE METHOD FOR THE PHILOSOPHERS PLATE

escription of area	% Cu	% Au	% Pb	% Bi	% Br	% Ag	% Fe	Mass balance	Trace elements
lished no. 1	3.3	0.43	0.66	0.11	0.8	93.1	nd	98.40	Cl
lished, scrubbed no. 1B	3.3	0.44	0.66	0.13	0.8	98.1	nd	103.40	Cl
ck, unpolished no. 2	2.2	1.50	0.66	0.23	1.5	96.5	<0.10	102.69	Cl,Ca
ck, unpolished no. 3	1.2	2.30	1.10	0.47	7.5	32.3	0.16	105.03	Cl,Ca
ont, unpolished no. 4	1.6	1.30	0.71	0.32	6.0	92.9	0.18	103.01	

Table 7.1: XRF analysis data of the Philosopher plate.

The X-ray fluorescence analysis data (XRF), shows the presence of bromine, often associated with silver artefacts from sea burial, as bromine has a strong affinity for both silver and gold. Bromine is also found in silver corrosion products from more usual burial environments as well, so the presence of bromine does not implicate a purely marine burial by itself. The presence of copper shows that the plates have been made with silver deliberately alloyed with a little copper, and the presence of gold and lead shows that the silver has been obtained from the cupellation of argentiferous lead ores.

It is interesting to compare the XRF data with the larger number of elements determined with Electron Microprobe Analysis (EPMA) analysis shown in Table 7.2. The XRF data was obtained with a Kevex 0750A X-ray fluorescence spectrometer operating at 50 KV, 50mA employing a Ba/Sr secondary target, in a lead-lined room. Note that this data is quantitative and that totals are not 100% as often seen, even today. This XRF data was obtained in 1990 with an instrument which is now completely obsolete, but it is a tribute to its sophisticated software, the secondary target system, the time spent by the operator on acquiring background counts, and peak deconvolution routines, that data this good could be obtained non-destructively over twenty years ago

The reasons for this impressive set of x-ray fluorescence data can be considered to be due to seven factors: First, the old Kevex instrument had to be operated in a lead-lined room, because of the great power of the x-ray beam compared with modern hand-held instruments; Second, the data was freshly acquired from a series of well-analysed silver standards and from silver chloride and silver bromide crystals; Third, the data was used to create a set of fundamental parameter ZAF corrections which were specifically made for this analysis; Fourth, the instrument is always calibrated for any flux variation by reference to a reference monitor standard, consisting of a block of Lucite; Fifth, the secondary target x-ray fluorescence system creates a background to the spectra which is practically flat; Sixth, peak deconvolution is carried out as the analysis proceeds, allowing complete control over the peaks remaining for XRF quantification.; Seventh, the entire data set takes many hours or even days of work to acquire and quantify.

In general, XRF is not to be trusted too heavily for the surface analysis of ancient silver alloys, as the results could be very misleading. This is because of the microstructural problems of altered surfaces, surfaces significantly enriched in silver by deliberate removal of copper, or effects of corrosion. The two-phase nature of many of the silver-copper alloys makes estimation of the bulk composition difficult on the basis of EPMA or other SEM techniques. One of the reliable ways to gain insight into the true composition of the silver-copper alloys is by metallography.

ELECTRON MICROPROBE WAVELENGTH DISPERSIVE ANALYSIS OF THE PHILOSOPHERS PLATE, POLISHED SECTION OF THE PLATE EDGE

Point	Ag-L	Au-M	Cu-K	Pb-M	Bi-L	Zn-K	Sn-L	As-L	Sb-L	Cd-L	Cl-K	Total
1	93.23	0.63	3.33	0.59	0.48	0.02	0.00	0.12	0.05	0.00	0.60	99.05
2	93.81	0.58	4.33	0.72	0.46	0.01	0.00	0.50	0.00	0.00	0.33	100.74
3	95.22	0.50	2.77	0.71	0.46	0.09	0.00	0.03	0.02	0.00	0.35	100.15
4	94.42	0.58	3.49	0.73	0.00	0.02	0.00	0.76	0.00	0.00	0.36	100.36
5	95.28	0.43	3.29	0.73	0.91	0.03	0.00	0.11	0.02	0.00	0.41	101.21
6	94.97	0.44	2.64	0.60	0.26	0.04	0.00	0.38	0.00	0.00	0.45	99.78
7	93.58	0.45	3.45	0.80	0.39	0.09	0.00	0.00	0.08	0.00	0.15	98.99
8	94.46	0.53	3.18	0.46	0.43	0.02	0.00	0.00	0.01	0.00	0.59	99.68
9	93.86	0.49	3.17	0.73	0.39	0.00	0.00	0.37	0.01	0.00	0.30	99.32
10	94.81	0.51	3.18	0.69	0.36	0.09	0.00	0.05	0.08	0.00	0.12	99.89
11	93.43	0.49	3.38	0.62	0.57	0.04	0.00	0.01	0.07	0.31	0.27	99.19
12	94.71	0.57	3.03	0.76	0.30	0.05	0.00	0.34	0.01	0.00	0.27	100.04
13	93.44	0.37	3.55	0.76	0.00	0.13	0.00	0.00	0.01	0.00	0.40	98.66
14	93.96	0.23	3.22	0.66	0.37	0.00	0.00	0.41	0.02	0.00	0.33	99.20
15	92.44	0.41	3.27	0.48	0.59	0.00	0.00	0.26	0.00	0.00	0.23	97.68

AVERAGED VALUES AND COUNTING STATISTICS FOR THE RESULTS GIVEN IN TABLE 2 FOR THE PHILOSOPHERS PLATE

	Ag-L	Au-M	Cu-K	Pb-M	Bi-L	Zn-K	Sn-L	As-L	Sb-L	Cd-L	Cl-K	Total
Ave.	94.11	0.48	3.29	0.67	0.40	0.04	0.00	0.22	0.03	0.02	0.34	99.60
+/−	0.82	0.10	0.38	0.10	0.22	0.04	0.00	0.23	0.03	0.08	0.14	0.88
Count Stats Rel%	0.50	14.11	1.88	9.41	25.41	25.46	25.50	25.50	25.50	24.63	4.29	

Table 7.2: Electron Microprobe Data for the Philosopher Plate.

The study was completed by semiquantitative inductively coupled plasma-mass spectrometry (ICP-MS). In the present study, bismuth, cadmium, chromium, gallium, germanium, gold, iodine, lead, palladium, samarium and silver were found. Note the absence of any detected zinc and platinum. Unpublished work by the Goldsmith's Company in England, used as data for determination of silver authenticity from the historical period, was used to compare with the data shown in this Table: the absence of zinc and platinum is more typical for ancient silver artefacts.

Neither set of analytical data suggests a modern origin. Gold levels of 4000 – 6000 ppm are unusually high for Renaissance silver, and the zinc levels are low for this period. Zinc was only detected by ICP-MS in the Fisherman plate at 31 ppm. The reason for the presence of traces of zinc in Renaissance silver is the use or reuse of brass. It is unusual to find zinc in ancient silver alloys.

SEMIQUANTITATIVE INDUCTIVELY COUPLED PLASMA/MASS SPECTROMETRY REPORT ON THE PHILOSOPHERS PLATE

	ppm	Detection limit		ppm	Detection limit
Aluminum	nd<3600	3600	Mercury	nd<8	8
Antimony	nd<17	17	Molybdenum	nd<0.9	0.9
Arsenic	nd<2200	2200	Neodymium	nd<0.9	0.9
Barium	nd<25	25	Nickel	nd<11	11
Beryllium	nd<4	4	Niobium	nd<0.9	0.9
Bismuth	371	1	Osmium	nd<0.9	0.9
Boron	nd<12000	12000	Palladium	8.7	0.9
Bromine	nd<350	350	Platinum	nd<3	3
Cadmium	4.4	0.9	Praseodymium	nd<0.9	0.9
Calcium	nd<2900	2900	Rhenium	nd<0.9	0.9
Cerium	nd<0.9	0.9	Rhodium	nd<0.9	0.9
Cesium	nd<1	1	Rubidium	nd<13	13
Chromium	450	250	Ruthenium	nd<0.9	0.9
Cobalt	nd<0.9	0.9	Samarium	1.3	0.9
Copper	27300	9	Selenium	nd<320	320
Dysprosium	nd<0.9	0.9	Silver	131000	0.9
Erbium	nd<0.9	0.9	Sodium	nd<20000	20000
Europium	nd<0.9	0.9	Strontium	nd<62	62
Gadolinium	nd<0.9	0.9	Tantalum	nd<0.9	0.9
Gallium	11	7	Tellurium	nd<4	4
Germanium	26	7	Thallium	nd<1	1
Gold	2560	0.9	Thorium	nd<0.9	0.9
Hafnium	nd<2	2	Thulium	nd<0.9	0.9
Holmium	nd<0.9	0.9	Tin	nd<13	13
Iodine	225	7	Titanium	nd<77	77
Iridium	nd<0.9	0.9	Tungsten	nd<88	88
Iron	nd<550	550	Uranium	nd<0.9	0.9
Lanthanum	nd<0.9	0.9	Vanadium	nd<2700	2700
Lead	4940	85	Ytterbium	nd<0.9	0.9
Lithium	nd<170	170	Yttrium	nd<0.9	0.9
Lutetium	nd<6	6	Zinc	nd<16	16
Magnesium	nd<390	390	Zirconium	nd<89	89
Manganese	nd<25	25			

Elements not analyzed: all gases, C, P, S, K, Si, Sc, In, Tb

nd = not detected

Table 7.3: ICP-MS Data for the Philosopher Plate.

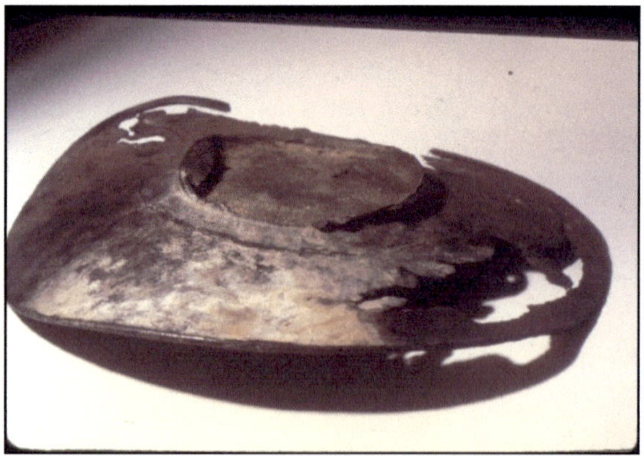

Figure 7.20 Fisherman plate. Late Antique, Eastern Mediterranean, 500-600 AD. Silver with foil gilding. 23 5/8 in. 83.AM.347. Diameter: 60cm. Photograph by courtesy of the J. Paul Getty Museum, Villa Collection, Malibu, California.

Figure 7.21. (opposite) Fisherman plate from the reverse, showing very fragmentary and heavily corroded condition of the plate before it was gap-filled with epoxy resin and partially reshaped during the process of conservation. Photograph by the author, circa 1985. The art historical critique that this object is a forgery is hard to maintain in the light of this evidence.

X-RAY FLUORESCENCE ANALYSIS OF THE SURFACE OF THE FISHERMAN PLATE

Description of area	% Cu	% Au	% Pb	% Ag	% Cl	% Hg	% Br
Front: hip area no. 3	0.64	0.96	0.73	96.4	nd	nd	0.60
Front: arm area no. 4	0.46	1.10	0.53	91.8	nd	nd	1.50
Front no. 5	0.64	0.94	0.69	92.4	nd	nd	1.10
Front no. 6	0.92	0.74	0.56	93.9	nd	nd	0.70
Back: main body polished no. 9	0.97	0.67	0.57	97.6	nd	nd	0.20
Back: base, rim no. 8	1.20	0.61	0.50	90.5	0.14	nd	0.12
Back: base, polished no. 10B	1.40	0.68	0.67	97.8	nd	nd	0.03
Front: gold no. 7	0.37	62.50	0.42	36.1	0.17	6.4	1.10

nd = not detected

ELECTRON MICROPROBE ANALYTICAL DATA FOR THE FISHERMAN PLATE BY WAVELENGTH DISPERSIVE ANALYSIS ON THE JOEL SUPERPROBE AT THE GETTY CONSERVATION INSTITUTE

Point	Ag-L	Au-M	Cu-K	Pb-M	Bi-L	Zn-K	Sn-L	As-L	Sb-L	Cd-L	Cl-K	Total
16	96.48	0.68	1.34	0.26	0.47	0.00	0.00	0.14	0.01	0.00	0.08	99.46
17	97.67	0.55	1.32	0.28	0.42	0.03	0.00	0.35	0.07	0.00	0.08	100.77
18	97.58	0.68	1.12	0.37	0.00	0.13	0.00	0.00	0.00	0.00	0.09	99.97
19	97.76	0.60	1.05	0.32	0.71	0.00	0.00	0.29	0.01	0.00	0.05	100.79
20	97.02	0.67	1.03	0.38	0.62	0.00	0.00	0.22	0.03	0.00	0.03	100.00
21	97.30	0.60	1.15	0.29	0.40	0.08	0.00	0.59	0.00	0.00	0.07	100.48
22	93.15	0.53	1.25	6.19	0.48	0.00	0.00	0.04	0.01	0.00	0.08	101.73
23	95.48	0.21	1.26	0.32	0.18	0.02	0.00	0.00	0.06	0.00	0.06	97.59
24	96.67	0.56	1.25	0.24	0.00	0.07	0.00	0.35	0.04	0.00	0.06	99.24
25	96.25	0.65	1.16	0.66	0.56	0.03	0.00	0.94	0.00	0.00	0.05	100.30
26	96.05	0.54	1.22	0.39	0.28	0.02	0.01	0.08	0.00	0.00	0.06	98.65
27	97.03	0.67	1.11	0.34	0.20	0.00	0.00	0.12	0.01	0.00	0.05	99.53
28	96.65	0.66	1.03	0.30	0.28	0.00	0.00	0.06	0.08	0.00	0.04	99.10
29	95.98	0.56	1.29	0.45	0.00	0.03	0.00	0.32	0.00	0.00	0.04	98.67
30	95.77	0.75	1.19	0.30	0.29	0.00	0.00	0.00	0.00	0.00	0.05	98.35

AVERAGED VALUES AND COUNTING STATISTICS FOR THE FISHERMAN PLATE DATA IN TABLE 6

	Ag-L	Au-M	Cu-K	Pb-M	Bi-L	Zn-K	Sn-L	As-L	Sb-L	Cd-L	Cl-K	Total
Ave.	96.46	0.59	1.18	0.74	0.33	0.03	0.00	0.23	0.02	0.00	0.06	99.64
+/-	1.15	0.12	0.10	1.51	0.23	0.04	0.00	0.26	0.03	0.00	0.02	1.09
Count Stats Rel%	0.50	10.93	2.05	17.60	25.50	25.50	25.50	25.17	25.50	25.50	23.51	

Table 7.4. X-Ray Fluorescence Data and Electron Microprobe Data for the Fisherman Plate.

SEMIQUANTITATIVE INDUCTIVE COUPLED PLASMA/MASS SPECTROMETRY REPORT ON THE FISHERMAN PLATE

	ppm	Detection limit		ppm	Detection limit
Aluminum	nd<1700	1700	Mercury	5	4
Antimony	nd<8	8	Molybdenum	nd<0.4	0.4
Arsenic	nd<1200	1200	Neodymium	0.6	0.4
Barium	nd<12	12	Nickel	22	5
Beryllium	nd<2	2	Niobium	nd<0.4	0.4
Bismuth	66	0.5	Osmium	nd<0.4	0.4
Boron	nd<5600	5600	Palladium	4.9	0.4
Bromine	nd<170	170	Platinum	3	1
Cadmium	2.6	0.4	Praseodymium	nd<0.4	0.4
Calcium	nd<1400	1400	Rhenium	nd<0.4	0.4
Cerium	nd<0.4	0.4	Rhodium	nd<0.4	0.4
Cesium	nd<0.6	0.6	Rubidium	nd<6	6
Chromium	230	120	Ruthenium	nd<0.4	0.4
Cobalt	nd<0.4	0.4	Samarium	1	0.4
Copper	7290	4	Selenium	nd<150	150
Dysprosium	nd<0.4	0.4	Silver	71300	0.4
Erbium	nd<0.4	0.4	Sodium	nd<9400	9400
Europium	nd<0.4	0.4	Strontium	nd<30	30
Gadolinium	nd<0.4	0.4	Tantalum	nd<0.4	0.4
Gallium	6	3	Tellurium	nd<2	2
Germanium	14	3	Thallium	nd<0.6	0.6
Gold	2190	0.4	Thorium	nd<0.4	0.4
Hafnium	nd<1	1	Thulium	nd<0.4	0.4
Holmium	nd<0.4	0.4	Tin	nd<6	6
Iodine	35	3	Titanium	nd<37	37
Iridium	nd<0.4	0.4	Tungsten	nd<42	42
Iron	nd<260	260	Uranium	nd<0.4	0.4
Lanthanum	nd<0.4	0.4	Vanadium	nd<1300	1300
Lead	2880	41	Ytterbium	nd<0.4	0.4
Lithium	nd<84	84	Yttrium	nd<0.4	0.4
Lutetium	nd<3	3	Zinc	31	8
Magnesium	nd<190	190	Zirconium	nd<43	43
Manganese	nd<12	12			

Elements not analyzed: all gases, C, P, S, K, Si, Sc, In, Tb
nd = not detected

Table 7.5 ICP-MS Data for the Fisherman Plate.

The figures on the Fisherman plate are very thick in some cases, 6mm thick. The section removed from an edge shows a large grain size, hammered-out cast dendritic structure of the original ingot from which the plate was made can clearly be seen orientated along the length of the section showing that a substantial amount of working and annealing took place in shaping the plate before it was carved to shape and partially gilded.

LEAD ISOTOPE DATA FOR THE GETTY MUSEUM'S SILVER PLATES AND DATA FOR SOME OTHER SILVER OBJECTS

Sample no.	Pb isotopic ratios 208/206	207/206	204/206	Museum	Object	Date	Museum Acc. No.
LE.230	2.0802	0.8390	0.05338	Hermitage Museum	Ewer (handle)	582–602 A.D.	omega 826
LE.174	2.0704	0.8368	0.05353	Hermitage Museum	Dish with Venus	c. 6th C. A.D. (c. 550 A.D.)	omega 3
BI.10	2.0741	0.8386	0.05363	Metropolitan Museum	"Cyprus" silver plate	c. 6th C. A.D.	17.190.394
BI.11	2.0719	0.8380	0.05368	Metropolitan Museum	"Cyprus" silver plate (rim of foot)		17.190.396
BI.13	2.0739	0.8383	0.05338	Metropolitan Museum	"Cyprus" silver plate	c. 630 A.D.	52.25.2
112.A.85	2.0736	0.83743	0.053503	Getty Museum	Philosophers plate	c. 4th–6th C. A.D.	83.AM.342
113.A.85	2.0804	0.84215	0.053703	Getty Museum	Fisherman plate	c. 2nd–4th C. A.D.	83.AM.347
188.G.87	2.10381	0.83404	0.05252	Getty Museum	Strozzi basin	c. 1625 A.D.	143.SB.82

Table 7.6. Comparison of Lead Isotope Data for the Fisherman Plate, the Philosopher Plate and the Strozzi (Fanelli) Basin in the Getty Museum.

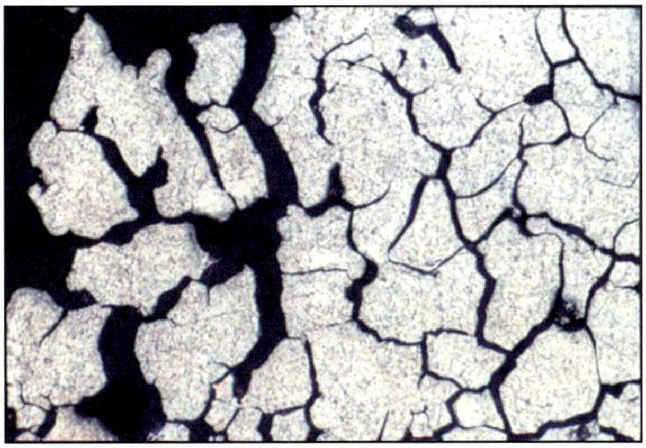

Figure 7.22 Photomicrograph of a section taken from the edge of the Fisherman plate showing the severe intercrystalline corrosion which has occurred in this plate, isolating the grains along the grain boundaries and gradually weakening the structure until the crystals have no coherence. Unetched, x160.

These grains have an ASTM number of 3 and prominent well-developed twinning passing across some of the large grains. These twin lines are predominantly straight, showing that the final stage of manufacture was an annealing operation. Evidence of the heavy working to shape of the Fisherman plate is apparent as well as the grain boundary embrittlement of the silver, which in the microstructure shown in Figure 7.22 has reached such an extreme state, that the silver has lost all coherence of the grains themselves.

Conclusion

Both plates have been shown beyond reasonable doubt to be ancient. They have been worked to shape from cast silver ingots, and have been carved to shape. This is not a Renaissance technique of silverworking, which refutes, by itself, the art historical view of the probable date of the Philosopher plate as being from the Renaissance, and the view that the Fisherman plate is a modern forgery. The scientific evidence suggests that the plates are much earlier in date, fully in keeping with a Byzantine origin.

Case Study Number 2. The Fanelli silver plate in the collections of the J. Paul Getty Museum, after a design by Bernardo Strozzi, dated to 1620-1625 AD. Metallographic examination was carried out because its authenticity was called into question by a Getty Conservator.

The Fanelli Silver Plate

The Fanelli silver basin, after an original design by Bernardo Strozzi silver plate, from Renaissance Italy, in the collections of the J. Paul Getty Museum, dated to between 1620-1625 AD, is shown Figure 7.23 and in a variety of microstructures in Figures 7.24, 7.25 and 7.26.

At one time this plate was thought to be an electrotype or later copy, due to the presence of silicon rubber in the crevices of the surface, the apparent content of traces of cadmium in the silver, and the observation of tiny bubbles of raised metal on the surface. Cadmium content is associated with recent silver objects because of the recycling of silver and other metals with traces of cadmium used in the earlier periods of the 20th century and the later 19th.

These were the reasons for the investigation of the Strozzi silver plate to try to resolve this mystery and to confirm its period of manufacture. The explanation for the small raised bubbles of silver on the plate, which had been noticed by the conservator during conservation cleaning, were in fact due to oxygen absorption as a result of slight over-annealing of the silver plate during working. Oxygen is quite soluble in hot silver. As the silver or silver-copper alloy cools down, some of the oxygen is released and an "orange-peel" effect may develop. The eradication of this surface imperfection may necessitate grinding part of the surface away to remove the oxidized skin. Needless to say, this was avoided at all costs by skilled silversmiths. Internal oxidation of copper could also be created by this oxygen absorption, useful for surface enrichment by pickling away the oxidized zone in acid, but not of any help in the fabrication of fine silver repoussé surfaces.

The explanation for the traces of silicon rubber, found by the conservator in the recesses of the plate were later discovered to be due to the previous Italian owner having taken a silicon rubber sectional mould from the plate, so that he could keep a copy of it himself before selling it to the Getty Museum

Figure 7.23. Silver basin. Perhaps modeled by Francesco Fanelli, after a sketch by Bernardo Strozzi, Italian, Genoa, 1620-1625. Probably made by an unknown Dutch or Flemish silversmith. The basin shows scenes from the life of Queen Cleopatra. Diameter 75.6cm. 85.DG.81. This Fanelli silver after a design by Strozzi was investigated in detail because at one time the plate was thought to be a silicon-rubber molded copy from an original. These assertions were to be proved false: the Getty plate is one of the masterpieces of Renaissance silver-working. Photograph by courtesy of the J. Paul Getty Museum, Brentwood, California..

The detection of cadmium in the silver composition was more problematic, since it had been carried out by the normally very reliable method of electron microprobe analysis (EPMA). Eventually, the author discovered what had happened: the scientist operating the electron microprobe had mistaken a silver sum
peak for a peak of cadmium. Sum peaks, where, for example two gold Lalpha peaks form an additional peak in the 22Kv region are a common problem and have to be accounted for by the very careful calibration of the x-ray microprobe, which in this case, had not been done correctly. The final proof of the genuine nature of the Fanelli plate was revealed by a series of microstructural studies.

The method of working of a complex silver plate such as this with many cast additions is described by Benvenuto Cellini (Ashbee 1967:93). After creating a clay model of the entire plate Cellini writes:

"I went straight at the job, with sheer ability of hammer work...striking from front and from back. Arms. Legs and body I hammered out in separate pieces ....when I had given them all their shape I soldered & fitted them together as before. The solder I used was *ottavo*, that is a solder composed of one-eight part of an ounce of copper to one of silver..."

The hammering of the Fanelli plate has been carried out with great skill, to the limit of the ability of the silver to take any further working, as a series of fine cracks can be seen on the reverse. The silver plate would first be cast and then worked and hammered to the required diameter and the design drawn on the back or scribed into the silver surface on the front so that faint impressions would be seen on the back as a guide to the shaping of the silver. Having undertaken this, the silver plate would then be rested on pitch or leather and hammered from the reverse, inspecting progress on the front. This is why the silver is badly cracked on the reverse, as the metal is extended to the limit of its ductility as it is pushed upwards.

The use of silver-copper solders for the attachment of separate parts was exactly what was observed in this plate. Instead of hammering out the small additions of some arms and horses lower legs to be attached to the repoussèd plate, the small clay parts which could not have been raised in silver were cut from the original clay model and then invested in small moulds so that the cast additions could then be exactly fitted to the worked plate with silver-copper solder. After filing and cleaning of the surface, these cast additions and the silver solder itself would be undetectable from a visual observation of the front surface. The plate shown in Figure 7.23 is a masterpiece of its kind and probably the most outstanding example of a Renaissance silver plate in the world. None of the other plates in collections around the world examined by the author have the complex three-dimensional representation of soldiers, horses and putti, which is achieved by the Fanelli plate with the same degree of artistic finesse and skill.

The soldered additions to the plate were examined with metallography. As can be seen in the etched section of Figure 7.24, the solder here is close to the eutectic composition of 79% silver, 21% copper, and is probably slightly higher in silver content than that as the second phase appears to be of silver-rich grains with an infill of the (α+β) eutectic.

Figure 7.25 shows the microstructure of a more copper-rich solder, with less than the eutectic amount of silver, so that the dendritic structure has copper-rich primary dendrites with an infill of the eutectic.

Figure 7.24. Section from the silver basin by Fanelli, in the collections of the J. Paul Getty Museum, showing the presence of a silver alloy, close to the eutectic composition for the solder used to join a part of an arm to the worked silver of the base. The silver alloy of the base is uppermost in this photomicrograph. Note the very good join between the silver-rich solder which has a cast structure and the worked silver grains of the bowl itself. Etched in acidified potassium dichromate, x360.

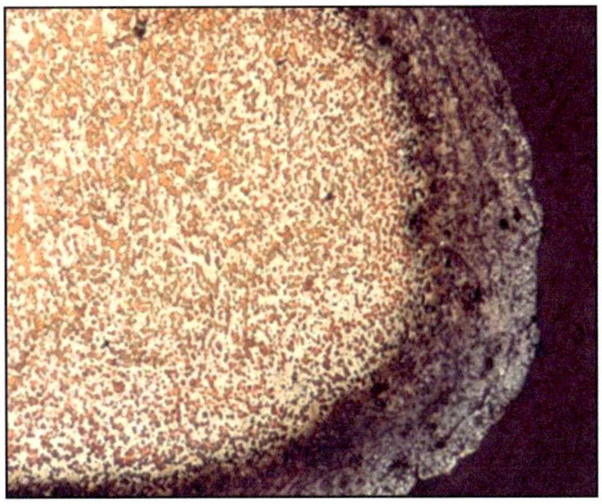

Figure 7.25 Photomicrograph of a section through a soldered join in the Fanelli plate, which was used to join a separately cast arm to one of the figures of the soldiers from the middle panel. Etched in potassium dichromate, x130.

Figure 7.26. Three microstructures of the Fanelli silver plate. Left, a polarized light view, showing the recrystallized grains of the surface zone with oxidized crust affected by heavy working, x 270. Top right, view of a solder join between an arm and the plate itself showing microstructure of the solder close to the eutectic composition, x 170. Bottom right, cross-section through the plate itself, revealing twinned grains and discontinuous precipitation of the silver. X 110. All etched in acidified potassium dichromate.

Conclusion

The Fanelli silver plate is a Renaissance masterpiece. The working of the silver has been carried to the limit of its ductility, while the addition of cast design elements has been carried out with great skill. Discontinuous precipitation within the silver grains reveals that the plate must be old. This precipitation is of interest since it shows that artefacts only 500 years old may show grain displacement. Several other silver plates dating from the same time period are in Museum collections, but few of these show the same degree of craftsmanship that the Fanelli plate manifests. Commonly, the degree of repoussé working of other silver plates is not as detailed and high, and many do not make use of additions of cast silver design elements which have to be cut off in wax or plaster from the original design and then modelled and cast in silver, being then soldered with silver-solder to the repoussé plate. The Fanelli plate is a masterpiece in silver of a high order.

Case Study Number 3. Ceremonial Sican Tumi Figure from the Middle Sican Period, Peru, 850-1050 AD. The metallographic examination was carried out to determine how the tumi had been fabricated and revealed a surprise, as a result of the metallographic study. The Sican kingdom on the northern coast of Peru, briefly existed as a powerful force in the region. Influence from the Sican area was felt into Ecuador and as far south as Pachacamac. Many high-quality artefacts in copper, silver and gold were produced. This was commented on by the Spanish conquistadors, Garcilaso de la Vega and Pedro de Cieza de Leon (1564), the latter writing:

> "Many valuable works were made by goldsmiths attached to the Temple, fashioned from gold and silver which arrived each year from many provinces of the Empire. These fine crafts arouse the highest admiration for it is amazing to see how few tools and instruments the smiths use and despite this how easily they produce things of the finest quality…."

Cieza de Leon was writing about Inca metalworking of the area, from about 1450 – 1532 AD, many years after the Sican presence on the northern coast. However, metallurgical techniques underwent little change in terms of the tools available to the metalsmiths. The upper part of the silver tumi is shown in Figure 7.27. It is 37 cm in height overall. The figure depicted surmounting the tumi knife is thought to represent a Sican lord, variously described as a depiction of Naymlap (Kauffmann 1973), the mythological founder of the Lambayeque dynasty, and as a personification of the Moon God. The proper left hand wing of the God is missing and, apart from tears in the silver sheet is in good condition.

The tumi itself has stepped cut-out design on the shaft, which is attached to a serrated-edged sheet or plate which, in turn, is soldered to a rectangular box made up of two parts, the rectangular box base and sides, closed carefully with a separate upper flat sheet. The legs of the lord are soldered to this base. The feet are made of two pieces, lower and upper parts, while the legs are made of hollow silver alloy tubes soldered down the length of both sides. The silver alloy figurine is soldered to these leg parts with the garment of the trousers covering the legs by partially wrapping them around from the front. The garment pieces are, in turn, joined to the body at the lower waist. The repoussé figure is made in two halves and joined around the outer edges along the length of the figure. The wings of the figure are made in two separate pieces and soldered onto the body. Both the tear-drop-shaped eyes and radially embossed ear spools are pierced, probably for attachments, now lost. Two small hollow sheet bird motif figures are attached as dangles from the extremities of the head-dress. These bird dangles have been made in two halves, soldered together. The construction of the scroll-work of the head-dress is based on a design employing four silver rectangular-sectioned hammered strips, which are actually made from two separate lengths, soldered together.

From the repoussé head outwards, this head-dress consists of ten individual soldered scroll designs, also made in silver alloy strip of rectangular cross-section. Forty-six remaining granulated silver alloy spheres are soldered between the second and third wire, followed by 39 small silver circles; finally, on the outer surface of the wire, 70 remaining tiny granulated silver spheres are soldered to the outer rim of the wire. The whole assemblage has been very skillfully made. There is an elaborate geometric design on the reverse of the figure which has a fringed tunic at the back and front. The trousers are decorated with rows of dots, zig-zag patterns and circles bounded in each case by a slightly raised border separating the designs. From the shape of the figure's nose, the lord appears to be wearing a crescentic nose-ornament.

The remaining wing is also decorated with back and front with banded designs and these is a row of raised ovoid dots at the outer edge. Other examples of this type of ceremonial tumi have been published in Sotheby's (1993), Gold of the Andes (1963), and Caredo Muro and Shimada (1985). The latter attempted to divide ceremonial tumis into a number of classes, or types, based on design elements. The one described here may roughly correspond with their type 2, showing a standing figure with wings, holding a cup with both hands at waist level and feet shown in plane view.

A section for metallographic examination was taken from the proper right side of the skirt. Extensive corrosion has occurred on both sides resulting in partial delamination of a surface crust. The thickness of the sheet at the edge of this skirt is 125 microns. The section is shown in Figure 7.28. The area has a lightly deformed cast structure. This is shown by the orientated, dendritic array of cellular regions of the silver-rich phase of the binary copper-silver system, surrounded by darker etching eutectic. The dendrites appear slightly deformed, but here the sheet metal has not been extensively altered in shape by working and annealing. If extensive working had taken place, then the dendritic structure of the material would have been distorted or extensively flattened out along the length of the sheet.

Figure 7.27. Partial view of the Sican Ceremonial Tumi. Photograph by the author.

The result of this metallographic examination was a surprise: it seemed impossible that a repoussé silver-copper alloy could have an essentially cast structure. In order to confirm this initial result, a region at the reverse of the fringe was chosen for study where the sheet appeared to be heavily bent. The resulting section is shown in Figure 7.27.

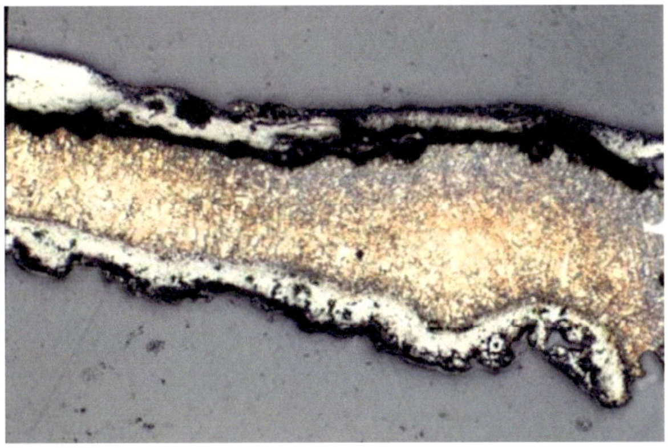

Figure 7.28 Unetched section through the silver tumi, revealing that the structure is principally as-cast, x87.

The distribution of the eutectic in Figure 7.28 is close to an equi-axed morphology. The dendrites are not apparent but the cellular array, especially the dark etching eutectic phase regions, shows that hammering and annealing has not been extensive. The structure is similar to that of an annealed cast morphology. The thickness of the sheet in this area is about 400 microns, or 0.4 mm, a very common thickness for ancient Peruvian sheet metalwork. The mounted and polished sheet fragments were analysed by electron probe microanalysis in a Joel superprobe employing pure element standards and wavelength analysis. The results are shown in Table 7.8. It is difficult to estimate the overall composition from electron microprobe analysis, due to segregation in casting and the two-phase nature of the alloy. The best estimate is probably given in analysis point 9, in which 92.3% silver 5.6% copper was detected. The concentrations of other impurities present are rather low, with percentages of lead less than 0.1%, no arsenic detected, and percentages of gold less than 0.1%.

Stos-Gale and Gale (1982) found the lead content of cupelled silver alloys from the ancient Mediterranean area to be in the range of 0.05% to 2.5%. Howe and Petersen (1994), in discussing silver-copper alloys from the Wanka cultural area in the Mantaro valley of the Peruvian central highlands, dating to Inca or immediate pre-Inca settlements, found lead contents between 0.2% and 1.64%, levels which are strongly indicative of the smelting of silver-lead ores or concentrate, followed by refining of the silver using cupellation. The lead concentration in the Sican tumi examined here averages 0.057%, perhaps at the lower possible limit of lead-silver smelting. The gold concentration, averaging 0.038% is also low compared with Old World silver alloys, where silver extracted from lead ores may contain anything from 0.3 to 3% of gold; however, in the New World, many silver ores of the central Andean region may contain relatively low levels of gold and the values found by Howe and Petersen (1994) are also fairly low, from less than 26 ppm to 2900 ppm.

The result of the metallographic study shows that Sican metalsmiths were pre-casting silver-copper alloys into thin blanks which could them be cut and shaped by repoussé working in areas where this was essential for producing the design. If these objects were mass produced, it could have been more efficient to use a technique of pre-casting of blank sheet to the approximate size required, rather than employing extensive hammering from a cast ingot. Until a substantial number of these figures have been analysed and metallographic studies carried out, there is no way of knowing if we are dealing with a unique piece, or whether we can infer from this examination any more far-reaching conclusions. Ingenuity and innovative methods for the working and shaping of metals is ever apparent in ancient South America: this is yet one more surprising example of such innovation.

Electron microprobe data for Sican tumi

Point	*Fe*	*Pb*	*S*	*Cu*	*Cl*	*Ag*	*As*	*Ni*	*Au*	*Total*
1	0.01	0.06	0.10	3.16	7.22	84.45	0.00	0.00	0.00	95.00
2	0.01	0.08	0.07	3.03	3.92	82.86	0.00	0.00	0.11	90.08
3	0.01	0.06	0.05	3.27	6.17	87.17	0.00	0.00	0.01	96.74
4	0.00	0.00	0.04	4.19	5.66	88.92	0.00	0.00	0.01	98.82
5	0.00	0.09	0.05	4.37	5.57	87.51	0.07	0.01	0.10	97.77
6	0.02	0.05	0.02	5.40	4.03	85.29	0.00	0.00	0.08	94.89
7	0.02	0.06	0.03	3.73	3.36	85.84	0.00	0.01	0.00	93.05
8	0.01	0.11	0.06	4.90	4.23	81.18	0.06	0.01	0.00	90.56
9	0.00	0.06	0.05	5.58	0.21	92.28	0.00	0.00	0.00	98.18
10	0.03	0.04	0.03	2.54	4.47	85.33	0.03	0.00	0.07	92.54
11	0.00	0.07	0.04	3.10	5.75	88.12	0.00	0.00	0.00	97.08
12	0.00	0.11	0.11	2.75	5.91	88.06	0.02	0.00	0.00	96.96
13	0.03	0.01	0.02	2.68	7.91	89.92	0.00	0.02	0.08	100.67
14	0.01	0.09	0.07	5.16	0.60	90.97	0.01	0.00	0.12	97.03
15	0.01	0.06	0.10	5.15	0.56	90.07	0.00	0.00	0.05	96.00
16	0.00	0.10	0.07	5.21	0.62	90.17	0.00	0.01	0.00	96.18
Stats. %	25.5	25.5	10.9	1.1	1.9	0.8	25.5	25.5	25.5	
MDL	0.026	0.078	0.010	0.034	0.011	0.050	0.064	0.028	0.101	

Stats. %: counting statistics (per cent). Low totals are due to porosity and possible light elements not detected. Conditions: 20 KeV, 40° take off angle, 12 nA beam current, 1 μm spot size.
MDL: mean detection limit (weight per cent). Values below this number are not statistically reliable.

Table 7.8 EPMA data for the Sican Tumi.

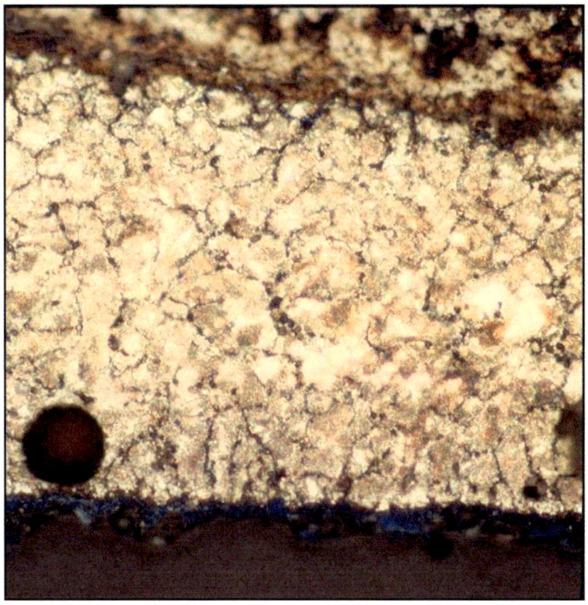

Figure 7.29. Metallographic evidence for the principally cast nature of the silver tumi: undistorted spherical porosity and a cast dendritic structure with very little subsequent alteration. Etched in acidified potassium dichromate, x160.

Figure 7.30. Silver Lion Rhyton, partially gilt. Parthian. Courtesy of the J. Paul Getty Museum, Villa Collection, Malibu, California. 86.AM.754

Silver artefacts are frequently gilded or partially gilded, inlaid with niello, glass, gold foil or other finishes. A magnificent example of gilded silver is the Parthian lion rhyton shown in Figure 7.30, whose microstructure will be discussed in a later volume in this series when gilding is described.

The microstructure shown in Figure 7.31 is from a coin of the Skandagupta kings, India, which is worked and annealed and then struck hot, in a silver-copper alloy with about 18-19% silver content. The alloys are then coated with a golden lacquer to make them appear golden.

The Gupta king Chandragupta II (Vikramaditya) was succeeded by his son Kumaragupta, who died in AD 455 and was succeeded by his son Skandagupta. During his reign, the invasions of the Huns became more frequent. Skandagupta repelled their early invasions and recovered most of the imperial provinces and produced a series of lacquered silver-copper alloy coins, whose golden surface still survives on many examples even today.

The microstructure shows the retention of the silver interdendritic phase as elongated stringers through the length of the section.

The interdendritic silver-rich phase has been elongated by working and despite the heavy working the alloy microstructure is still not in full equilibrium according to the phase diagram, because of dendritic segregation. The surface of the coin was then enriched in silver by pickling away some of the copper and finally coated with a golden lacquer to make the coins appear to be made of gold, probably composed of a coloured shellac.

From the relative proportions of the silver-rich phase and the copper-rich phase it is possible to estimate the approximate composition of the alloy. For the Scandagupta coin, this is about 18% silver, and 82% copper. Examination of different areas of the section shows that the coin has been hot-worked to shape, as the two-phase alloy has been distorted around the little depressions of the coin surface.

Figure 7.31. Photomicrograph of a cross-section of the silver coin of the Skandagupta kings, colour etched in ammonium persulphate followed by Klemm's No. 1 reagent. x240.

The Lead-Tin System

Another important eutectic alloy is that of the lead-tin system which produces a range of microstructures rather similar to that of the copper-silver system. Examples of the alloys themselves will be dealt with in Volume IV of this series.

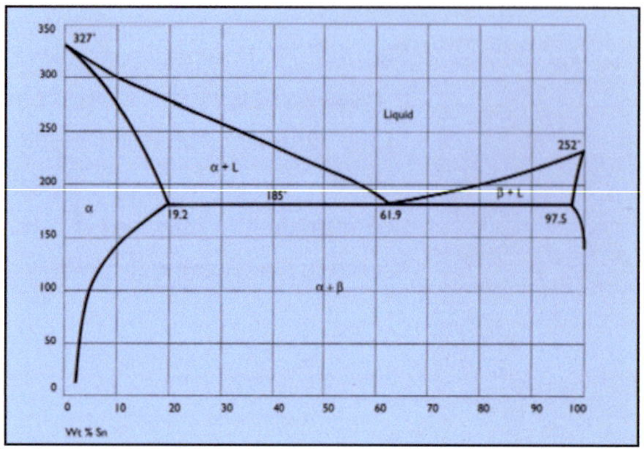

Figure 7.32. The tin-lead binary system. The type of microstructures to be seen in these alloys is similar to the copper-silver binary eutectic type.

In the diagram shown in Figure 7.33 the microstructures resulting from the cooling of an alloy with 40% tin, 60% lead are illustrated. The system is often dendritic and segregated in the same way as the silver-copper alloys.

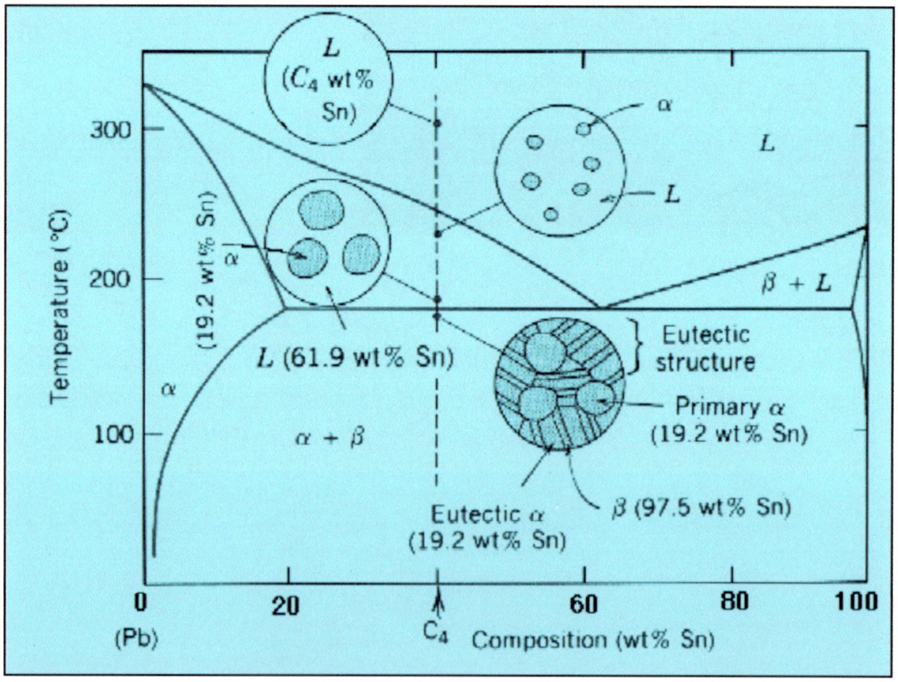

Figure 7.33. The tin-lead system in which we are following the cooling of an alloy with 40% tin, 60% lead.

As the alloy cools down it reaches the liquidus curve and small grains of alpha phase will precipitate of initial composition 19.2% tin and 81.8% lead. The composition of these crystals change as the alloy cools down until, when the line at 185 °C is reached, the remaining liquid breaks down into a fine eutectic mix. The alpha phase will have the composition of 19.2% tin 81.8% lead and the beta phase, 97.5% tin and 2.5% lead. However, in cast lead-tin alloys casting segregation is often observed instead of the equilibrium phases predicted by the diagram, in an analogous manner to that of the silver-copper alloys. Lead-tin alloys, soft solders as they are also called, are very useful for soldering materials together, but the silvery colour of the solder can be distracting for precious metalwork in gold and even Pliny offers caution in its use for this purpose.

Chapter Eight: Copper, Copper-Arsenic, Copper-Antimony, Copper-Nickel and Copper-Iron Alloys.

Bronze vessels that have been interred under the earth a thousand years appear pure green the colour of kingfisher feathers . . . those that have been immersed in water a thousand years are pure emerald in colour with a jade-like lustre. Those that have not been immersed as long as a thousand years are emerald green but lack the lustre . . . those that have been transmitted down from antiquity, not under water or earth but through the hands of men, have the colour of purple cloth and a red mottling like sand, which protrudes when excessive, and looks like first-quality cinnabar. When boiled in a pan of hot water the mottling becomes more pronounced.

Zhao Xigu, circa 1100 AD

During the Song dynasty (960–1279), when the connoisseur Zhao Xigu was writing, ancient Chinese bronzes were already being excavated and studied. Indeed, these antecedents of Chinese art were a source of inspiration and fascination long before an interest in the appearance of ancient bronzes arose in the West. This early appreciation of patinas is shown by the fact that Zhao Xigu was prepared to undertake experimental work on them, boiling them in water to observe the slightest reaction or change in appearance.

Bronze gave its name to an entire epoch of human history. Native copper could be hammered into useful artefacts by many different cultures, but bronze was the alloy that made so many significant advances possible. Before the advent of the Bronze Age, the utilization of native copper began very early, in the 7th or 8th millennium BC.

Native copper and Extracted Copper

Native copper is easy to cold-work into small, useful objects such as pins or needles, or into small ornaments such as necklace beads, dangles, or rings. Native copper use is sporadic and often difficult to determine. Copper in the native state may show similar trace impurity patterns to copper smelting from oxide ores or malachite. There is a tendency for native copper to show metallographic features that are distinct from smelted copper, but much of these differences may be lost if the copper is extensively worked and annealed. Figure 8.3 is a sample of native copper which has been worked and annealed, in which folding, characteristic of native copper use in some artefacts, can still be seen.

Folding and elongated twinned grains are the characteristics of native copper. Growth features may also be seen in which impurities in the native copper are segregated to certain planes or discrete regions of the copper. Metallography combined with careful determination of trace elements may still afford clues to the origin of the copper to allow some conclusions to be reached. Figure 8.1 shows the variation in mechanical properties of pure copper with percentage reduction in area by rolling. The grains present at 0% reduction have a Rockwell F scale hardness of about 45, which is comparable with the Vickers number. The copper grains become deformed and squashed on rolling to 68% reduction as seen in the photomicrograph in Figure 8.1.

The Rockwell hardness is now increased to 95, the tensile strength goes up but the elongation drops greatly after only 30% reduction in thickness, as the copper becomes brittle.

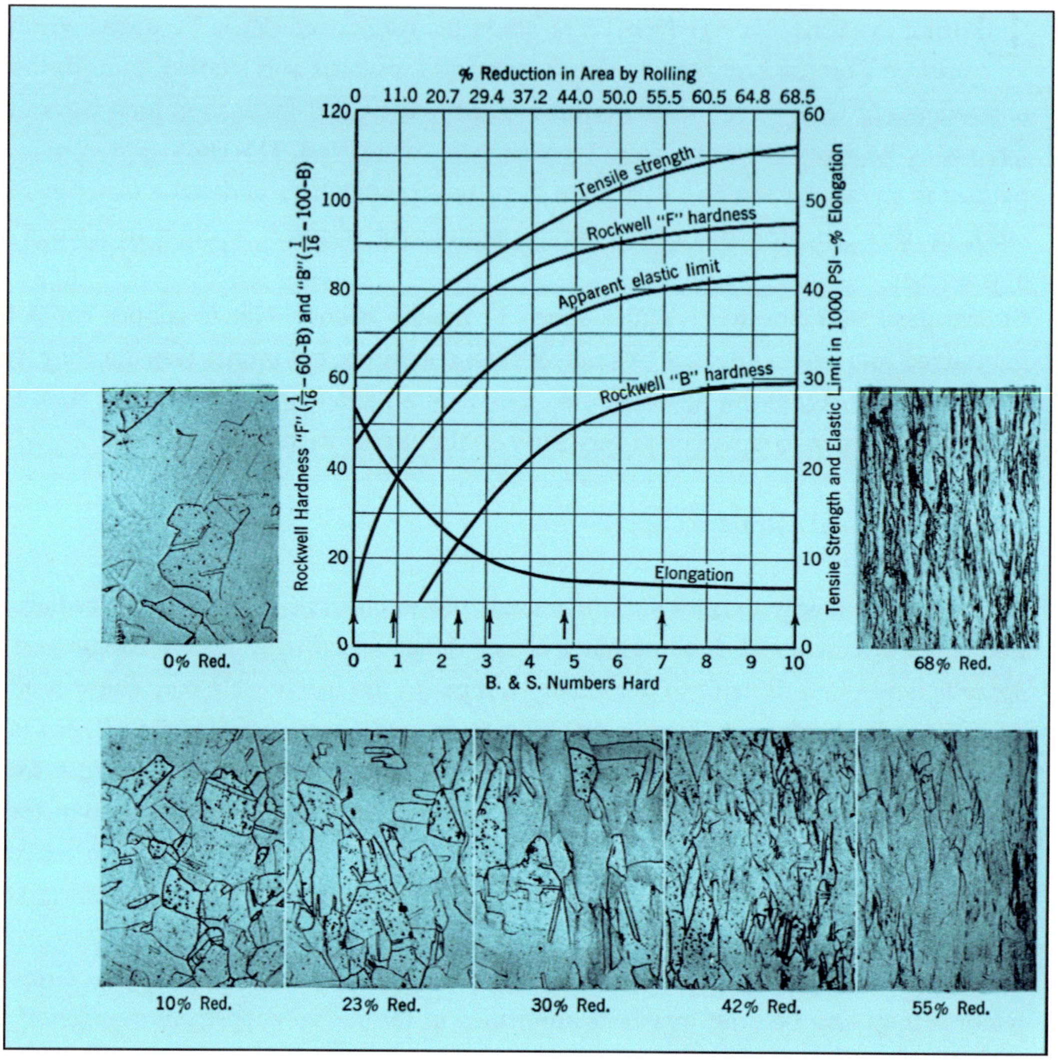

Figure 8.1 After Samans (1963). Mechanical properties and metallographic appearance of copper after various stages of cold-working by rolling. The reduction percentages shown range from 0% to 68%. Note that the grains of copper have a fibrous aspect to them at 68% reduction on working. In ancient copper objects, such working is found, particularly in small awls or pins or other objects which have been extensively cold-hammered to shape them. Note how the tensile strength and hardness increase as the extent of cold-working rises, but the elongation falls. As copper is cold-worked, the brittleness of the alloy also increases as it has less ability to move under an applied load. The hardness increases to about 120 Hv and the elongation falls off quite quickly, changing relatively little from 44% reduction in area to 68% reduction. Annealing of heavily worked copper will easily re-form recrystallized grains, the grains being twinned.

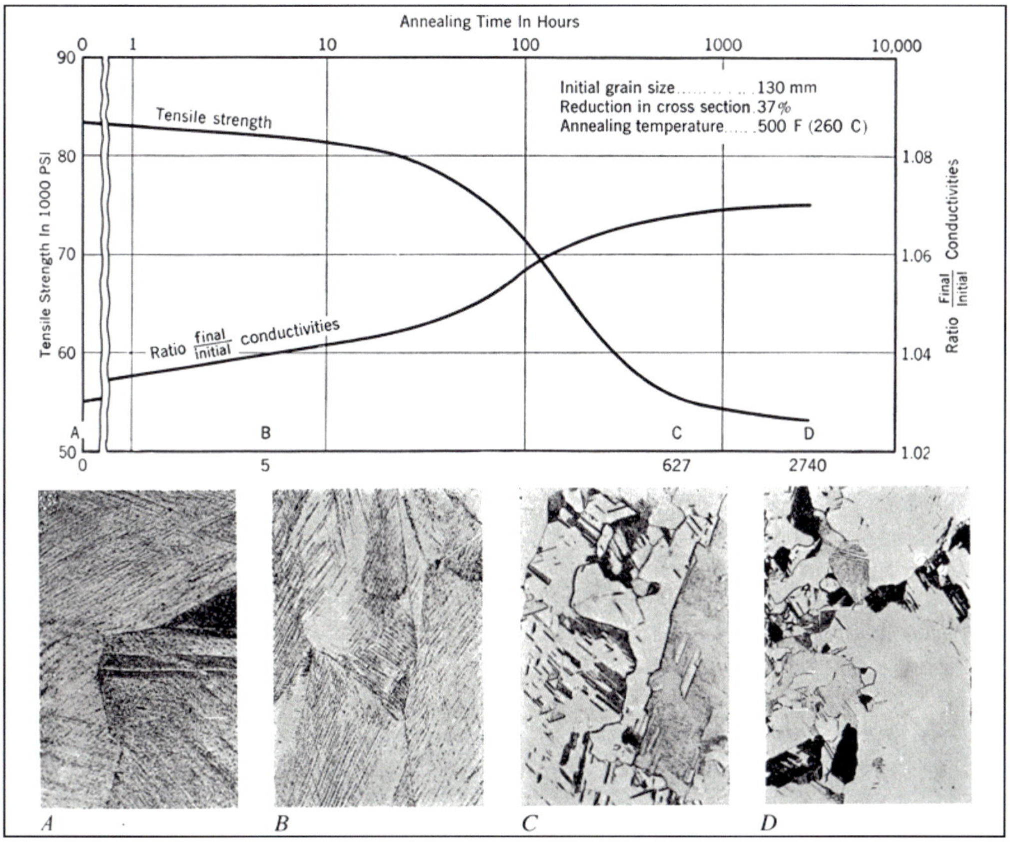

Figure 8.2 After Samans (1963), showing the change in properties of annealed copper at 260°C. The tensile strength drops and conductivity increases on annealing.

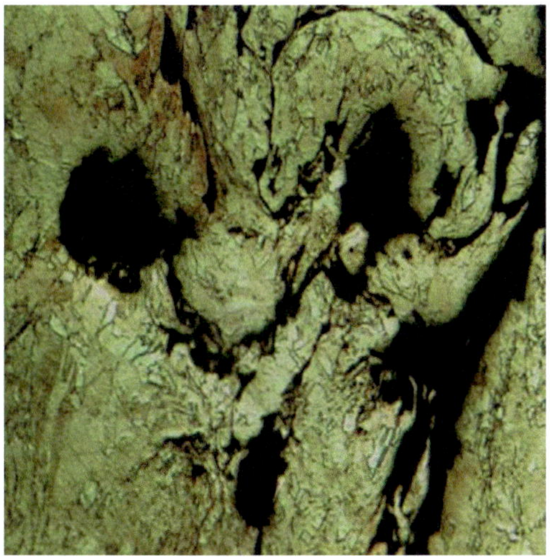

Figure 8.3. Photomicrograph of a worked and annealed sample of native copper from the Great Lakes Region, USA. The hammering and annealing of this native copper has created twinned crystals with straight twin lines and extensive folding of the native copper has occurred on working, a typical feature of native copper artefacts. Etched in alcoholic ferric chloride, x 150, with slightly oblique illumination.

Case Study Number 4. Neolithic copper artefacts from the ditched Neolithic enclosure of Makriyalos, Greece. Metallographic examination was carried out to ascertain whether native or smelted copper was in use in this site which dates back to the 3rd Millennium BC. Metallographic study often can provide clues to the use of native copper, especially if folding or other unusual features are seen in the resulting sections. The site of Makriyalos had been known for some time, but as a result of a road-widening scheme between Athens and Thessaloniki, rescue excavations were undertaken, as seen in Figure 8.4, when the true extent of the site was. discovered. It is the largest ditched Neolithic enclosure in the whole of Europe. It covers some 50 hectacres. About 20% of the site was destroyed in 1970 during roadworks but enough remains for good excavation data. The Neolithic occupation was dated primarily by pottery.

The present chronological estimate places Makriyalos I and II between the limits of the Late Neolithic, which lasted from ca 5400/5300 BC to ca 4700/4500 BC. Further details of the site are given in the paper by Pappa and Besios (1999). Very little is known about metallurgical practices during the Late Neolithic, mainly because of the small number of artifacts found and analysed from stratified deposits (McGeehan-Liritzis and Gale 1988, Demoule and Perles 1993). A number of recent studies have added to our knowledge of the region and the early dates of copper metallurgy in the Balkan region. For example, Pernicka et al (1997) discuss the prehistoric copper from Bulgaria and its composition. Gale et al (2003) also look at the early copper metallurgy in a wider context, as do Hauptmann et al (1999) and Bougarit (2007). Sixty-five objects were found in Makriyalos II of which nearly half are small cylindrical beads made of curved sheet copper. There are also pins, awls, pieces of thin wire, a small chisel, a sheet in the shape of a 'ring idol' and other unidentified objects. 21 of these have been studied by the author.

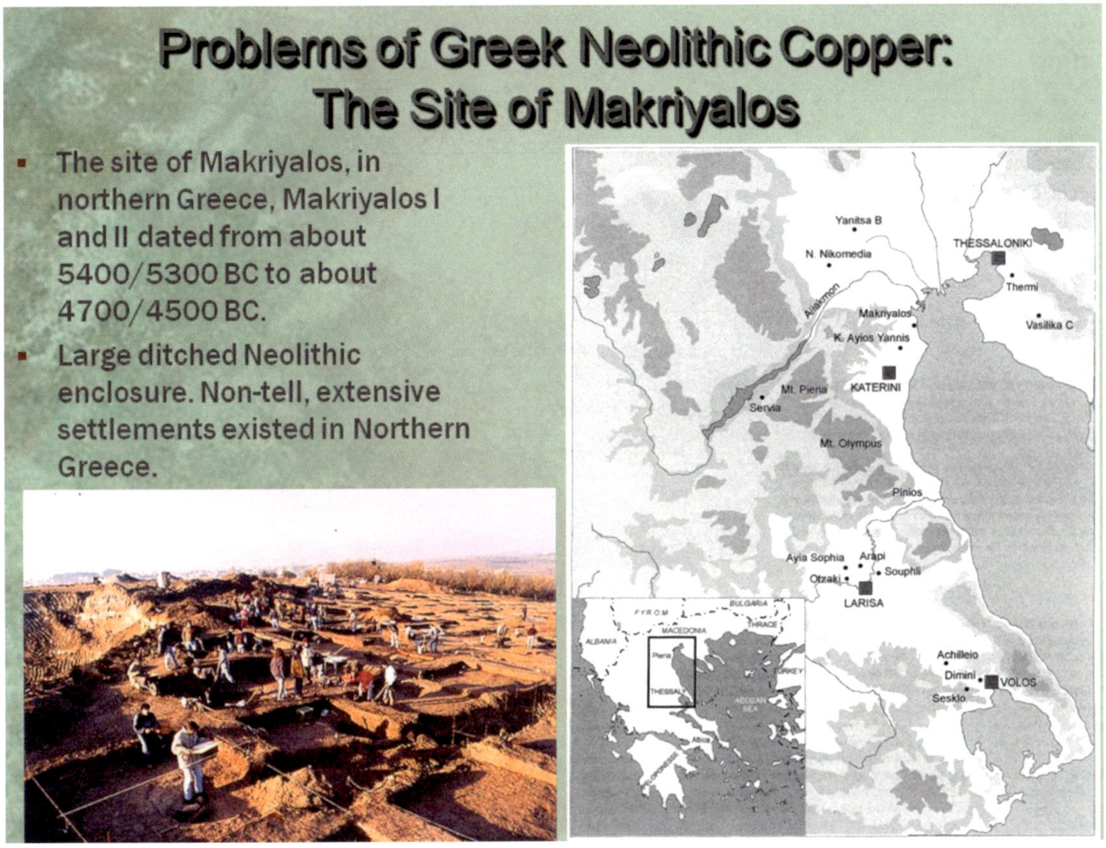

Figure 8.4 The location of the site and a view of the rescue excavations. The beginnings of the site have been dated to the Late Neolithic, from about 5400-4900 BC by Pappa and Besios (1999).

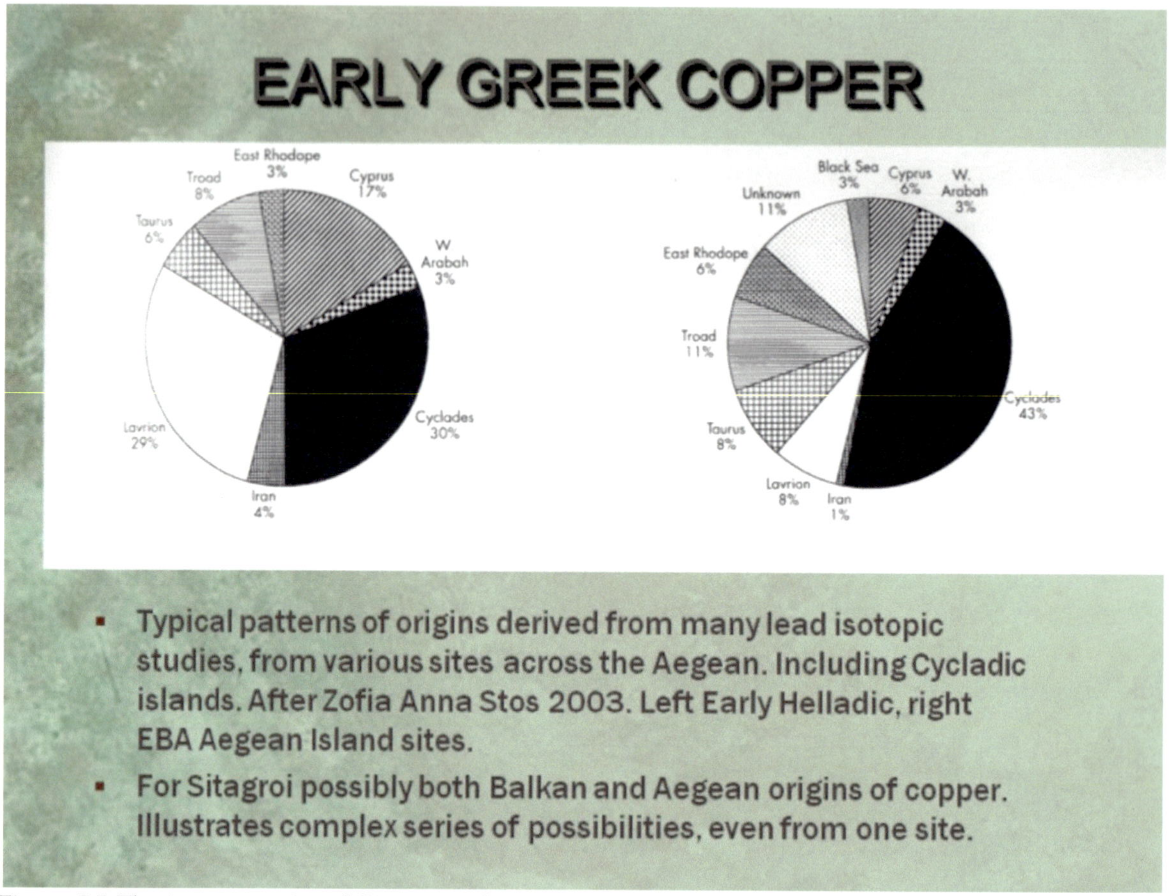

Figure 8.5 The important site of Sitagroi and Makriyalos may be linked by the type of copper utilized. No lead isotope work has yet been done on the Makriyalos material, but these charts show the typical distribution of sources.

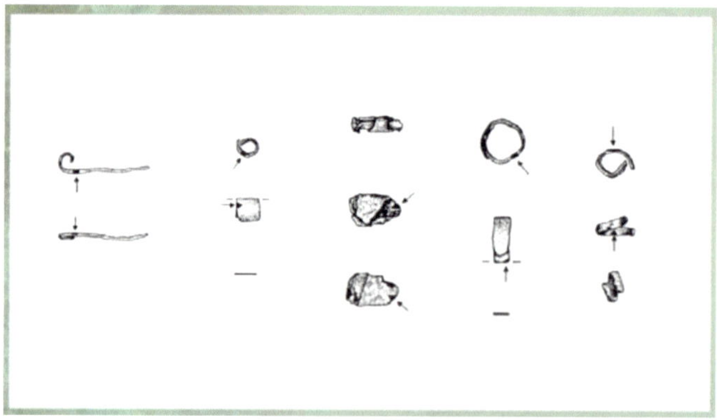

Figure 8.6 Some of the small excavated fragments, pins and rings from the Makriyalos II excavations. Shown at half their actual size.

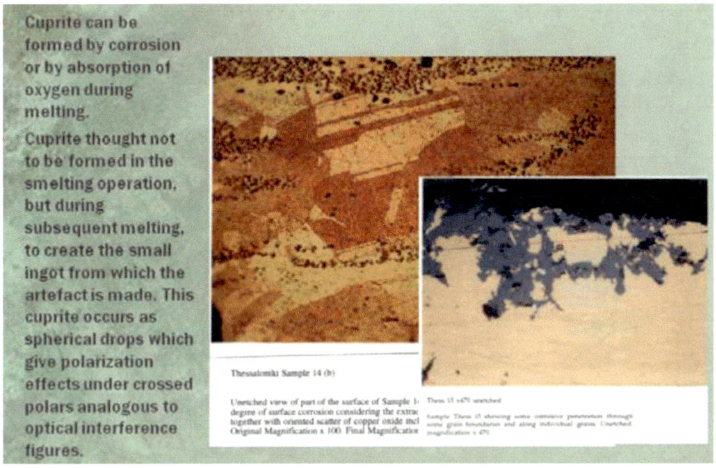

Figure 8.7 Two of the Makriyalos objects showing typical microstructures with extensive cuprite inclusions. Note that in this early metalwork there are no sulphide inclusions. The cuprite, which I term primary cuprite, to distinguish it from cuprite formed by subsequent corrosion processes, or larger cuprite globules from surface oxidation. Small spherical droplets of cuprite give a stationary cross under crossed-polars, corresponding to the optical indicatrix of these cubic minerals. The indicatrix is usually applicable to transmission effects, but the pleochroic reaction of cuprite provides the same optical phenomenon.

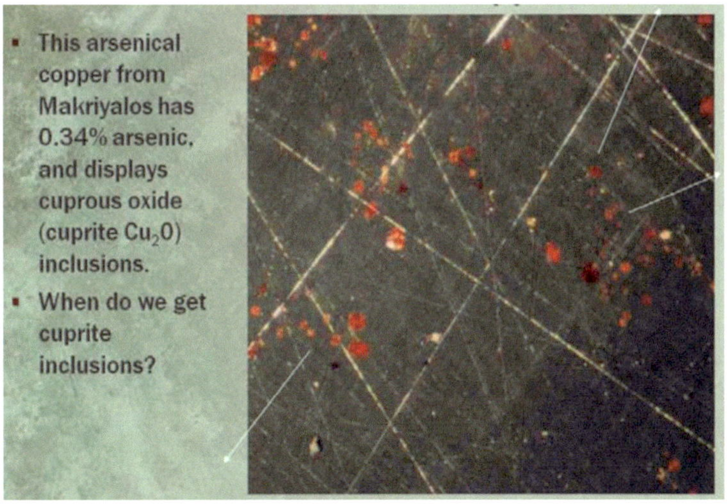

Figure 8.8 Many of the objects from the site show that smelted copper was used, some with small traces of arsenic. There are no sulphide inclusions showing that copper sulpharsenide ores were not used for the smelting operation. It may well be that at this early stage some of the oxidized copper ores available contained a little arsenic.

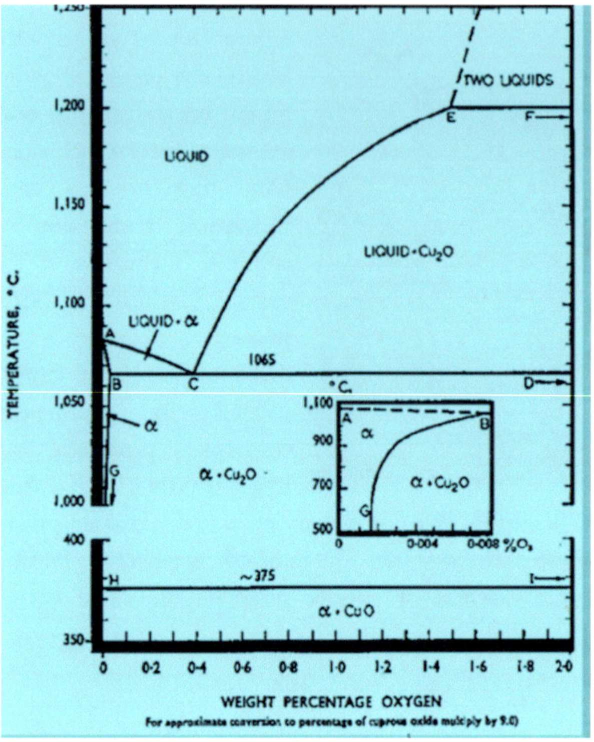

Figure 8.9 The copper-oxygen system. Oxide eutectic formation and copper oxide inclusions are important in discussions of early copper extraction.

The copper-oxygen system has an interesting eutectic point, C, which lies below the melting point of pure copper. On cooling below the liquidus temperature, shown on line AC, as the temperature falls, copper nuclei grow in the liquid, gradually increasing their size, causing the liquid to become richer in oxygen. The composition of the liquid, follows the line AC until, at the eutectic point C, the liquid remaining between the primary grains solidifies, to form copper and cuprite. During fire-refining, air may be picked up by molten copper. Oxygen can therefore be absorbed by this copper. Hydrogen is also picked up in the furnace, particularly during reducing or 'poling' operations, and co-exists in equilibrium with the oxygen. On solidification, this equilibrium is disturbed, the oxygen and hydrogen reacting together to form steam, which can become entrapped in the casting. By carefully reducing the oxygen to a controlled level, the volume of steam cavities may be made to counteract the natural contraction on solidification of the cast metal. This can produce castings with a level top surface rather than a concave casting surface. The oxygen in tough-pitch copper can be converted to steam when the metal is heated in a reducing atmosphere such as may be used in annealing and heating operations. This is known as "gassing" because hydrogen diffuses into the metal and reacts with the cuprous oxide to form insoluble water vapour. The expansion forces the grains apart and can embrittle the metal as a result. Hauptmann et al (2002) maintain, in an opposite argument to that of Tylecote, that cuprite cannot be present in smelted copper directly, that even in oxhide ingots this is evidence for the remelting of the copper. They refer to Figure 8.10 for their argument. We always suspect that copper with cuprite eutectic has picked up this oxygen in a subsequent melting operation. Most of the Makriyalos copper artefacts have cuprite inclusions from the primary melting phase of fabrication.

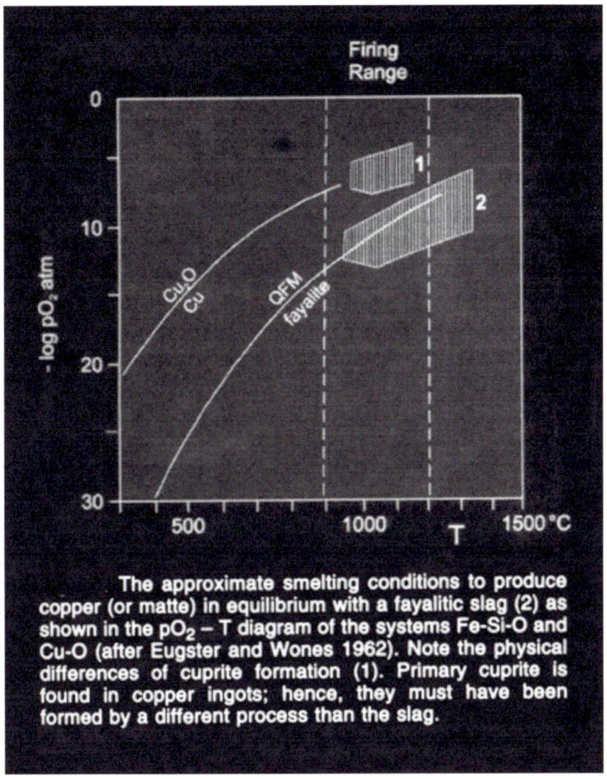

Figure 8.10 The occurrence of oxides in copper can be linked to firing temperatures and the partial pressure of oxygen. After Hauptmann et al (2002).

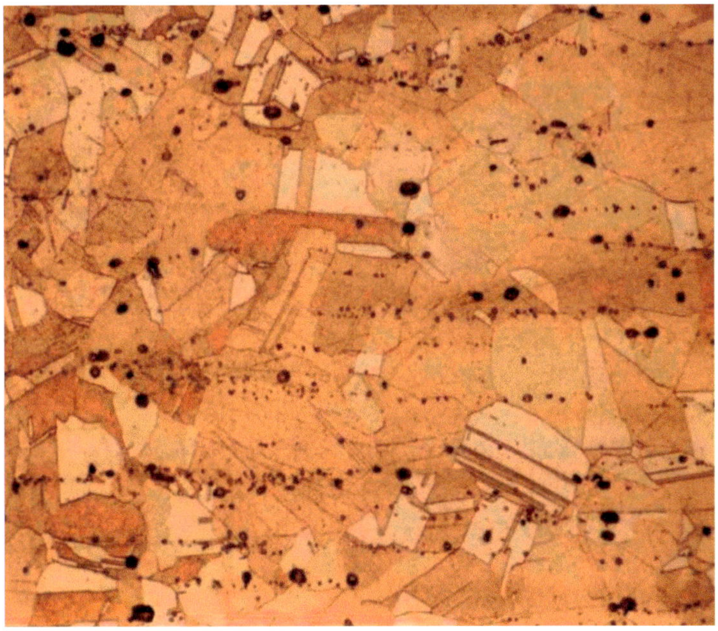

Figure 8.11. Sample 11. Not a typical native copper microstructure. Shown here etched in alcoholic ferric chloride, x150. The cuprite globules are dissolved away, leaving small dark holes in a worked and annealed grain matrix of twinned crystals, showing annealing twins with straight twin lines, showing that annealing was the final working operation

A paper of relevance here is that of Wayman (1999) who discusses the microstructure and non-metallic inclusions in native copper which should be consulted for further discussion..

Sample: Sample No. 11, Neolithic Greek copper fragment

	ppm	Detection Limit		ppm
*Aluminum	ND	40	*Mercury	ND
*Antimony	265	7	*Molybdenum	ND
*Arsenic	1940	10	Neodymium	ND
*Barium	ND	1	*Nickel	59
*Beryllium	ND	1	Niobium	ND
*Bismuth	55	1	Osmium	ND
*Boron	ND	40	Palladium	ND
Bromine	ND	900	Platinum	ND
*Cadmium	ND	2	Praseodymium	ND
*Calcium	ND	10000	Rhenium	ND
Cerium	ND	1	Rhodium	6
Cesium	ND	1	Rubidium	ND
*Chromium	ND	50	Ruthenium	ND
*Cobalt	ND	2	Samarium	ND
*Copper	MATRIX		*Selenium	ND
Dysprosium	ND	1	*Silver	294
Erbium	ND	1	*Sodium	ND
Europium	ND	1	*Strontium	ND
Gadolinium	ND	1	Tantalum	ND
Gallium	ND	20	Tellurium	ND
Germanium	ND	1	*Thallium	ND
*Gold	ND	2	Thorium	ND
Hafnium	ND	1	Thulium	ND
Holmium	ND	1	*Tin	3
Iodine	ND	900	*Titanium	ND
Iridium	ND	1	Tungsten	ND
*Iron	9500	600	Uranium	ND
Lanthanum	ND	1	*Vanadium	ND
*Lead	1740	2	Ytterbium	ND
*Lithium	ND	2	Yttrium	ND
Lutetium	ND	10	*Zinc	ND
*Magnesium	ND	70	Zirconium	ND
*Manganese	ND	10		

Date Analyzed: 1-16-97
MATRIX = Major Element * = Quantitative analysis
Elements Not Analyzed: All Gasses, C, P, S, Si, Sc, In, Tb

Figure 8.12 ICP-MS data, confirming the metallographic interpretation. With small amounts of lead, antimony and arsenic, this artefact does not appear to have been made from native copper.

Sample: Sample No. 12, Neolithic Greek copper fragment

	ppm	Detection Limit		ppm	Detection Limit
*Aluminum	ND	40	*Mercury	ND	6
*Antimony	ND	8	*Molybdenum	ND	6
*Arsenic	ND	10	Neodymium	ND	1
*Barium	3	1	*Nickel	112	8
*Beryllium	ND	1	Niobium	ND	1
*Bismuth	ND	1	Osmium	ND	1
*Boron	ND	40	Palladium	ND	1
Bromine	ND	1000	Platinum	ND	1
*Cadmium	ND	2	Praseodymium	ND	1
*Calcium	ND	10000	Rhenium	ND	1
Cerium	ND	1	Rhodium	6	1
Cesium	ND	1	Rubidium	ND	1
*Chromium	189	60	Ruthenium	ND	1
*Cobalt	16	2	Samarium	ND	1
*Copper	MATRIX		*Selenium	ND	90
Dysprosium	ND	1	*Silver	19	1
Erbium	ND	1	*Sodium	159	70
Europium	ND	1	*Strontium	ND	1
Gadolinium	ND	1	Tantalum	ND	1
Gallium	ND	20	Tellurium	ND	1
Germanium	ND	1	*Thallium	ND	1
*Gold	ND	2	Thorium	ND	1
Hafnium	ND	1	Thulium	ND	1
Holmium	ND	1	*Tin	ND	1
Iodine	ND	1000	*Titanium	ND	40
Iridium	ND	1	Tungsten	ND	80
*Iron	29300	700	Uranium	ND	1
Lanthanum	ND	1	*Vanadium	ND	2000
*Lead	57	2	Ytterbium	ND	1
*Lithium	ND	2	Yttrium	ND	1
Lutetium	ND	10	*Zinc	ND	60
*Magnesium	ND	80	Zirconium	ND	1
*Manganese	66	10			

Date Analyzed: 1-16-97
MATRIX = Major Element * = Quantitative analysis
Elements Not Analyzed: All Gasses, C, P, S, Si, Sc, In, Tb

Figure 8.13 ICP-MS results for sample number 12, showing that the copper is relatively pure. Iron, chromium and manganese may very well be corrosion penetration interference in this case, as the samples available for the ICP-MS study were very tiny. Notice that tin, arsenic and antimony are all not detected in this analysis, which is fairly typical for the copper artefacts from this site.

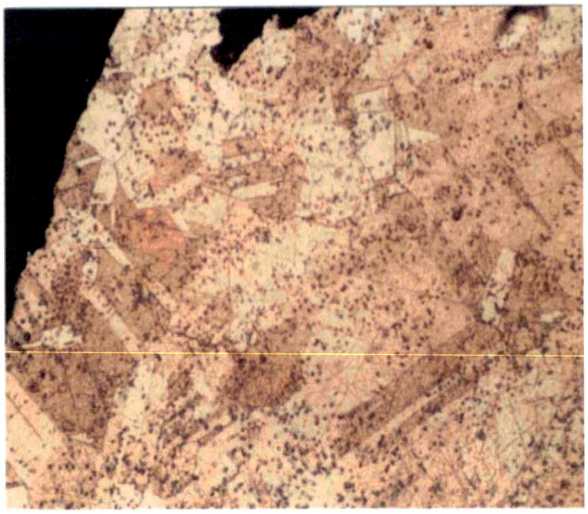

Figure 8.14 Another example of a copper object from the site which is permeated with extensive copper oxide globules from melting of the copper to create the ingot from which the copper object was made. Etched in alcoholic ferric chloride x 170.

Figure 8.15 Makriyalos sample 17. Etched in alcoholic ferric chloride and under quarter-wave plate. Same details can be seen as for the last photomicrograph: a worked and annealed microstructure with rather less cuprite globules than in some of these copper artefacts. X130

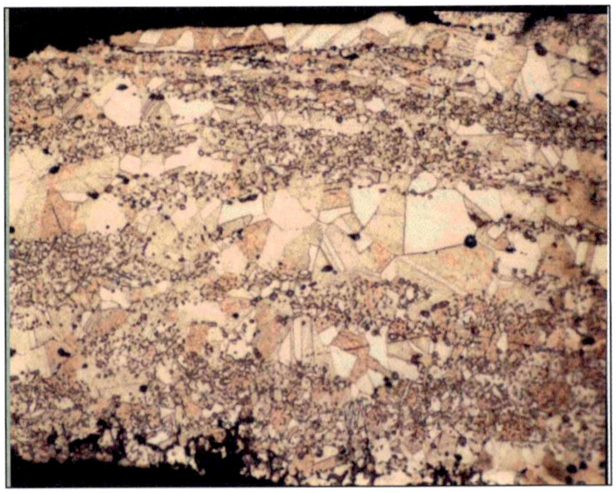

Figure 8.16 . Makriyalos Sample 20. View of the microstructure showing extensive stringers of copper-copper oxide eutectic, due to considerable oxygen absorption into the copper when melting the ingot. The grains are recrystallized and twinned showing that artefact had been shaped by working and annealing. Etched in alcoholic ferric chloride x120.

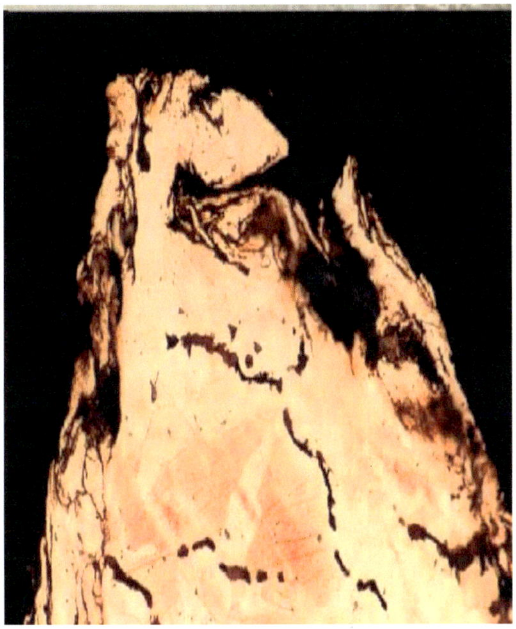

Figure 8.17 a

Figure 8.17b

Figure 8.17a and 8.17b Two small objects from the site of Makriyalios sectioned and etched. The section in 8.17a shows a strange arrangement of porosity, large grains and folding of the copper towards the surfaces of the objects. These are all features associated with native copper. The section at the right shows small cuprite inclusions and well-worked and recrystallized grains with all the appearance either of native copper which has been completely molten and worked and annealed into a finished object, or a copper derived from already smelted metal rather than native copper. Objects are etched in alcoholic ferric chloride, x95.

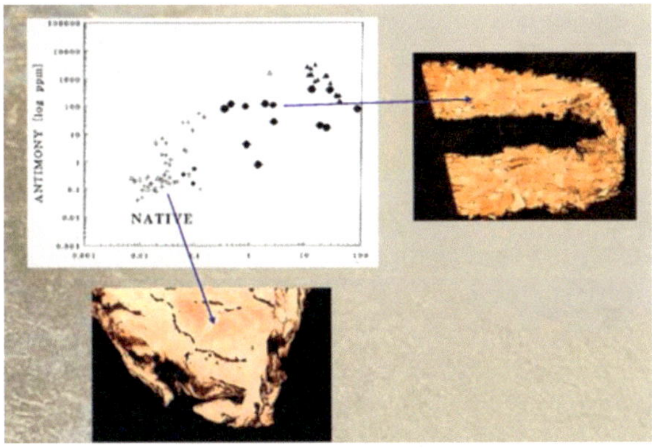

Figure 8.18 The low gold (bottom x axis) and low antimony (y axis) are more commonly associated with native copper, while higher gold and antimony content may be associated with smelted copper. This distinction may be useful but in association with the microstructure of the two artefacts from this important Greek site, are strong evidence for the argument that the site provides evidence of the shift from native copper utilization to smelted copper in the Neolithic period. In general most of the metalwork appears to be of smelted copper, but a few pieces are probably of native copper.

The alloys of copper and arsenic: arsenical copper.

The gossan, the weathered cap of copper mineral deposits may be leached by weathering and enriched in arsenic and sometimes antimony as well. The extraction of copper from these kinds of deposits and the smelting of them naturally results in the production of copper-arsenic alloys, and when the weathered zone is exhausted, the arsenic concentration may drop to levels unsuitable for alloying purposes. There are other reasons why copper-arsenic alloys were utilised and why they may have been replaced or supplanted by tin bronzes in later periods. Arsenic fumes are highly toxic, although since many smelting operations were carried out in specialized areas away from living spaces, this may have not been a sufficient reason to give up arsenic in favour of tin. In some regions it is possible that speiss may have been of benefit in allowing alloying of copper with speiss containing some arsenic. The full story of arsenical copper production in some regions of the world has yet to be unraveled.

Tin bronzes are somewhat more controllable than arsenical copper in a number of respects, although how important these issues of control are is hard to assess. Arsenic is easily lost from copper alloys by continued melting and casting, even though it is hard to get rid of all of the arsenic entirely by this means. But a drop in arsenic content means less desirable mechanical properties. Remelting of tin bronze results in very little loss of tin in comparison to the loss of arsenic. We also have to consider technological preferences of certain societies. If the control of tin is an important facet of use bronze production, then some cultures may have exercised state or imperial control over alloy production, which is hard to achieve with arsenical copper, because the ores are not necessarily distinct.

Tin deposits may occur in very different geological settings to copper deposits and trade may be an important aspect of the distribution of tin in the form of ingots. There may be many different reasons or different kinds of explanation for the changes which occurred in the early Bronze Age in the Old World, or in early alloying in South America, particularly in Peru, Argentina, Chile, Ecuador and parts of Mexico, where metallurgy developed independently from the Old World, but with the same kind of transitions from arsenical copper to tin bronze in regions of the continent which utilised smelted copper.

Cosmelting experiments to produce copper-arsenic alloys were carried out by Lechtman in 1984. The experiments included crucible and furnace techniques utilizing ores collected in the Peruvian Andes. Smelting charges consisted of mixtures of copper oxide ore with either copper sulpharsenide or iron sulpharsenide ore. The experimental furnaces were constructed to resemble furnaces excavated at Batán Grande, a large ore smelting site on the Peruvian north coast where copper-arsenic alloys were produced during the Middle Horizon and later. The cosmelting experiments yielded coherent copper-arsenic alloy ingots over a wide range of oxide to sulpharsenide ore mixtures. Crucible and furnace charges containing ratios of between 2:1 and 4:1 oxide:sulpharsenide mineral produced clean metal, fully separated from slag or matte byproducts. The sulphide ores were not roasted prior to smelting and in the experiments, no flux was added to the charges. The copper-arsenic alloys used in ancient Peruvian artifacts could have been made easily, deliberately or accidentally, by cosmelting procedures, which were in all probability, sought by the metalsmith for copper-arsenic alloy production. For further reading on these issues, which continues to be a source of considerable interest in the early development of metallurgy, see Hosler (1988,1994); Lechtman (1996); Charles (1979, 1980); Tylecote, (1976); Muhly (1988), Ravich and Ryndina (1995) and the recent issue of World Prehistory already referenced in Chapter One.

Copper-arsenic alloys are now considered much more important than they once were. There is still debate over whether they were deliberately produced from the smelting of copper ores with added arsenic minerals, or whether they began as accidental arsenical impurities in copper ores themselves. Probably both explanations are correct in different times and places. Because it is very hard to control the percentage of arsenic produced in copper-arsenic alloys as compared with tin bronzes, the percentages may vary considerably, and can be very low, for example, 0,3% arsenic, in which case the inadvertent presence of the metal may be more likely to be responsible, or the content may be 3.7%, in which case a deliberate alloy may be involved, especially if part of a coherent culturally based arsenical copper alloy utilization.

Copper-arsenic alloys are very difficult to homogenize compared with tin bronzes, and this is due to the steep fall between the solidus and liquidus curves shown in the phase diagrams of Figure 8.19 and 8.20. There are differences between the two, and the latest diagram, Figure 8.20 is probably preferable. The alloys of arsenic usually range from fractions of a percent, up to about 8-9% of arsenic. Alloys with higher arsenic contents than this are quite unusual.

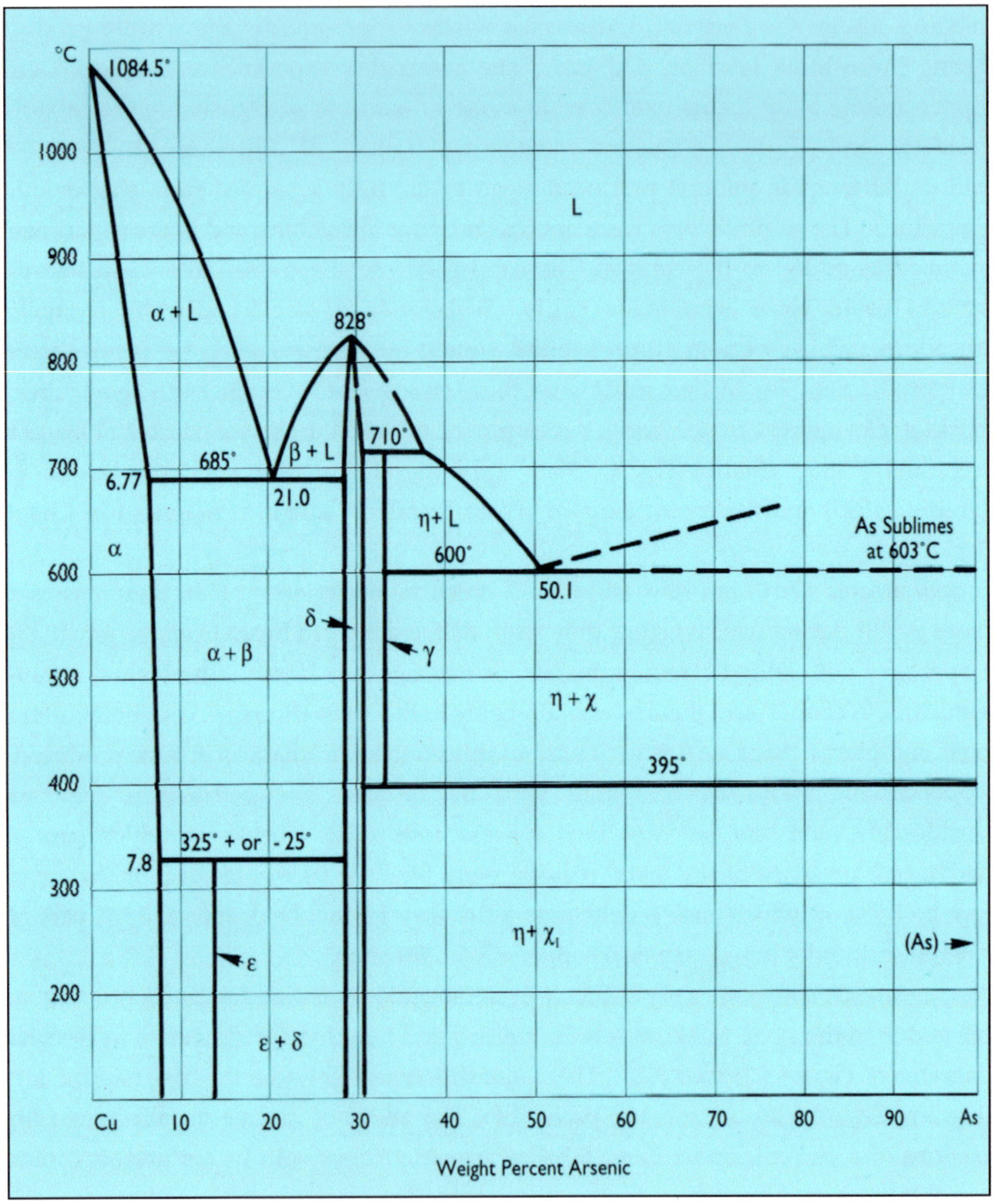

Figure 8.19 The copper-arsenic phase diagram (After Scott 1991)

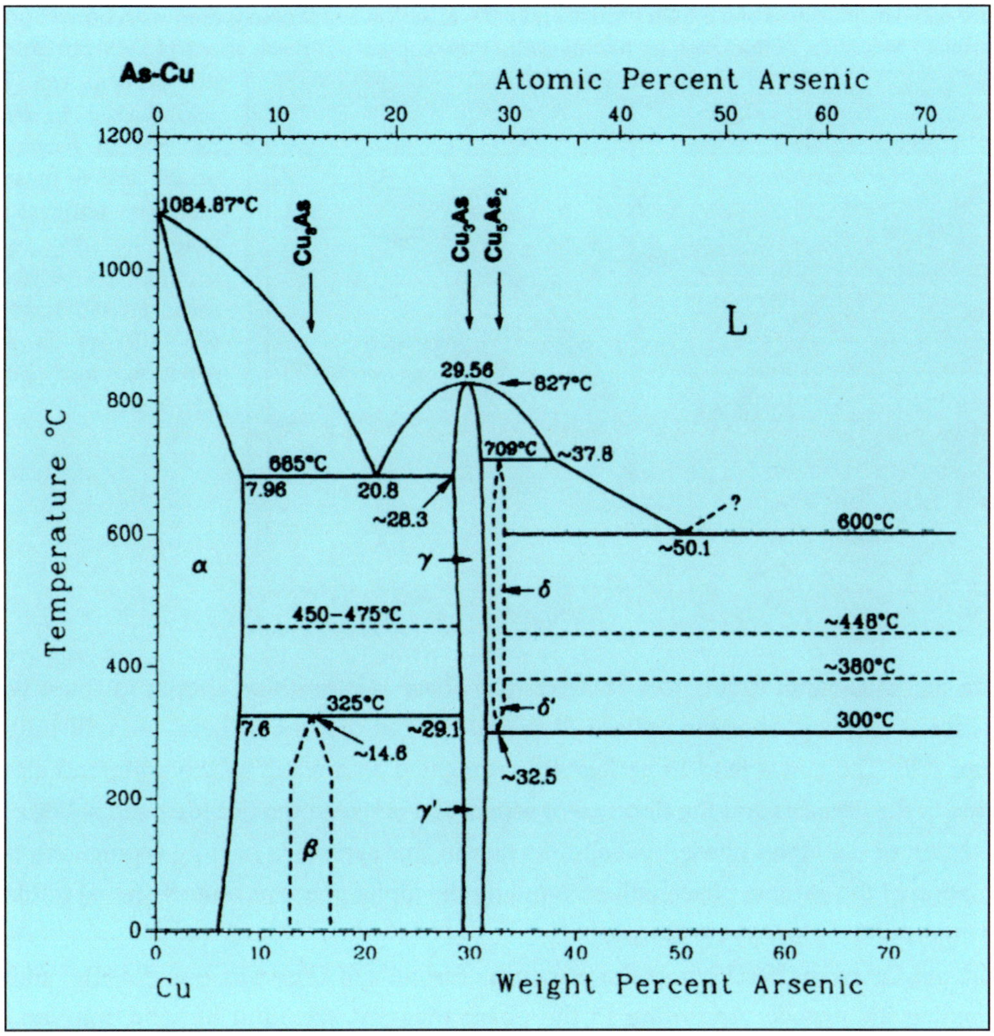

Figure 8.20. The copper-arsenic phase diagram according to Subramanian and Laughlin (1988). This diagram is the currently preferred version of the labelling of this system which has undergone some improvements by Subramanian and Laughlin.

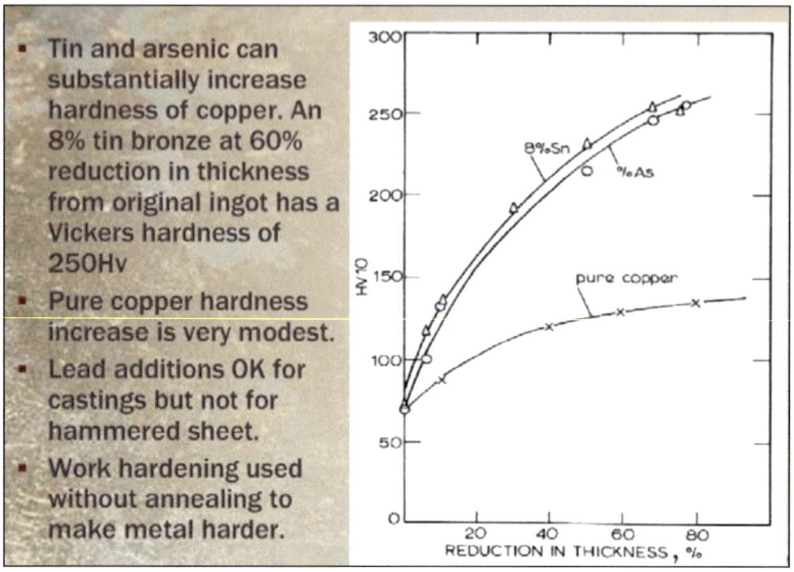

Figure 8.21 The hardness of a 8% tin bronze and a 5% arsenical copper compared with pure copper in terms of reduction of thickness on hammering. Pure copper can reach about 140 HV but an 8% tin bronze can be cold-worked to over 250 HV. A hardness of 250 HV is greater than the hardness of iron in its soft condition. Carbon steel can often reach 600 HV so is far superior, but pure iron is not.

There are substantial differences between the phase relationships shown in these two versions of the copper-arsenic phase diagram. For the discussion here, we will utilize Figure 8.20. The alloys tend to be heavily segregated as a result of the wide and steep incline in the liquidus and the distance of separation between the liquidus and solidus in the region of the alpha phase. As solidification in this system is heavily segregated, the formation of the gamma phase will accompany the alpha phase as eutectic ($\alpha+\gamma$) infill in cast ingots with as low as 2% arsenic.

Budd and Ottaway (1991) report the presence of islands of Cu_3As in experimental alloys containing 2% arsenic. According to the phase diagram, the limit of solid solution of arsenic in copper is 7.9%. Between 7.9% and 20.8% of arsenic, the second phase, gamma, forms which is an ordered phase of composition Cu_3As. These ordered phases have essentially no ductility, so if the arsenic content of copper-arsenic alloys becomes too high (circa 13%) then they may crack badly on hammering. Many alloy microstructures with lower arsenic contents are characterized by the presence of primary dendrites of the alpha phase, with an infill of the ($\alpha+\gamma$) eutectic. Since the alloys tend to be heavily cored from the cast state, the eutectic may be somewhat divorced and appear as principally gamma phase precipitates surrounded by cored alpha grains or dendrite

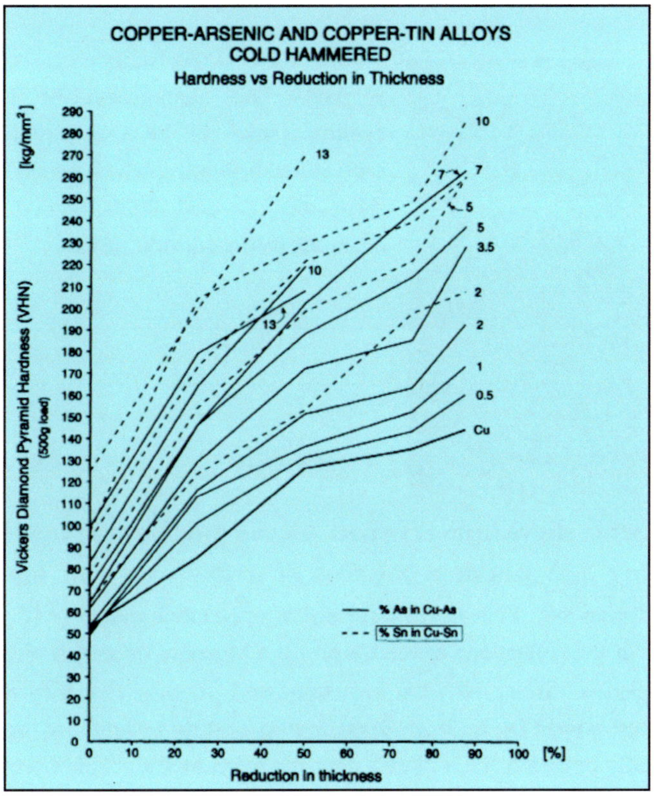

Figure 8.22 After Lechtman and Kline (1999) showing the variation in hardness for both copper-tin alloys and copper arsenic alloys. The copper arsenic alloys, with relevant percentages by weight are shown in this diagram. A 7% arsenical copper can reach 280HV with 90% reduction in thickness on hammering. A 10% tin bronze can reach 290HV.

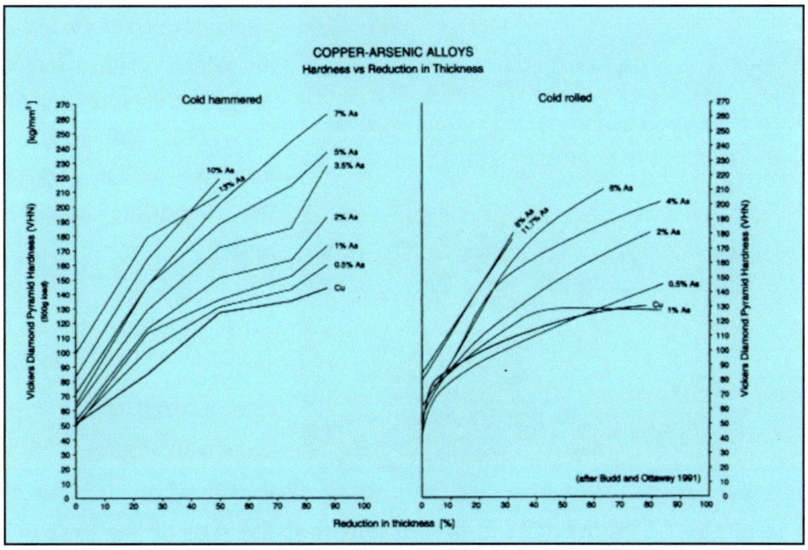

Figure 8.23 The hardness of copper-arsenic alloys compared with percentage reduction in thickness, showing the very high hardness values which can be attained on cold hammering. A 3.5% arsenic bronze, for example, can reach 230 Hv after 90% reduction. Other compositions of interest can be seen in these two figures. After Lechtman and Kline (1999)

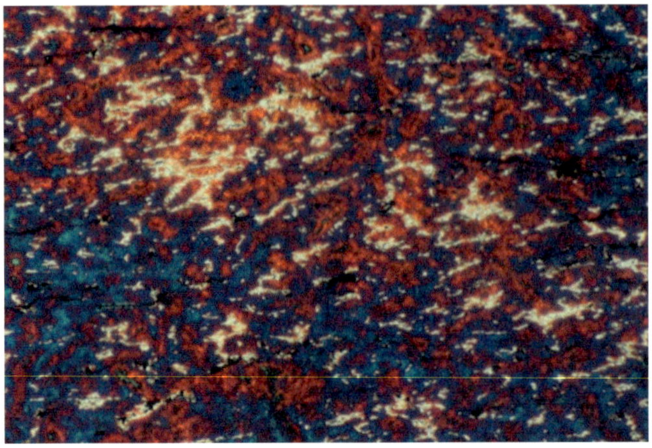

Figure 8.24 8As92Cu cast, cold worked. The segregation of the epsilon phase can be seen together with extensive coring which occurs in these cast arsenical alloys. Etched in alcoholic ferric chloride, x85.

Case Study Number 5. Copper-arsenic alloys from Ecuador. Metallographic examination was carried out to assess casting and working practices of a few axes and other implements Figure 8.25 Cast Ecuadorian axe. These axes were cast in open stone moulds, often made in some numbers. This axe is in the collections of the Cambridge Museum of Archaeology and Anthropology and the section also confirms, from the inclusions and porosity that they are indeed cast flat in the mould, as they would be from an open mould casting technology, or a modification of this in which a partially enclosed lid is placed over the open mould. This we can see from the directional solidification and the presence of oxide slag to the top of the axe section, corresponding to the upper surface of the axe in the mold. The arsenical content creates a heavily segregated cast structure from the molten state as the dendritic structure forms.

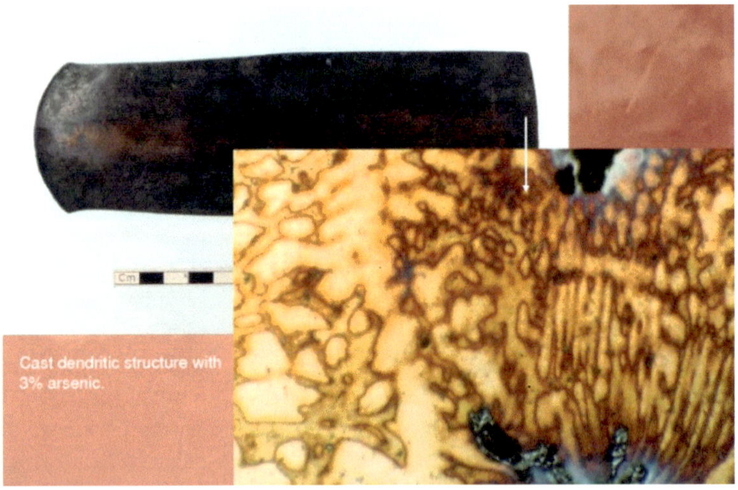

This penannular nose ring whose microstructure is shown in Figure 8.26, is 17 mm in internal diameter and is made in a circular copper alloy rod, about 3.1 mm in diameter. The grain structure of Figure 8.26 is unusual in that the size of the grains is much smaller on one side than the other. This may be due to differential working of the arsenical copper on one side of the nose-ring than the other, but such features are rarely seen in worked and annealed arsenical copper.

Figure 8.26 Ecu1294 From the site of La Compañia, a penannular nose ring of solid construction in arsenical copper. Etched in 10% aq. potassium peroxydisulphate, followed by saturated sodium thiosulphate and 1 gram of potassium metabisulphite in 100ml of solution. X140

Figure 8.27 From the site of La Compañia, a penannular nose ring of solid construction in arsenical copper. Etched in 10% aq. potassium peroxydisulphate, followed by saturated sodium thiosulphate and 1 gram of potassium metabisulphite in 100ml of solution. X200.

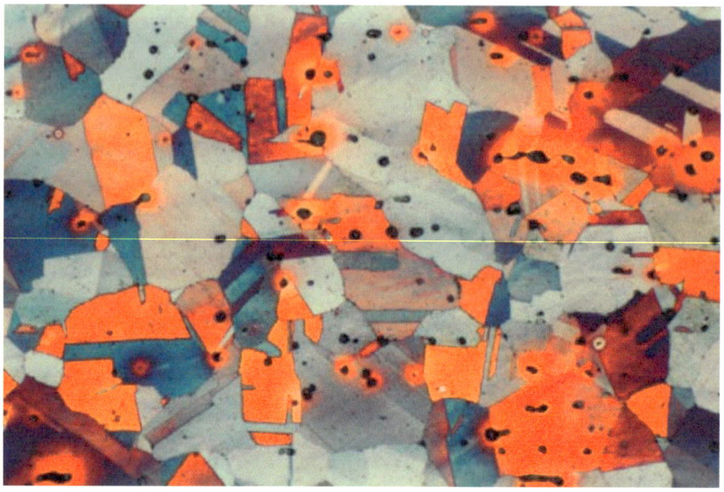

Figure 8.28 From the site of La Compañia, a penannular nose ring of solid construction in arsenical copper. Etched in 10% aq. potassium peroxydisulphate, followed by saturated sodium thiosulphate and 1 gram of potassium metabisulphite in 100ml of solution. X400.

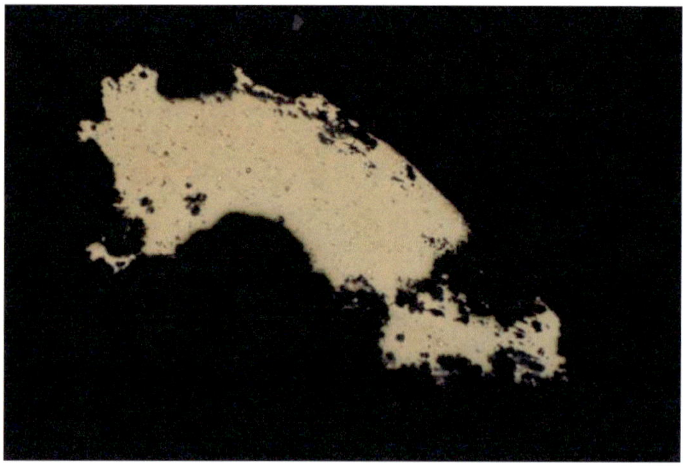

Figure 8.29 Chinese copper-arsenic alloy hoe excavated from the site of Hejiahsan, in west Yunnan, late Spring and Autumn period to middle Warring States period. x50. The sample is from the blade of this heart-shaped hoe. Unetched.

These samples are from the research of Yali Yun (Yali 2010) into the metallurgy of the ancient west Dian region of Yunnan province in the Bronze Age from the 12th century BC to 5th century BC. Erhai is located in the eastern region of Yunnan province, connecting inner China with southeast Asia. Erhai lake in the centre of this region was the communication hub of the Southwest silk road. The objects from the excavations numbered over 3000 artefacts of which 2600 were bronzes of different kinds. The metals can be divided into a number of groups (a) unalloyed copper, with iron tin or arsenic in very low concentrations (b) tin bronzes, from 2.6% to 22.4% tin, with decorative metalwork having higher tin contents and weapons with less than 10%. (c) Copper-tin-lead ternary alloys with lead between 6 and 13% (d) copper-arsenic alloys with arsenic between 2.7% - 7.7%, and one piece with 23% arsenic for decorative use. The normal range of arsenical content of these alloys are shown here in Figures 8.29-8.34 and show the working and annealing to shape of these alloys, together with persistent arsenic segregation and the presence of small sulphide inclusions from the smelting of the copper from sulphide ores. The inclusion level is very modest, the grain size is moderate, showing well-developed copper extraction metallurgy and relatively clean alloy production.

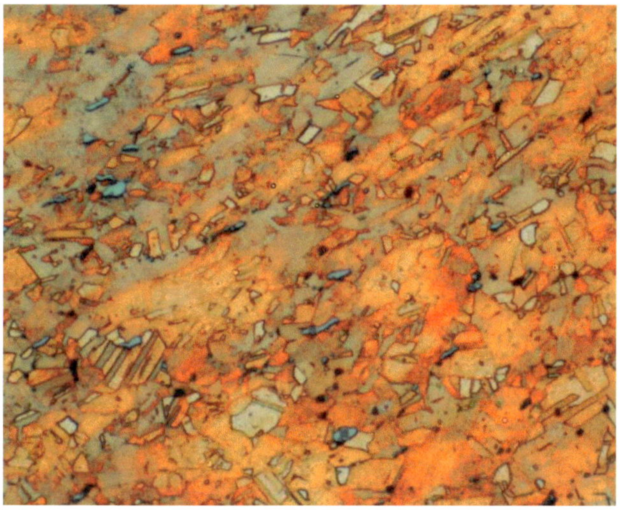

Figure 8.30 Chinese copper-arsenic alloy hoe excavated from the site of Hejiahsan, in west Yunnan, late Spring and Autumn period to middle Warring States period. Photomicrograph of the blade showing blue copper sulphide inclusions along the length of the hoe from working to shape, with recrystallized grains and segregation, producing variable green and orange colours from unequal arsenic distribution. X 240. Etched in Klemm's reagent number 2.

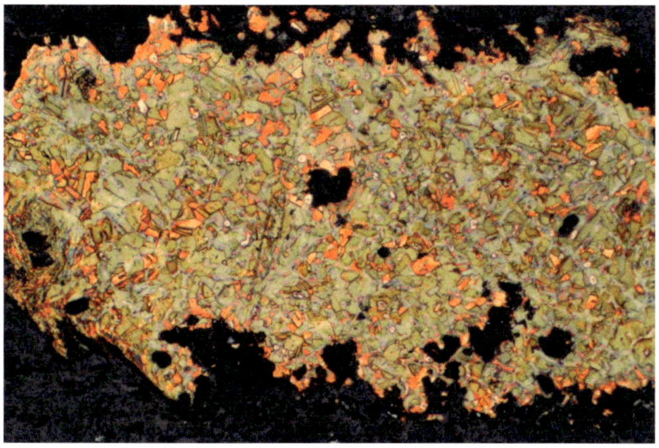

Figure 8.31 Sample 9862 Yun Yali Chinese copper-arsenic alloy hoe excavated from the site of Hejiahsan, in west Yunnan, late Spring and Autumn period to middle Warring States period. Photomicrograph of the blade showing blue copper sulphide inclusions along the length of the hoe from working to shape, with recrystallized grains. Microphotograph by Yun Yali. The sample was from the blade of leaf-shaped hoe.x100.

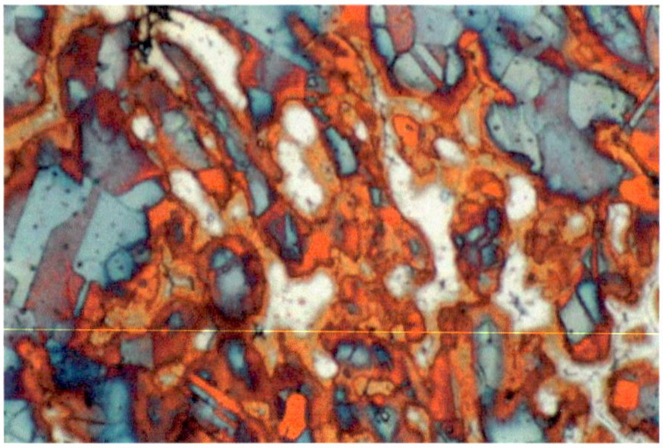

Figure 8.32 Sample 9862 Chinese copper-arsenic alloy hoe excavated from the site of Hejiahsan, in west Yunnan, late Spring and Autumn period to middle Warring States period. Photomicrograph of the blade showing blue copper sulphide inclusions along the length of the hoe from working to shape, with recrystallized grains. Microphotograph by Yun Yali. X400

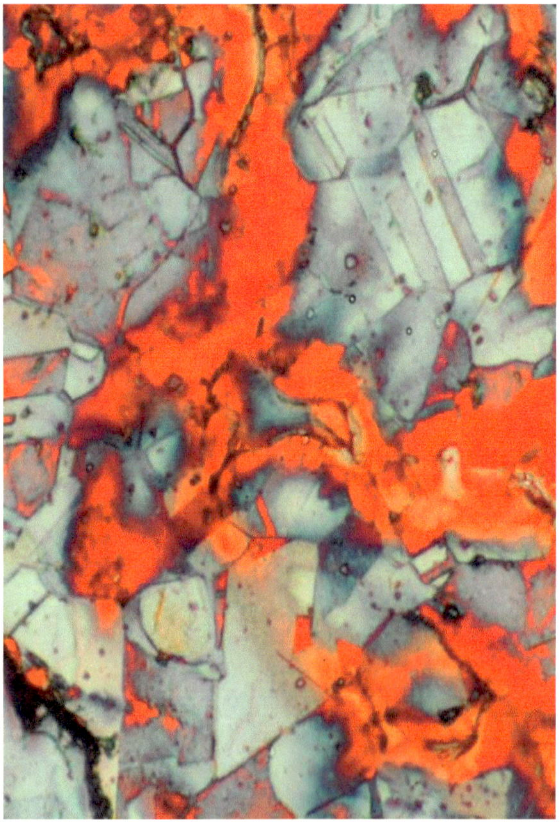

Figure 8.33 Sample DY1 by Yun Yali. Metallographic sample from the blade of a bronze dagger which was unearthed from the Yongzhi site, Deqin county, northwestern Yunnan. Late Warring States period. Etched in 10% aq. potassium peroxydisulphate, followed by saturated sodium thiosulphate and 1 gram of potassium metabisulphite in 100ml of solution X400

Figure 8.34 Sample DY1 Yun Yali Sample DY1. Metallographic sample from the blade of a bronze dagger unearthed from the Yongzhi site, Deqin county, northwestern Yunnan. Late Warring States period. Etched in 10% aq. potassium peroxydisulphate, followed by saturated sodium thiosulphate and 1 gram of potassium metabisulphite in 100ml of solution X400. The microstructure shows that the casting segregation from the original melt still persists in the copper-arsenic alloy here, despite the working and annealing, producing a twinned and recrystallized grain structure.

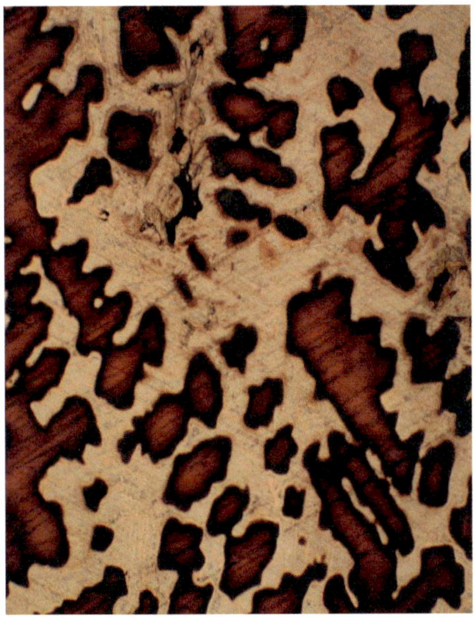

Figure 8.35 An 18% arsenical copper as cast showing a heavily segregated cast dendritic structure. X 80. The darker regions are the copper-rich alpha grains and the infill is a two-phase arsenic-rich matrix. Etched in alcoholic ferric chloride.

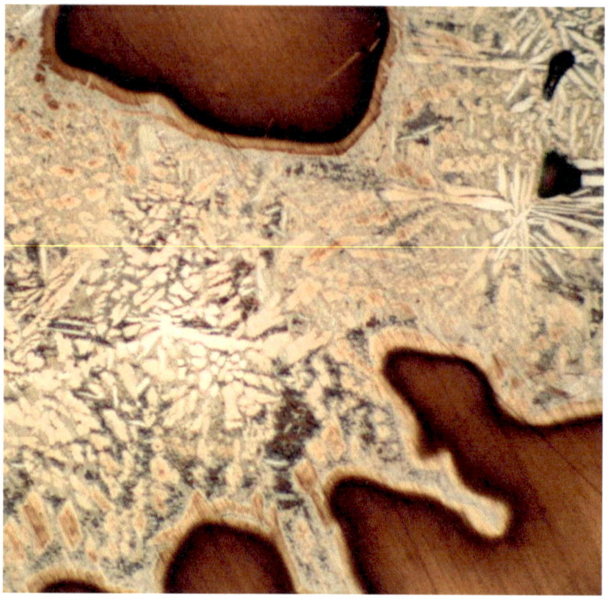

Figure 8.36 The 18% arsenical copper alloy showing the two-phase arsenic interdendritic matrix. Etched in alcoholic ferric chloride, x180.

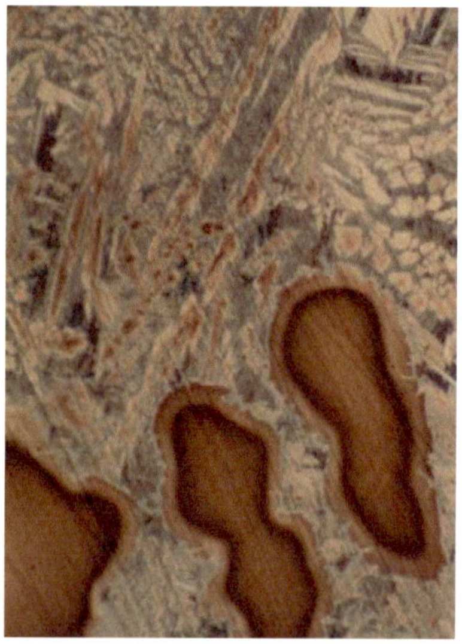

Figure 8.37 The 18% arsenic alloy at x 400 magnification showing complex structure of the acicular precipitation of the arsenic-rich phases.

Copper-Antimony Alloys

Copper-antimony alloys were used in many different regions of the world where antimony ores are to be found, such as in Hungary, where antimonial copper was locally available. The Urnfield culture of Hungary, from about 1300 BC - 750 BC was a late Bronze Age culture of central Europe. The name comes from the custom of cremating the dead and placing their ashes in urns which were then buried in fields. The Urnfield culture was already connected with the appearance of iron. The extent to which antimony was used in different cultures is still not fully accounted for: there is strong evidence that it was already in use at that time in Switzerland (Schwab 2010). Antimony-rich copper can be found in ingots from the early Iron Age and the alloys were widely and systematically used by the Celts for coins and jewellery during the late Iron Age. The Celts employed antimony, often with arsenic-rich Fahlore copper to imitate silver and to reduce the tin content (Northover 1992; Lutz and Schwab 2008 and Schwab, forthcoming). Relevant publications on the topic include: Northover (2004; 1998). In other regions of the world, Scott (1998) reported on the presence of copper-tin-antimony alloys used for a La Aguada style plaque from South America, shown in Figure 8.39, and as analyses become more sophisticated we can expect several other occurrences to be reported in the future on this exciting topic.

The properties of copper-antimony alloys was investigated by Archbutt and Prytherch (1937: 72). Hampe, whose work is quoted by Archbutt and Prytherch in their 1937 compilation, reported that the limit at which hot shortness occurs is lower for antimony that in for arsenic and that with 0.5 per cent antimony in the absence of oxygen there is already a tendency to hot shortness. An alloy with 1% antimony was found to be very hard and hot short but could withstand a fair amount of cold-working. The tensile strength of copper is improved by antimony both at room temperature and 250 ^{o}C. When more oxygen is present, antimony behaves like arsenic, interacting with the cuprous oxide and breaking up the copper-cuprous oxide eutectic into globules differing in colour from cuprous oxide and sometimes appearing duplex in character (Archbutt and Prytherch 1937: 77).

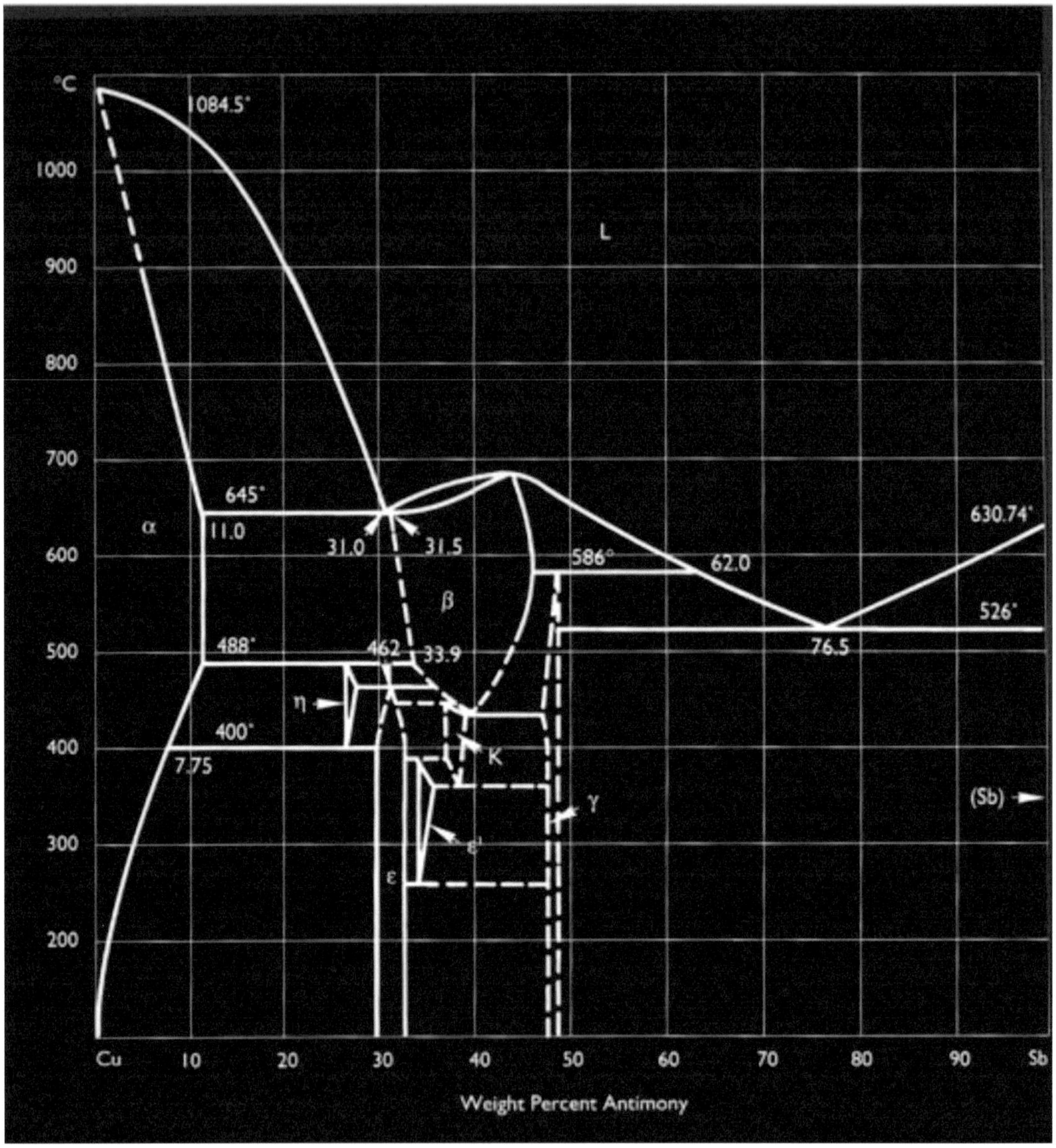

Figure 8.38 The copper-antimony phase diagram, which has similarities to the copper-tin phase diagram in the region of the copper-rich alloys, with four successive phase transitions in the 10-15% antimony region. Higher antimony alloys will have alpha + epsilon phases where the epsilon is very brittle. The alloy was used in the ancient Urnfield culture in Hungary.

The earliest use of antimony-containing alloys comes from the early 4th millennium from the astonishing Nahal Mishmar hoard found in a cave in the Judean desert, made in a ternary copper-arsenic-antimony alloy (Shalev and Northover 1993). The Caucasus is known to have some ancient antimonial bronzes, from the period of 1500 BC – 1100 BC, but its occurrence is still rare (Chernykh 1992, Tylecote 1987). As with many alloying elements, the reddish or bronze colour of copper is gradually lost as alloying element content rises, and the copper-antimony alloys with more than 10% of antimony may be golden or silver-white. In the Nahal Mishmar material, antimony was found to be present often as globules of pure antimony oxide surrounded by an outer rim of copper sulphide. The copper contained about 4% of antimony.

Antimonial copper is uncommon in ancient China. One example from the Hanzhong site had 72.4Cu, 9.53Sb, and 0.9As with 0.04Sn. Chen et al (2009) report on some alloys from the Shang Kingdom, a group of Hanzhong bronzes from southwest Shaanxi. A scepter-shaped object had as much as 9.5% antimony with 0.9% arsenic and a trace of tin. Metallographic examination showed an annealed microstructure of recrystallized equiaxed grains. Antimony often occurred as globules of antimony oxide, surrounded by copper sulphide to form small spherical structures surrounded by a crazed copper sulphide. This structure is unusual and the authors think represent hot oxidation of antimonial bronze during oxygen diffusion along grain boundaries into an area containing both copper sulphide and the intermetallic phase Cu_3Sb, the epsilon phase as shown on Figure 8.38.

Figure 8.39 La Aguada style bronze cast plaque, 84mm x 51mm, dated to about 650 AD - 850 AD from Argentina or Bolivia. Cast in a copper-tin-antimony alloy. Probably cast by the lost-wax process (Scott 1998).

Alloys of Copper and Nickel

Arsenic and tin are not the only alloying elements of importance in antiquity, either intentionally or unintentionally. For example a small Near Eastern cylinder seal which was once described by Reynold Higgins in his book on Minoan and Mycenaean art as being made of haematite is a cupro-nickel alloy. The seal was once in the Marcopoli collection and is now held in a private collection in the USA. The seal was probably cast, although the figures and inscription are directly engraved. The inscription is in Hittite and reads *Ta-ka-na-ni*. The crudity of the style and the inscription (which is upside down) suggest a provincial origin to Beatrice Teisser who examined the seal. She writes that a northern Syrian or southwestern Anatolian origin is likely and that the seal probably dates from the late fourteenth to early twelfth century BC, a time when the Hittites were active in northern Syria. XRF analysis of the cylinder seal proved that it was not made of haematite, but a copper-nickel alloy containing small amounts of cobalt, iron and lead.

A very fine copper-alloy bull's head, analysed at the Institute of Archaeology, London, and said to come from Anatolia, was shown by electron microprobe analysis to have 77.45 Cu, 20.93 Ni, 1.37 Co, 0.25 Fe by weight percentage. Such compositions are quite startling if one fails to realize that widespread use was made of cupro-nickel alloys in the manufacture of decorative objects in the Near East, and that objects containing nickel have been found from such important sites as Ur, Kish, and Tell Asmar. For example, a square pin from Ur was found to contain 94.41 Cu, 1.77 Ni, 0.03 Fe, tr, Pb, with no tin or sulphur detected (Elam 1932). It is not certain from where the nickel used to make these alloys came, since nickel and copper ores are not usually closely related.

Copper-nickel alloys have an important history as coinage alloys. The Bactrian kings Agathocles and Pantaleon, along with their contemporary Euthydemus II, are unique in the ancient world, in that they were the first in the world to issue copper-nickel coins. These coins may be indicative of the existence of trade links with China at that time. Local Fahlore copper of the eastern Alps in the region of Mitterberg can contain several percent of nickel (Stöller et al 2009). Schwab (2010) notes the existence of early Iron Age brooches from Austria made of nickel-rich bronze which suffer from the aging effects that these ternary alloys are prone to.

Copper and nickel are soluble in each other over the entire compositional range which means that, unlike copper-antimony or copper-iron alloys, that the alloys are easily worked and very ductile. As copper and nickel are mutually soluble, there is no second phase present in this system and therefore the alloy is more resistant to corrosion as there is no galvanic difference between phases in this system. If the grain size can be refined to produce small grains, then the alloy is ideal for coinage. It is a remarkable fact that it should have been used for coinage so long ago by the Bactrians.

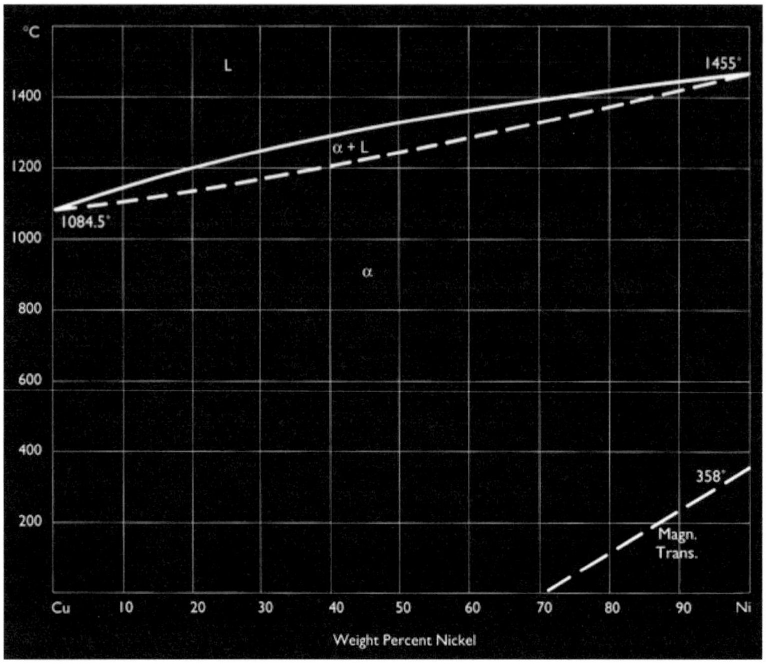

Figure 8.40 The copper-nickel phase diagram (after Scott 1991). Alloys of cupro-nickel, as they are often called are commonly used for coinage alloys today, but were used for the same purpose 2,000 years ago by the Bactrians.

Cheng and Schwitter (1957) assert that one source was China, while Camman (1958) suggests Persia and other possibilities. The reliable analyses of nickeliferous Bactrian coinage goes back to the work of Dr. Walter Flight (1868, 1882) at the British Museum. Flight had been sent some Bactrian coins of Euthydemus II of the 2nd century BC from India by Major-General Alexander Cunningham. Flight determined the composition to be 77% copper and 20% nickel. Besides nickel alone, there are also alloys of copper-arsenic-nickel in the ancient Near and Middle East mainly in the 4th – 3rd millennium BC. In Bronze Age Europe such alloys are very rare. Copper high in arsenic and nickel are found in the Wadi Arabah (Hauptmann 2007). The copper-arsenic-nickel alloys are common in the area between northern Syria, southern Anatolia and along the Upper Euphrates (Hauptmann 2007). Nickel as an alloying element has been found in Hittite –related objects from North Israel dating to the end of the 2nd millennium BC (Northover 1998). Copper-arsenic-nickel alloys are obtained in some regions where the local ores provide this kind of smelted metal or mix of locally available metals. Chen et al (2009) found an intermetallic phase with an approximate composition of $(Cu,Ni)_3As$ to be present in the ternary alloys. The nickel was found to be heavily enriched in the intermetallic phase compared with the alpha solid solution. The alloys of copper-arsenic-nickel are known from China from several sickle-shaped artefacts from Hanzhong (Chen et al 2009). These alloys have from 2.3% to 10.5% nickel, showing significant alloy groupings. One example, rather high in sulphur, had 70.2Cu, 13.0As, 5.01Ni, 1,77Fe, and a lot of sulphur and oxygen. These high arsenic-nickel-copper alloys will have a range of silvery colours.

Needham et al (1975) argue that cupro-nickel was known as a unique alloy by the Chinese during the reign of Liu An in 120 BC in Yunnan. Probably the Chinese copper-nickel alloy known as *paktong* was unknown at that stage, but the naturally occurring Yunnan ore which produced cupro-nickel alloys was probably a valuable internal trade commodity. The scholar Ho Wei describes the process in circa 1095 AD, which suggest the Chinese were not aware that nickel was a metal in its own right. The *paktong* alloy was described as being made from adding small prills of naturally-occurring *"Yunnan"* ore to a bath of molten copper. The late Ming and Ching literature have very little information about *paktong*. However, it is first mentioned specifically by name in the *Thien Kung Khai Wu* of circa 1637:

"When *lu kan shih* (zinc carbonate, calamine) or *wo chhein* (zinc metal) is mixed and combined with *chih thung* (copper), one gets 'yellow bronze' (ordinary brass). When *phi shang* and other arsenic substances are heated with it, one gets 'white bronze' or white copper: *pai thong*. When alum and niter and other chemicals are mixed together one gets *ching thung*: green bronze." [Needham et al 1974].

During the European importation of Chinese white-copper during 1750 to 1800, increased attention was made to discovering its constituents. Peat and Cookson found that: "the darkest proved to contain 7.7% nickel and the lightest said to be indistinguishable from silver with a characteristic bell-like resonance when struck and considerable resistance to corrosion, 11.1%". Another trial by Fyfe estimated the nickel content at 31.6%. Guesswork ended when a Dr Dinwiddie in 1793 brought back, at considerable personal risk (smuggling of *paktong* ore was a capital crime, a law imposed by the Chinese Emperor) some of the ore from which *paktong* was made. Cupro-nickel was now widely understood and published by E. Thomason, in 1823, in his submission later rejected for not being new knowledge to the Royal Society of Arts.

Efforts to duplicate exactly the Chinese *paktong* failed in Europe due to a general lack of requisite complex cobalt-nickel-arsenic ores, except in Germany where such ores existed, which led to the production of copper-nickel alloys, known as German silver. In 1829, Percival Norton Johnston persuaded a Dr. Geitner to establish a foundry at Bow Common, behind Regents Park Canal in London, and obtained ingots of nickel-silver of 18% Ni, 55% Cu and 27% Zn. Between 1829 and 1833, Percival Norton Johnson was the first to refine cupro-nickel on the British Isles. Johnsons' most serious competitor, Dr. Benson devised greatly improved methods of cobalt and nickel suspension and marketed their own brand of nickel-silver: known as *British Plate*.

Form	Temper		Tensile Strength	Yield Strength	Elongation in 50mm	Hardness
	Standard	Former	MPa (N/mm^2)	MPa (N/mm^2)	%	Rockwell B
Seamless Condenser Tubes B111M-98	O61	Annealed	275	105	-	-
	H55	Light drawn	310	240	-	-
Plate Sheet Strip and Rolled Bar B122/122M-95	M20	As hot rolled	275-425	-	-	-
	H01	Quarter hard	350-460	-	-	51-78
	H02	Half hard	400-495	-	-	66-81
	H04	Hard	490-570	-	-	76-86
	H06	Extra Hard	505-585	-	-	80-88
	H08	Spring	540-605	-	-	83-91
Plate and sheet for pressure vessels, condensers and heat exchangers B171M-96	M20 and O25	(80mm and under)	275	105	30	
		(>60 to 140mm)	275	105	30	
$1N/mm^2$ is equivalent to 1MPa						

Table 8.1 Some mechanical properties of copper nickel alloys, with tensile strength, yield strength and some Rockwell B hardness scale measurements.

Lechtman (2003), describes bronze alloys of the Middle Horizon, during the period from 600-1000 AD, of both Wari and Tiwanaku (Tiahuanaco) cultural sites in the Andean cultural area, with particular attention to their elemental composition and site distribution. Alloys of copper-arsenic and copper-tin were in use until the Spanish entrada, but a ternary alloy of copper-arsenic-nickel was confined to the Middle Horizon period. Bronze production in the Middle Horizon coincided with a period of long-distance movement of materials and goods in the Andes, which also spread new technical information. In spite of this, uniform production of bronze alloys within cultural areas did not occur, which is not surprising, given the complexity of cultural developments.

Wari areas manufactured only arsenic bronze alloys in the Middle Horizon. In the area dominated by Tiwanaku, two alloys were used simultaneously, the ternary alloy of copper-arsenic-nickel and copper-tin bronze. There is no evidence of the use of the ternary alloy following the collapse of Tiwanaku. It is interesting that the same type of copper-arsenic-nickel alloys should also have been reported by Kunlong et al (2009) from

the remote frontiers of the Shang Kingdom: these are the Hanzhong bronzes from southwest Shaanxi, China. More than 200 bronze objects found in Hanzhong, during the Shang dynasty, were analyzed for their composition and microstructure. Forty-three typologically distinct, and probably culturally indigenous, items were found to be compositionally distinctive as well. Three special alloys, arsenic-bronze, antimonial bronze, and copper-nickel-arsenic ternary alloys were found to have been utilised.
The story of the use of nickel and nickel-containing alloys is clearly an interesting one. With further data and analyses we shall learn more, but the Hittite alloys with up to 20% of nickel remain some of the most remarkable.

Alloys of Copper and Iron

Attention to the issues of iron within smelted copper alloys was first drawn by Cooke and Aschenbrenner in 1975, who examined the cooling of a 6% iron in copper alloy. As the 94% Cu 6% Fe alloy cools, it reaches the liquidus at about 1215 ^{o}C. At this stage solid gamma iron phase begins to separate out. This gamma iron will contain about 8.3% copper in solid solution. As the temperature falls, more of the gamma iron-rich phase separates until at 1095 ^{o}C the precipitating iron contains about 8.5% copper. At the copper-rich side of the diagram, the composition of the still-liquid copper follows the liquidus until at 1095 ^{o}C, the copper will contain something like 3% iron in solution. At 1084.5 ^{o}C, a *peritectic* reaction will occur between the liquid and the precipitated gamma phase to give a solid solution of 96% copper and 4% iron. This means that given a very slow cooling rate the alloy should consist of a solid solution eta phase of 96% copper and 4% iron with residual gamma iron particles. As the temperature falls, the copper is gradually precipitated out of the alpha iron and at the same time the eta phase loses iron. In ancient specimens, so far no evidence has come to light to suggest that the peritectic reaction has occurred. Because of the presence of alpha-phase iron as ferrite, the copper alloys containing iron are often ferromagnetic and can be picked up or attracted with a magnet, which has long been known for some ancient coinage alloys.
Increases in iron content in copper may be linked to successive stages of technological innovation. In Europe, this may be associated with the spread of slagging processes in the middle of the 3rd millennium BC (Craddock and Meeks 1987). Here, iron may mix with the furnace charge through the intergrowth of ore with gangue or the use of various fluxing agents. In more primitive extractive processes, under poor reducing conditions, iron minerals that may have been present are not sufficiently reduced to get incorporated as iron in the copper (Craddock 1995). In highly reducing conditions, some of the iron mineral impurities could react with copper draining through slaggy deposits, to form

copper-iron alloys. Iron is soluble in molten copper, especially if sulphides are present (Craddock 1995: 138), resulting in copper sometimes showing several percent of iron, as found with some primary production sites such as Timna (Craddock 1988). Tylecote and Boydell (1978) showed that the amount of iron present in the crude copper cake or ingot could be reduced from about 2% to 0.5% by remelting the copper in a crucible. Iron contents in copper are usually an indication of higher temperatures and stronger reducing conditions in the furnaces. Hauptmann (2007) reports that the level of iron concentrations in Faynan copper, from the Bronze Age of Jordan can be as low as 0.001% and as high as 76%. Metallographic studies showed that some of the iron content was due to inclusions of copper-rich ferrite as well as iron phosphides.

In the later Mameluk period, Hauptmann (2007) found two-phased copper-iron alloys, one with more than 50% Fe. Analyses showed that both copper and iron contained dissolved content of the other metal: iron usually had 5-7% copper, with phosphorus greater than 1%, while copper contained 1.5-3% iron as small inclusions. This corresponds to some extent to the iron-copper phase diagram derived by Hansen and Anderko (1958) shown in Figure 8.41.

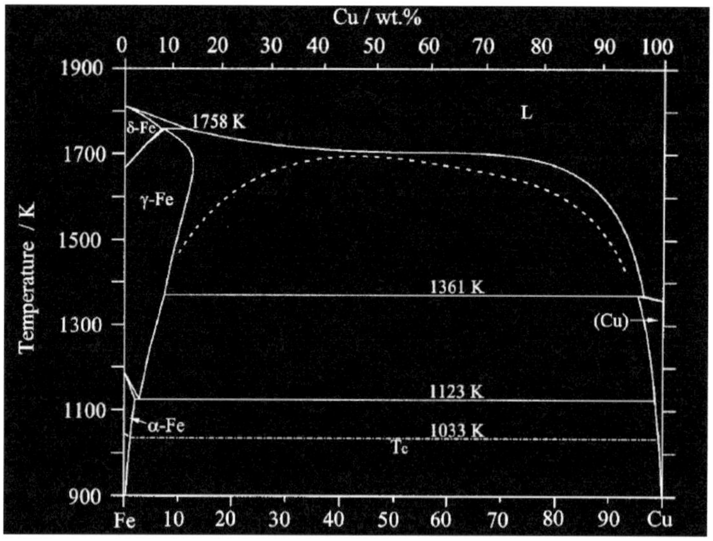

Figure 8.41 The iron-copper phase diagram (after Hansen and Anderko 1958). Note that here the percentages at the bottom of the chart are given in atomic percentages, with weight percent at the top and that the authors give the temperature scale in degrees absolute rather than degrees centigrade. The low solubility of iron in copper means that most of the iron is precipitated out as fine dendritic or rounded masses.

Iron is dissolved in copper at very high temperatures, such as 1450 °C but iron is precipitated out of the copper alpha solid solution on cooling, initially as austenite, which transforms at 835 °C to ferrite. The most common occurrence of iron in copper is as small dendritic rounded shapes, as the solidified iron formed during the copper solidification. Some of this iron content can simply be eliminated by remelting the copper in a crucible and skimming off the oxidized dross, taking most of the iron with it (Tylecote and Boydell 1978).

Although Hauptmann (2007) reports the presence of iron phosphides in copper, these are not common. They occur in all periods in Faynan copper from Jordan and are characteristic of this region. They can be explained by phosphorus bearing ores, and by charcoal which in the Arabah may include up to 10% phosphorus as the pentoxide in the ash (Merkel 1990). The system copper-iron-phosphorus-oxygen is discussed further by Hauptmann (2009: 209); one point of interest is the discovery of a eutectic intergrowth of Fe_3P with iron phosphate whose morphology is very similar to ledeburite, discussed in Volume II in this series. The same texture was reported for inclusions of cast iron in ancient copper from Sa Sedda, Sardinia (Tylecote 1983a and 1983b).

Alloys of Copper and Lead

Copper alloys are frequently leaded. Leaded bronze, brass and arsenical copper alloys are all common. Highly leaded alloys were quite popular in the European Iron Age, Hellenistic and Roman period. Segregation is quite common in these alloys. It is unusual for the lead to sink down through the alloy, in many cases it remains inside the alloy and in German is known as "Blocksaigerung" as the lead rich phases follow the heat. In the Mahdia bronze study Schwab (2010) illustrates an Iron Age alloy which is fully segregated into lead-copper and copper-lead. Lead does not usually occur with copper ores, so the lead content usually represents a deliberate addition of lead. The equilibrium diagram shows that the microstructure will consist of two distinct phases and that the copper grains will contain globules of lead.

Practically all of the copper will solidify before the lead-copper eutectic forms. This lead-copper eutectic is, for all practical purposes, pure lead, as it consists of 99.9% lead and 0.1% copper. This means that the lead is segregated while the solidification process is taking place. Ordinarily, the separation of lead globules would be expected to result in massive segregation and an unusable material would result. There is a monotectic reaction at 955 °C which occurs when the liquid from which the copper is separating out reaches a composition of 36% lead. At this point, a new liquid forms which contains about 87% lead.

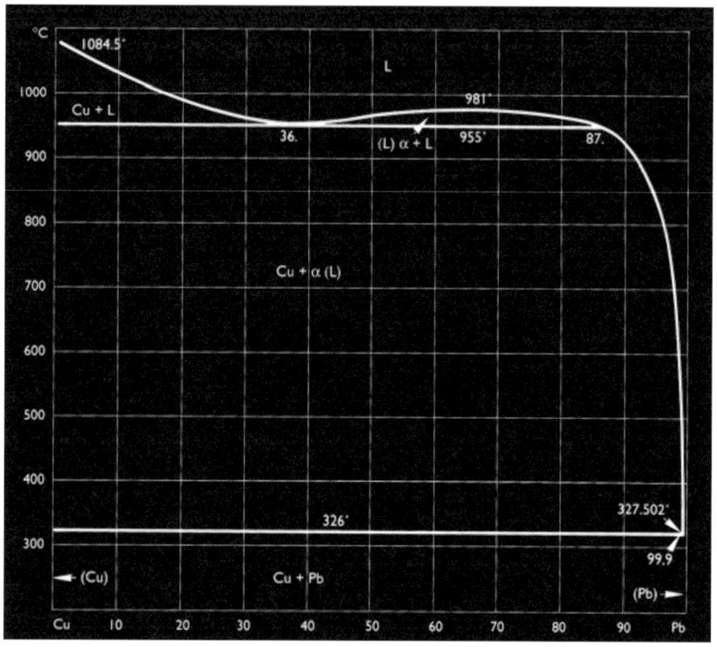

Figure 8.42 The copper-lead system. After Scott (1991)

This new liquid is heavier than the first liquid, and so it tends to sink under gravity. However, in practice the gross segregation is limited by the formation of a dendritic structure upon casting in the copper-rich alloys and, with a high cooling rate, the lead is finely dispersed among the copper grains or dendrites. With very high lead content alloys, the two liquids that separate out form an emulsion when they are cooled from about 1000 °C (Lord 1949). This emulsion results in a division into very fine droplets so that gross separation cannot occur. With leaded copper, brass or bronze alloys the lead usually occurs as small, finely dispersed spherical globules scattered at the grain boundaries and within the grains themselves. Lead has the effect of making the alloys of copper easier to cast; it can, for example, improve the fluidity of the alloy in the melt. Lead also makes copper difficult to work above levels of 0.5%, but is useful in the localized filing or smoothing of cast artefacts and aids machinability.

Chapter 9: Tin Bronzes

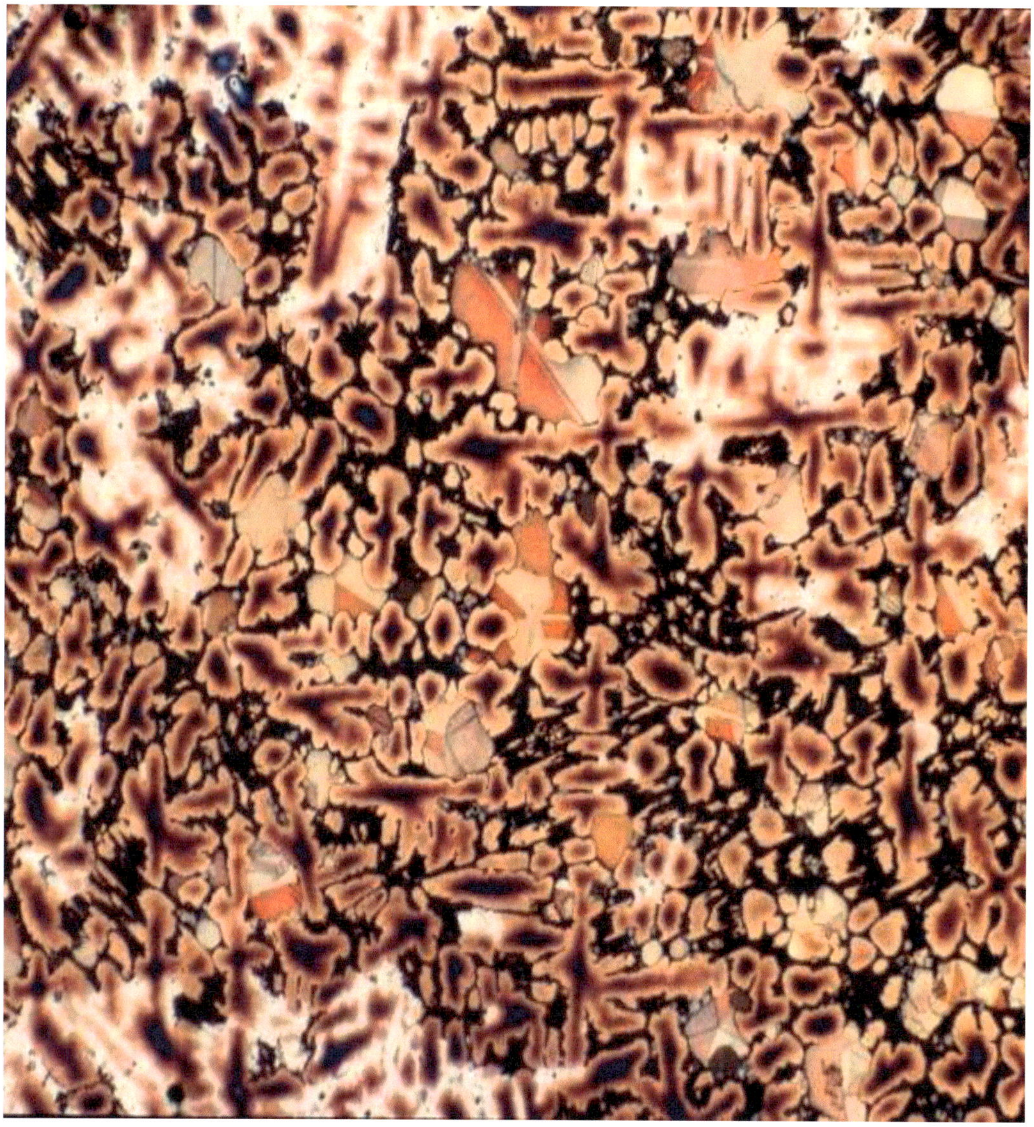

Frontispiece 1. Photomicrograph of a Luristan dagger handle from 1500 BC showing fine cast structure of the handle with about 17% of tin. Heavily cored with redeposited twinned crystals of metallic copper as a result of corrosion. Colour etched in lead thiosulphate, x 280.

Bronze in its traditional sense, is a binary alloy of copper and tin. It is one of the most employed alloys over the past four thousand years. The full phase diagram is shown in Figure 9.1. The melting point of tin is very much lower than copper and tin is a body-centered tetragonal metal, unlike copper in terms of lattice size and shape. This mismatch creates a variety of complexities in the phase diagram. In general, the bronze system can be divided into the low-tin bronzes, from 1-18% tin and the high-tin bronzes from 18%-30% tin.

The usual range of complete solid solubility of tin in copper is about 14%, in terms of practical use of the lower tin bronzes. In some Bronze Age artefacts, care was taken to try to anneal the bronze palstaves and axes to take as much tin into solid solution as possible up to the practical limit, which would make the tools as hard as possible without the brittleness of the delta phase resulting in cracking on impact. In normal casting in moulds of clay, stone or metal, the transition between the liquidus and solidus in the α-bronzes is a period of dendritic growth. The low-tin bronzes are, strictly speaking, those with up to 18% of tin, because this is the edge of the alpha solidus as can be seen from the diagram in Figure 9.1.

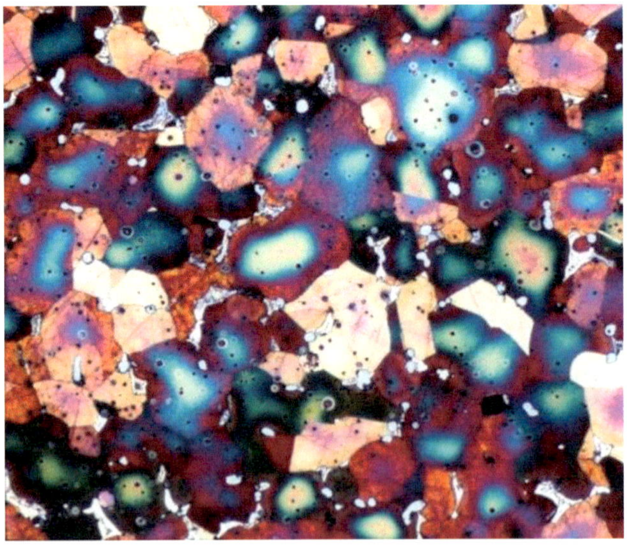

Frontispiece 2: A heavily cored, cast tin bronze toggle pin from Iran, Middle Bronze Age, with prominent coring. Few twinned crystals can be seen, showing that the cast bronze has received a limited amount of working and annealing to shape it. Colour etched in lead thiosulphate etch after pre-etching in potassium peroxydisulphate, x150.

Alloys with 18% of tin, at equilibrium, cool through the (α + liquid) region which extends from the melting point of pure copper at 1064 °C to 799 °C. At 799 °C, the entire mass changes from a solid + liquid phase to a solid α phase. The epsilon phase becomes important in the processes related to tinning where it does appear. In tinned surfaces there is a general progression from the higher tin phase, through to the lower content phases as one passes into the alloy interior. This means that the entire diagram is relevant to ancient metallurgy and is very useful in helping to explain what is seen metallographically. However, Figure 9.2 is also of great practical use, because it shows the nature of the phase relationships without the epsilon phase being shown in the lower tin containing regions where the alpha + delta phases tend to dominate.

The delta phase is a light blue-grey colour, easy to recognise in etched sections of the copper-tin alloys. As the delta phase is an *intermetallic* compound, it tends to behave as a crystalline solid without ductility. This can harden the alloys of copper and tin substantially, but the delta phase has to be kept under control. If there is sufficient delta to coat the grain boundaries, then the alloy becomes extremely brittle, and so is of little practical use for axes, halberds, spears, and daggers. This is why the tin content for practical purposes was always less than 14%.

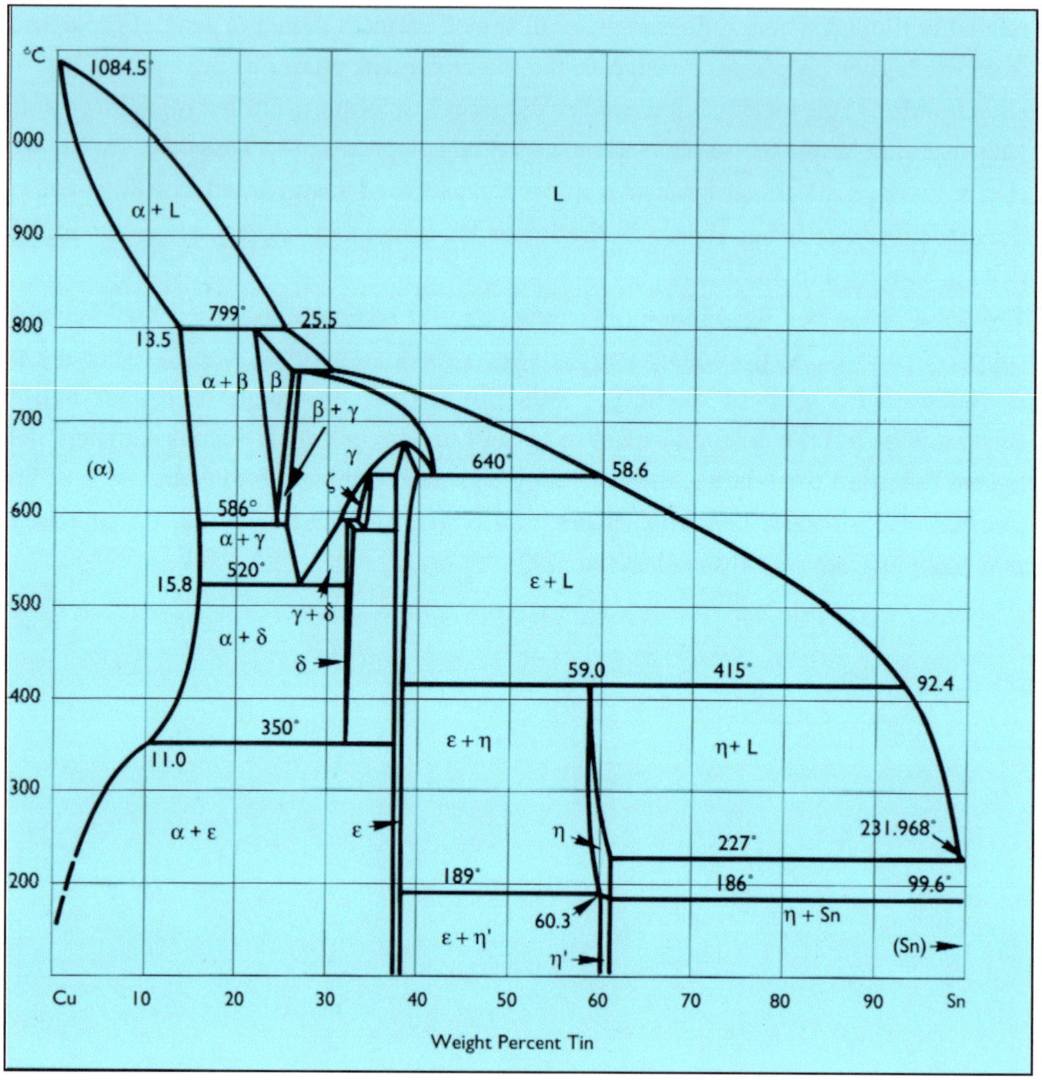

Figure 9.1 The copper-tin equilibrium diagram at full equilibrium. The overall diagram is useful for explaining both the phases which form at the high-tin end of the diagram, in the tinning of copper, bronze or iron artefacts, and at the low end, representing the composition of the bronzes below 18% of tin. However, the epsilon phase at the lower tin concentrations does not occur in as-cast alloys, but only after thousands of hours of annealing, so is essentially fictional for applications to ancient and historic bronze alloys with low tin concentrations.

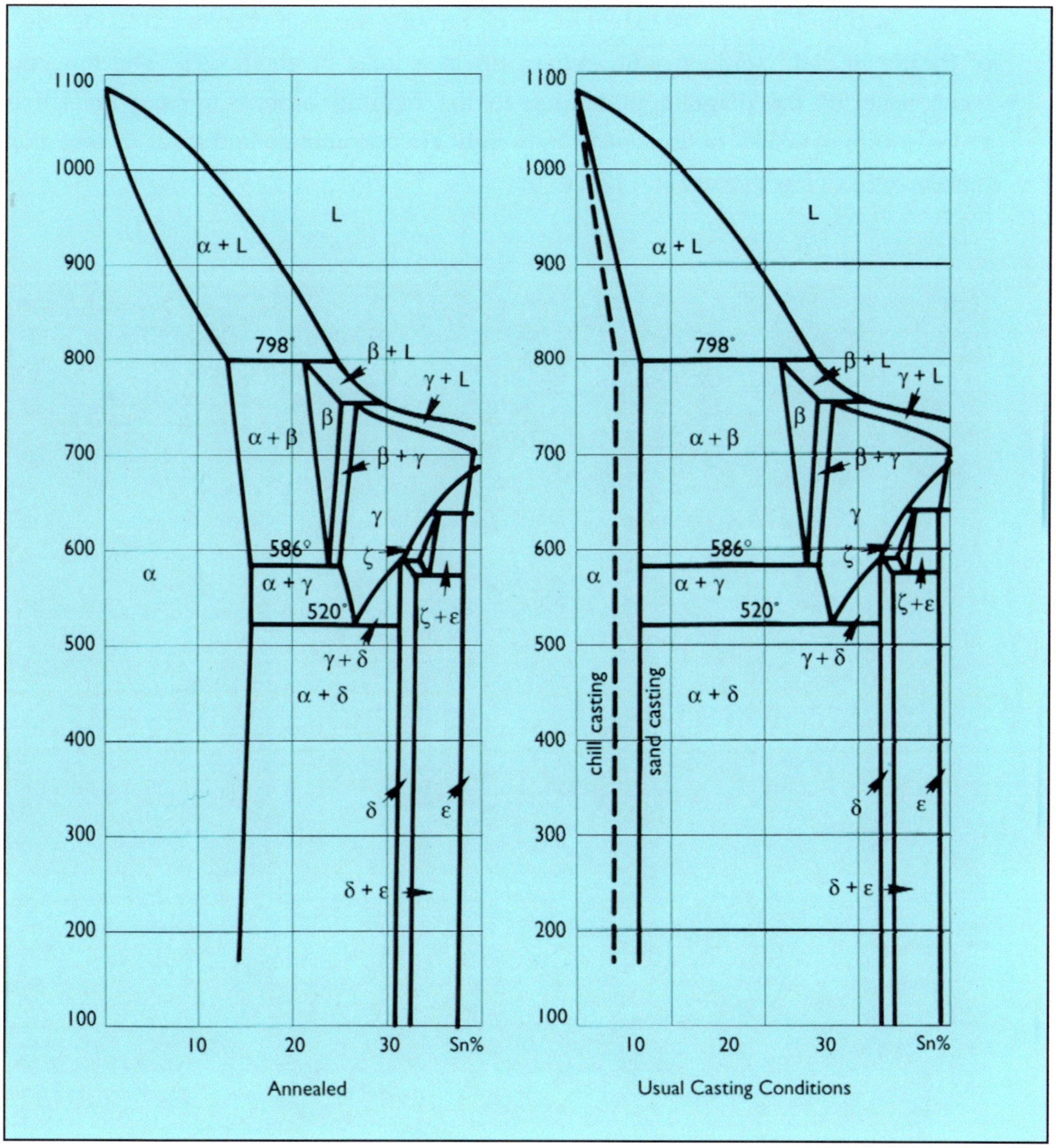

Figure 9.2 Realistic phase equilibria in the copper-tin system based on observation of laboratory cast alloys. Note that the epsilon phase still occurs at higher tin concentrations, but that the alpha + delta eutectoid phase, along with grains of alpha, dominates the low-tin alloys as they cool to room temperature. After Hansen and Pell-Walpole (1951).

Other versions of the copper-tin phase diagram have been drawn up by Hansen and Pell-Walpole (1951), which are shown in Figure 9.2 for the low-tin side of the diagram. These two diagrams show the phase relationships for bronzes which have been fully annealed in the left hand chart, and bronzes either chill or sand cast in the right hand version. In the annealed tin-bronzes, the extent of the alpha solid solution is enlarged to include all of the tin in solid solution with copper up to a limit of about 14%. The remaining components of the diagram pertaining to the high-tin bronzes remain unchanged. Bronzes with over 30% of tin content by weight are uncommon, and most ancient alloys have between a few percent and 12% of tin.

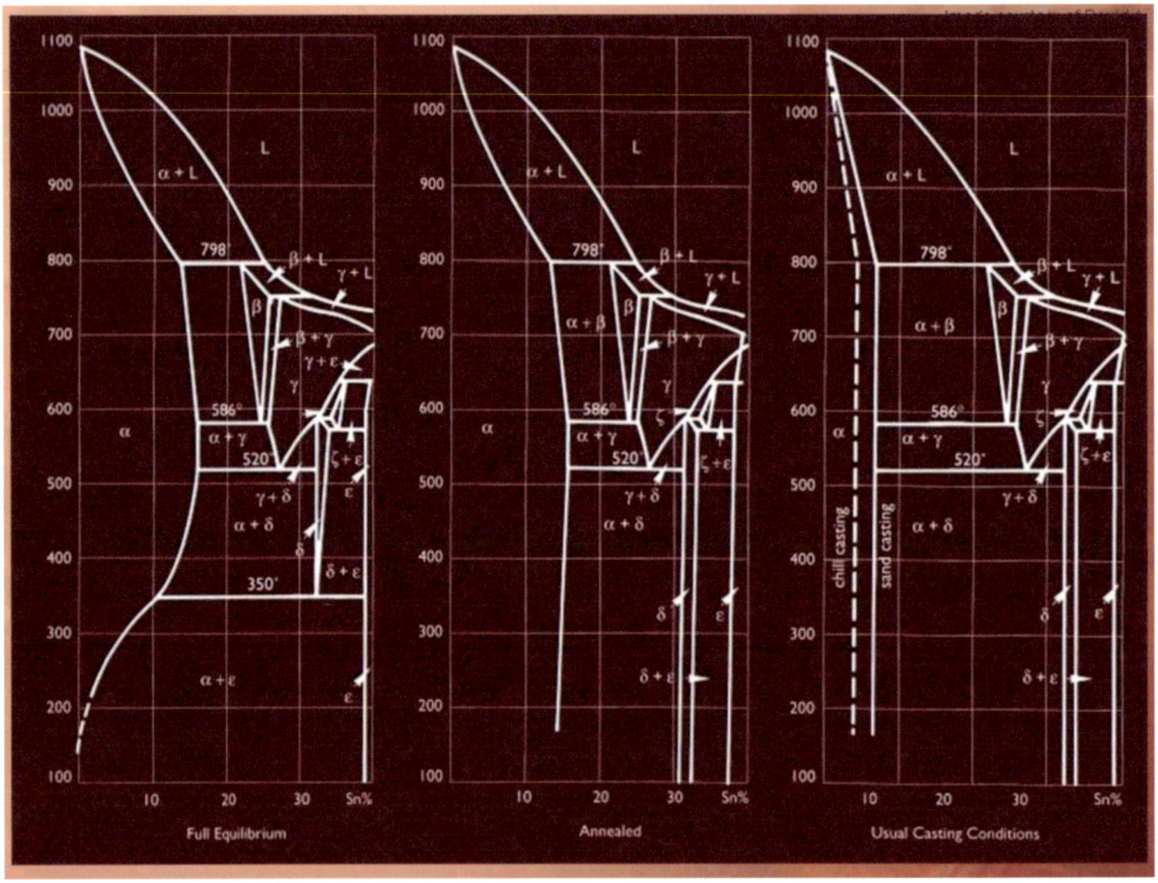

Figure 9.3 The three versions of the copper-rich side of the binary phase diagram. After Hansen and Bell-Walpole (1932).

The dotted line in the right hand figure indicates the extent of the alpha solid solution in the chill-cast bronzes. Because less time is allowed for equilibrium, there is a correspondingly greater amount of the delta phase present and less tin in solid solution in the alpha grains. In the sand-cast tin bronzes, the cooling is slower, so more of the tin can be absorbed into the solid solution alpha grains. The initial dendritic structure, illustrated well in Figure 9.9, will be lost if the alloy is hammered and annealed, especially if this hammering and annealing occurs in repeated cycles, which is the normal *modus operandi* of cold-working of a metal or alloy. The mechanical properties of the alloys can be rationalized on the basis of the phases formed, and accounts for the dramatic variations seen in the chart illustrated in Figure 9.4. Looking at the Brinell hardness, which is roughly comparable with the more usual Vickers DPN hardness, the hardness rises rapidly from around 100 to reach 300 at 25% tin. A Brinell hardness of 300 is very hard and difficult to cut with a jeweller's saw in terms of sampling of such artefacts.

On the other hand, the elongation which reaches 70 mkg/cm^2, falls away to nothing as soon as the delta phase begins to be an issue, at around 18% tin in as-cast alloys. Both the tensile strength and reduction in overall thickness are equally affected by this 18% limit in the as-cast alloy system.

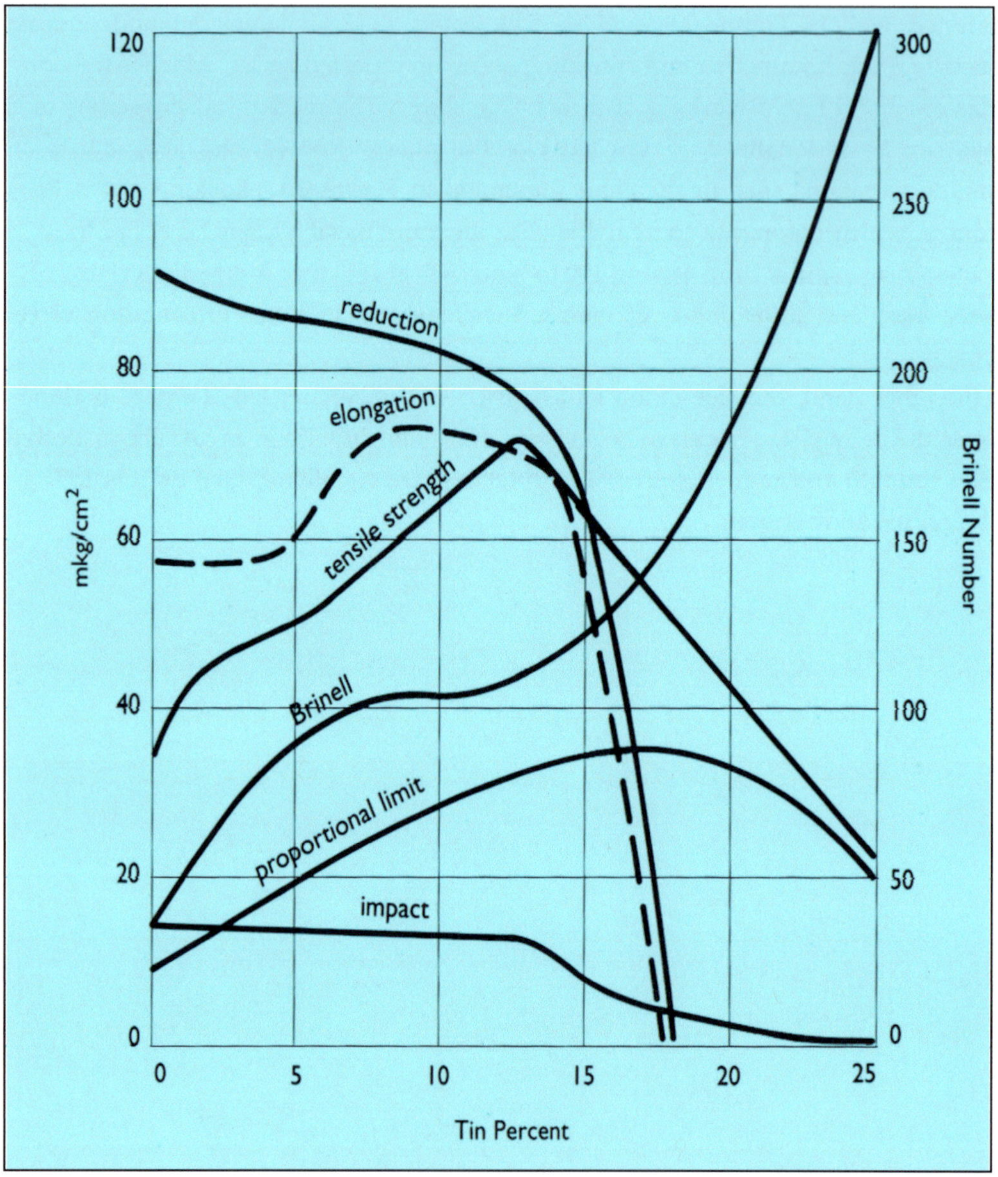

Figure 9.4 Impact hardness, Brinell hardness, elongation and reduction ability in tin bronzes up to 25% tin. After Hansen and Bell-Walpole (1932). Note that the Brinell hardness escalates to 300 Brinell at 25% tin content and that the tensile strength rises to about 14% tin, before undergoing a precipitous decline above 15% of tin. The impact hardness falls gradually up to 13% of tin and then declines more slowly than the tensile strength loss as the delta phase increases. The reduction on working falls from about 90 for pure copper to 40 for a 16% tin bronze and reaches zero around 18% of tin.

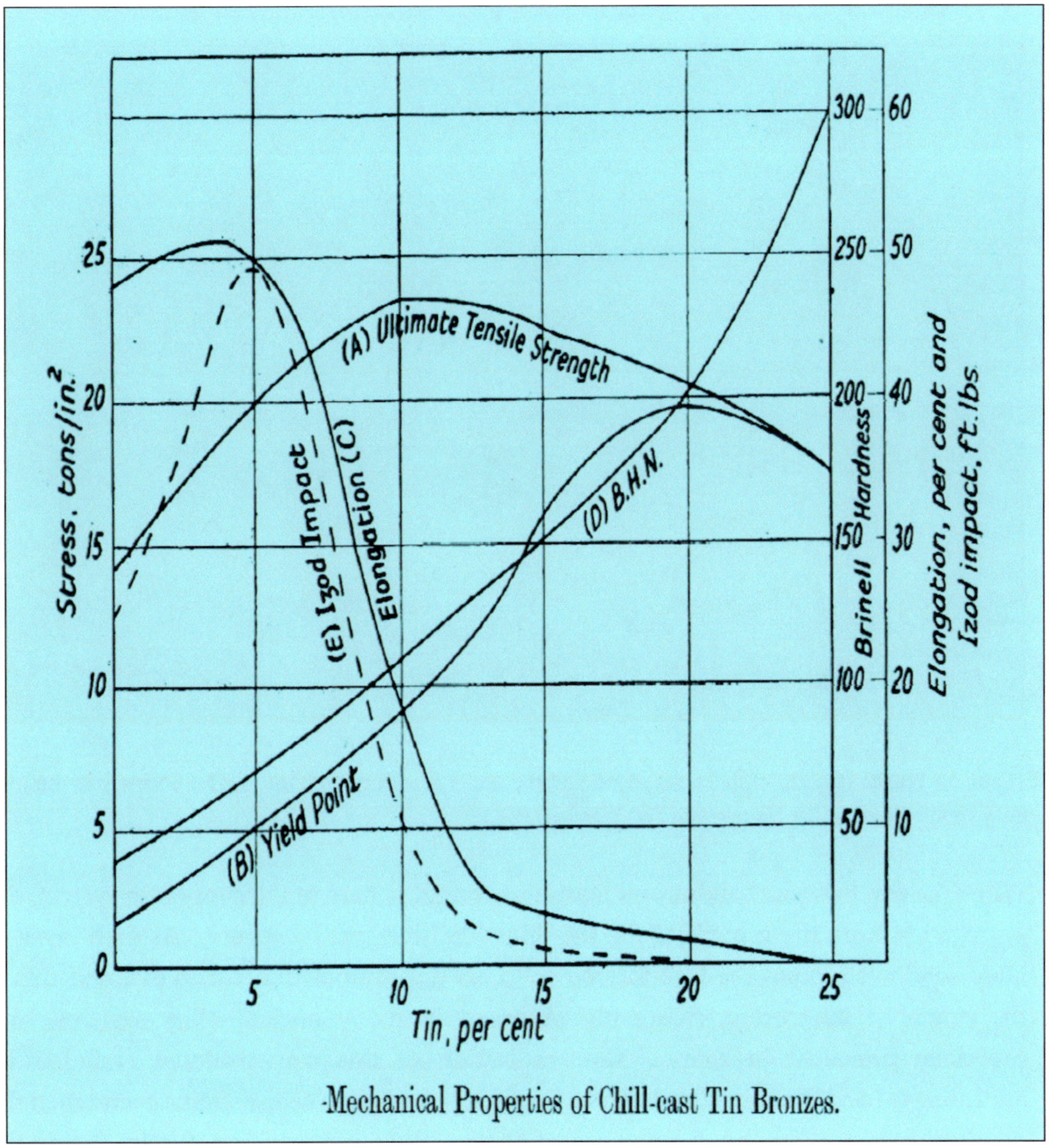

Figure 9.5 The ultimate tensile strength, elongation, and yield point for chill cast tin bronzes, up to 25% tin. After Hanson and Pell-Walpole (1951).

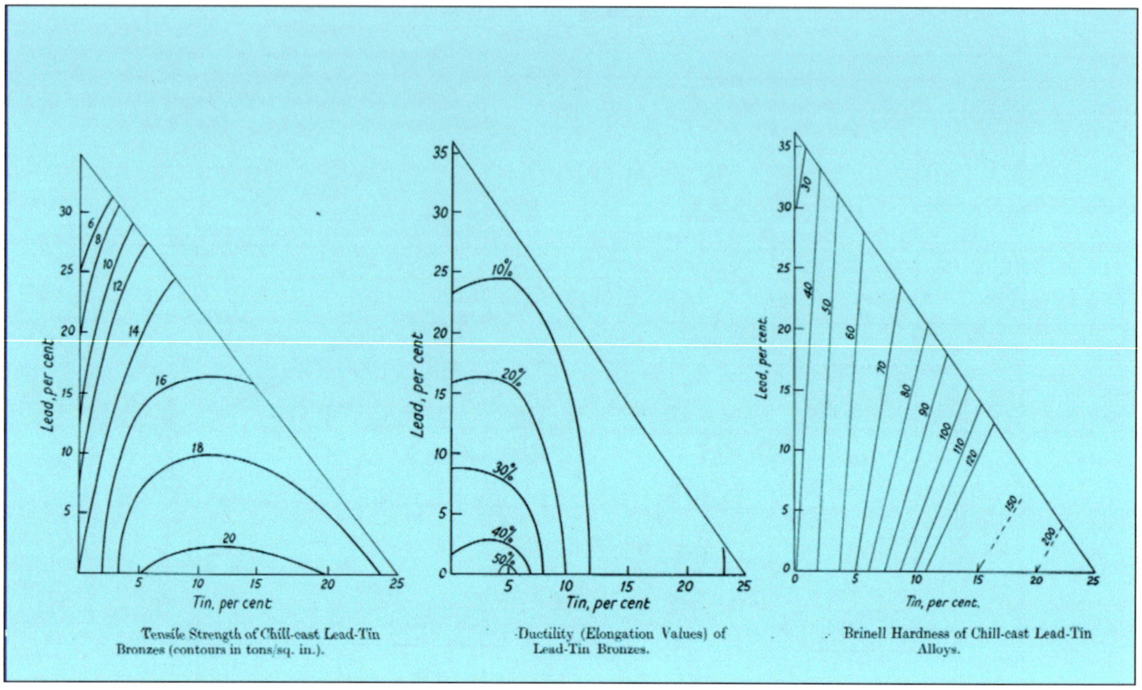

Figure 9.6 Tensile strength of the ternary lead-tin bronze alloys with percent loading for a variety of different alloy compositions. After Hanson and Pell-Walpole (1951).

When the gap between solidus and liquidus is large, as here in the copper-tin system, the segregation from the growth of the tree-like dendrites can be severe. As each layer of alloy solidifies around the first-formed solid, so the composition varies too, and this is the origin of the coring, frequently observed in many ancient alloy systems, and especially prevalent in some. The eradication of this compositional variation by annealing is hard to achieve and coring frequently continues across grains even when the dendritic cast structure has been removed by annealing and working. Notice that in the full equilibrium diagram, that the alpha solid solution is supposed to decompose to an alpha + epsilon (α+ε) phase towards room temperatures.

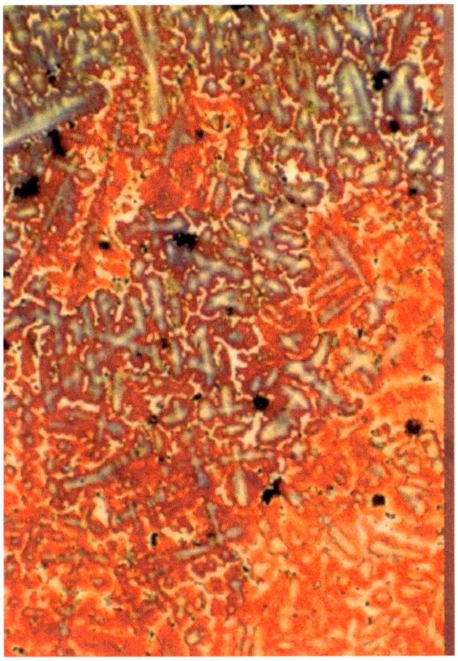

Figure 9.7 Bronze axe from Austria, colour etched in acidified thiosulphate etch, accentuating the dendritic coring. The sutectoid phase appears white and the alpha phase orange or red. Cored dendrites are lighter in colour, 10% tin, 90% copper. One of the most common ranges of the copper-tin system.

In practice, thousands of hours of annealing are necessary in order to create this transition and it is never seen in the low tin bronzes in ancient or historic samples. In fact, the copper-tin phase diagram at full equilibrium, for low-tin bronzes, is something of a theoretical model: it is a convenient model on which we can superimpose the reality of actual events when a tin bronze is cast. The reality is that the bronze will consist of alpha solid solution with an infill of the eutectoid of (α+δ) phase. Note that it is a *eutectoid* phase and not a *eutectic* phase, because the delta phase arises from the decomposition of a prior solid phase, rather than being derived from a liquid melt. The delta phase, $Cu_{31}Sn_8$, is a crystalline solid of a definitive chemical structure: it cannot vary in its composition such as the alpha phase which can absorb more or less tin depending on circumstances. These compounds of fixed composition are referred to as *intermetallic* compounds, and they tend to be very hard with no ductility. This is because the kind of covalent bonding which exists in these intermetallic compounds is the same essentially as in other crystals, which cannot be hammered into different shapes without breaking up. For ease of reference they are treated in the discussion as a phase like other components, but they are invariant in their properties. Whenever a phase diagram contains one of these straight vertical lines passing down into the solid regions then it is a sure sign of an intermetallic compound existing in the system. Note that there is another smaller straight line region in the copper-tin phase diagram, for the eta phase (η), which is the compound Cu_5Sn_6. The eta (η) phase and the epsilon phase (ε), become important in any discussion of tinned surfaces and higher tin content alloys which will be illustrated later in this chapter.

Notice in the diagram that at 799 ^{o}C above 18% of tin we enter regions where (α+β), and β solid solutions exist. Therefore, if we have an alloy which contains 20% by weight of tin, 80% copper, it will melt at a lower temperature than an alloy of 5% tin, and will form α dendrites and liquid, which then enters another solid solution region, that of (α+β), which decomposes at 586 °C to form a mixture of (α+γ), which lasts until 520 °C, before breaking down to (α+δ).

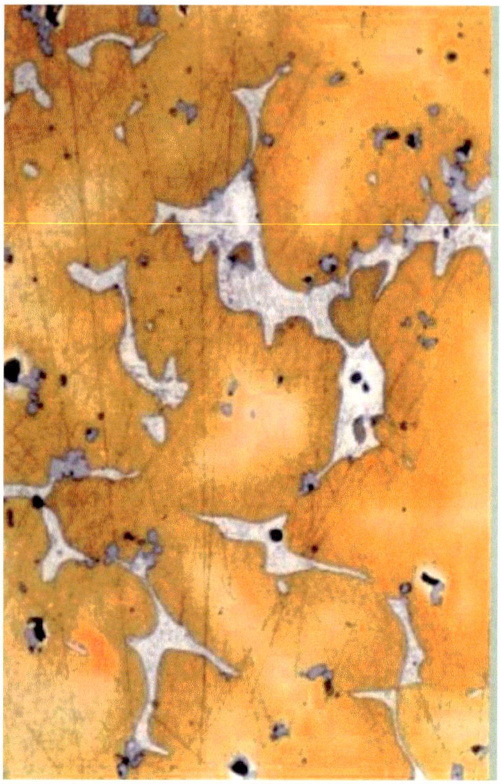

Figure 9.8 9% tin bronze showing the typical structure of the eutectoid phase. Note the presence of small copper sulphide inclusions (blue) and coring shown by the variaitions in colour on etching. If fully annealed the eutectoid phase would be absorbed by the alpha phaase. This may be seen in some Bronze Age axes especially, to reduce brittleness.

These potential series of phase changes which the high-tin bronzes can undergo means that the morphology of the resultant alloy at room temperature is very different from the dendritic morphology of the lower tin bronzes. The low-tin bronzes with tin up to 18% have too much of the eutectoid phase present to be workable, as this phase has no ductility and is hard but brittle.

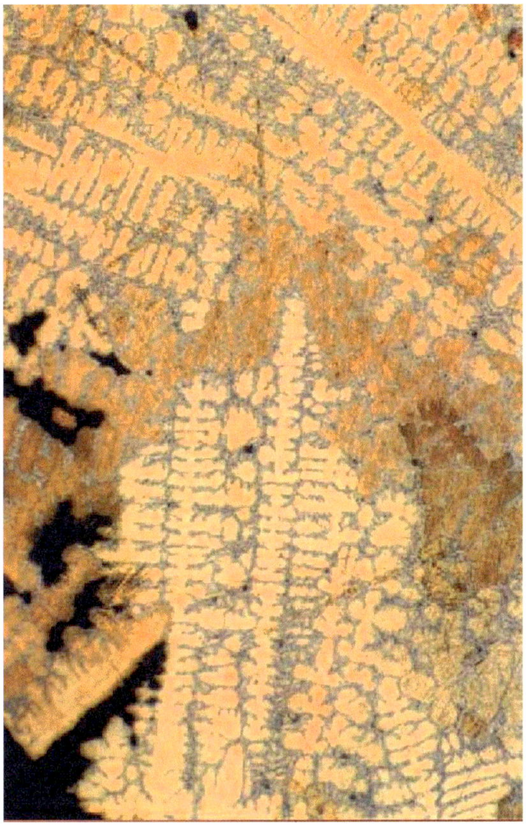

Figure 9.9 16%Sn, 84% copper laboratory ingot, slowly cooled. Coarse dendrites of alpha phase with an infill of the eutectoid. Some ancient bronzes have this composition such as bells, mirrors and handles.

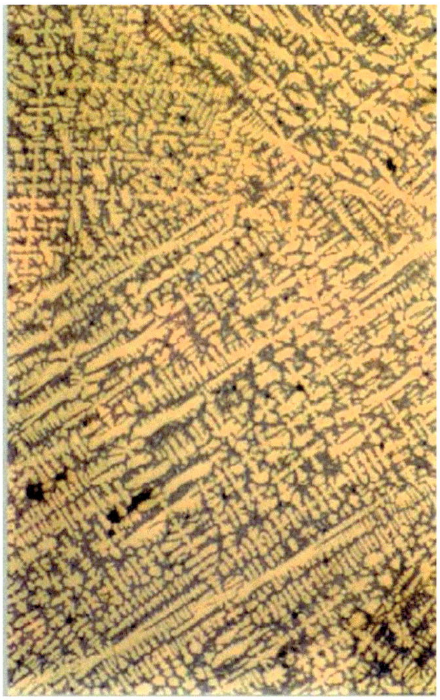

Figure 9.10 18% tin bronze, with 82% copper as cast. Finely developed alpha dendrites with an eutectic infill. This kind of alloy was not used for worked sheet bronze but only for ancient castings as it is too brittle to be hammered out.

The cellular morphology of the bronzes with about 28% of tin in the as cast condition are shown in Figure 9.11. The structure of this alloy is derived from the higher temperature gamma phase field which has an eutectoid transformation at 520 °C, in which the former solid will break down to a gamma and delta phase, which then breaks down to a delta and alpha + delta phase, in which the brittle delta phase occurs at the grain boundaries throughout the cast ingot. Consequently, these alloys are very brittle and if dropped, may shatter as the forces through the delta phase simply result in grain boundary failure.

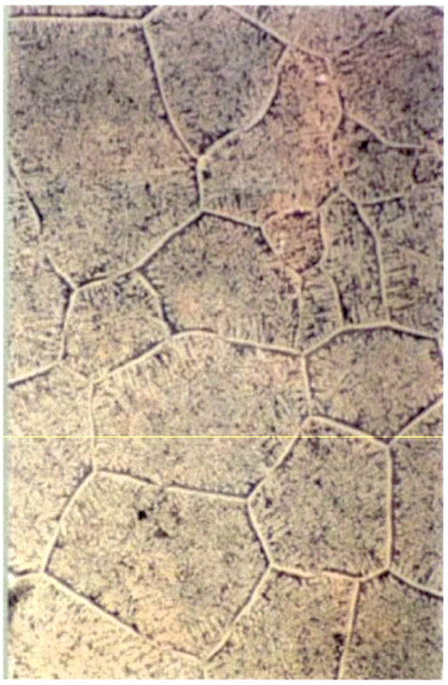

Figure 9.11. broze with 28% tin, 72% copper. These alloys have a cellular structure from the cast condition, not dendritic. Used in some ancient mirrors as speculum metal for telescope mirros as takes a very hard polish. Of decorative use too in historic times. Alloy is too brittle to be worked at all.

The brittle nature of these alloys can be overcome, firstly by reducing the tin content to get free of the gamma eutectic range, and secondly by quenching the alloy to hold one of the higher temperature phases in metastable equilibrium. In fact, the quenched alloy is very stable at room temperature, and the quenched structures survive for thousands of years. South-East Asia, India and Thailand were the pre-eminent centres of this extraordinary metallurgical development. The bronze can be made with sufficient tin content to pass into the β or $(\alpha+\beta)$ region of the diagram, when the alloy can be quenched from about 700-650 ^{o}C, within this phase field.

Figure 9.12: Polished and etched photomicrograph of part of an arrowhead from Palestine, dating to the Middle Bronze Age, about 1500 BC. The microstructure shows that the arrowhead has been cast without further working in a tin bronze with no lead addition. The large dendritic structure is clearly visible. Under partially polarized light, x80, etched in alcoholic ferric chloride.

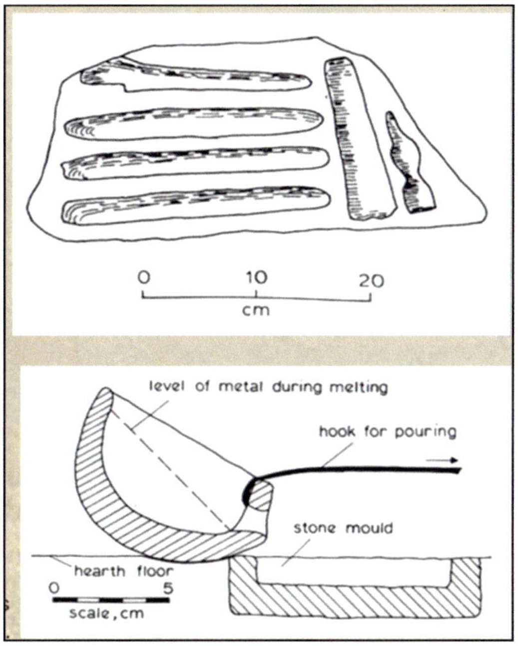

Figure 9.13a (After Tylecote 1976)., Late bronze Age crucible pouring, Keos, Greece) Hollow lost-wax casting already in use in 5th millennium BC in the Nahal Mishmar hoard. More typical are flat axes cast into an open mould. On heating in a crucible we tend to get reduction of water vapour by copper and the formation of cuprite and release of steam.

$2Cu + H_2O = Cu_2O + 2H$

If hydrogen enters the hot metal and the alloy then cools, gases might ruin the casting.

Early casting techniques often involved the use of stone moulds, or clay moulds. The problem with many castings under these conditions is that the bronze or copper alloy reacts with water vapour to create cuprite and hydrogen. The operation of poling the molten copper with a green twig, released this hydrogen as steam, leaving a certain amount of copper oxide behind in the alloy. In ancient alloys this helped to counteract shrinkage on casting and creates what is called tough-pitch copper, which has 0.4-0.6% oxygen, and is a well-known modern equivalent of this type of ancient alloy.

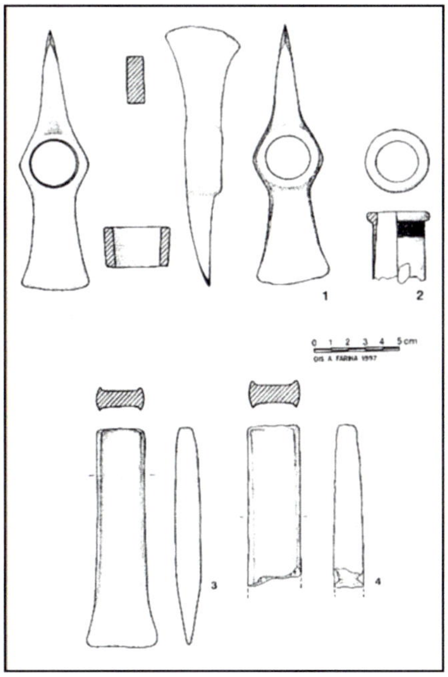

Figure 9.13b Cast bronze tools of these types were widespread in the Bronze Age and could be cast into open stone moulds, clay moulds or bronze moulds.. After Tylecote (1976). We can get rid of the hydrogen by 'poling' with green twigs. This helps to get rid of some of the hydrogen as steam. The presence of some cuprite helps to counteract the shrinkage of copper on cooling. The modern equivalent is touch-pitch copper which has a little oxygen content. We get some shrinkage allowed for in feeders and risers or by the surfaces of the casting itself contraction on solidification. These are some flat axes from Lawhead, Pentland Hills, Scotland, 12.5 inches long and weigh 6 lbs. The top hafted axes are cast from Sardinian oxhide ingots, from Pattada (Sassari), Sedda Ottinnera hoard.

In a series of experimental bronze castings, Ottaway and Wang (2004) maintained that some of their results did not accord with published phase diagrams. We have seen how chill-cast bronzes may differ from the theoretical phase equilibria as a result of the work of Hansen and Pell-Walpole in 1951 and, later in this chapter we will review the photomicrographs from original work by Heycock and Neville (1904) This review will show that the structures do, in fact, possess a logic that is dependent on the full phase diagram, or the diagrams proposed by Hansen and Pell-Walpole.

Ottaway and Wang (2004) found that bronzes cast in bronze moulds had by far the best surface and least flashing, requiring less finishing to prepare the artefact for use. However, the bronze moulds were troublesome to make and lasted for only nine castings before cracks appeared. Clay moulds were easier to make, even though requiring a lengthy drying and firing regime. They produced bronze artefacts with a good surface but flashing remained a problem and the moulds were not reusable. Sand moulds, though not reusable either, were easier to make and did not produce much flashing, but the bronze artefact had a rather rough surface. A comparison of tin or leaded tin alloys made clear that no one process produces the best bronze. For a bronze that can be well reduced in thickness by cold-working and annealing, a low tin bronze and no lead, cast in a sand mould and water quenched would give the best material. For maximum hardness, a high-tin bronze cast in a bronze mould and air-cooled would give the best result (Ottaway and Wang 2004). However the latter effect is only achieved with great brittleness and the beta-quenched bronzes in the high-tin category are mechanically superior to the air-cooled examples. The section is taken from a small bronze arrowhead from Palestine, showing that the entire arrowhead was made by casting into a mould with no further working to shape, and illustrates many of the features to be found in ancient castings.

There is substantial dendritic segregation, visible in Figure 9.14 as orientated patches of fernlike growth, with a second phase, the (α+ δ) eutectoid phase of the copper-tin system occupying the interdendritic spaces in the cast solid. The arrowhead has therefore been cast to shape in a mould without any further working. Figure 9.15, below, shows the cast structure of a Luristan ceremonial axe from Iran of the 5th century BC in which the (α+ δ) eutectoid phase can be clearly seen as a blue-grey jagged precipitate in a matrix of segregated α grains, marked with green-etching centres with haloes of blue and red.

Figure 9.14 Small Bronze Arrowhead from Palestine. The dendritic structure is undistorted, showing that these arrowheads were simply cast into moulds with no further working, x 180, etched in alcoholic ferric chloride. Dark regions are corroded (α + δ) eutectoid surrounding α phase dendrites.

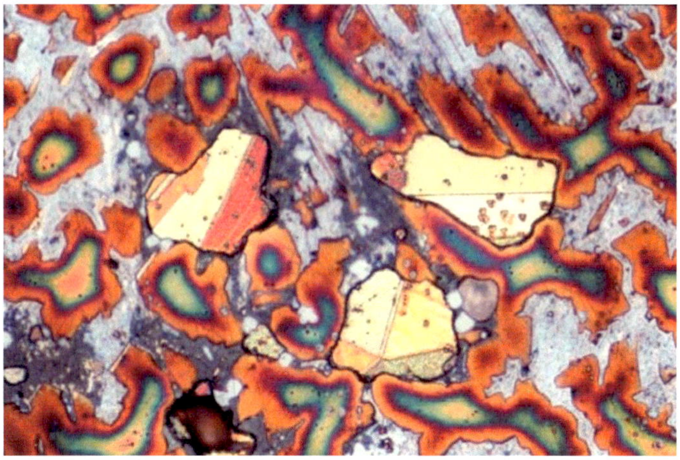

Figure 9.15. etched photomicrograph of a Luristan ceremonial axe, showing eutectoid phase with cored alpha and twinned crystals of redeposited copper. Etched in lead thiosulphate colour etch, x 360.

The section is colour etched and reveals the segregation well. The etching also shows the presence of twinned grains of copper which have displaced the (α+ δ) eutectoid phase, reprecipitating copper in the process as tin is lost.

The grains of copper are twinned naturally from deposition from solution, as seen in Figure 9.15. In some copper alloys, the processes of annealing and working can be combined into one operation, known as hot-working. Hot-working allows plastic movement of the metal to be attained so that the shape is altered and, when cooled, retains the new deformation. Some metals, such as iron and gold, can be welded together when hot to produce objects of different shape and size, although with most metals the presence of an oxide film makes it difficult to combine smaller pieces of metal together without recourse to melting them together into a new cast ingot or joining them with the use of solder.

Some alloys are hot-short and cannot be worked at elevated temperatures. For example, bronzes with about 17% of tin may be hot-short as the brittle delta phase surrounding the grains may result in shattered fragments being produced. These fragments are useful as chaplets or spacers and were much used in ancient Chinese bronze castings. Many brass alloys with high amounts of zinc are also hot-short and must be shaped by working and annealing rather than hot-working.

An example of twinned structures found in copper alloys that have been cast, cold-worked and annealed is shown in Figure 9.16. The section is taken from an early Bronze Age palstave dated to about 2700 BC from County Clare, Ireland. Although the axe was cast in an open mould and might be expected to show a cast, dendritic structure, these open-mould Bronze Age axes were often finished by working and annealing of the cast form to shape the final product. The cutting edge could be work-hardened by selective cold-working in this area, increasing substantially the hardness of these bronze tools. Both corrosion and some porosity are present in this microstructure, but the essentially annealed nature of the axe can be seen in the twinned grain structure in Figure 9.17: blue-etching regions are of redeposited copper, twinned, as a result of corrosion.

Figure 9.16. Section through a Palstave from County Clare, Ireland dated to 2700 BC. Colour etched in acidified lead thiosulphate etch. The recrystallized grains show twin lines and crystals of redeposited copper as a result of corrosion appear blue. Some corrosion has also occurred along stain lines showing extensive working to harden this Palstave, made in an 8% tin bronze, to the maximum extent.

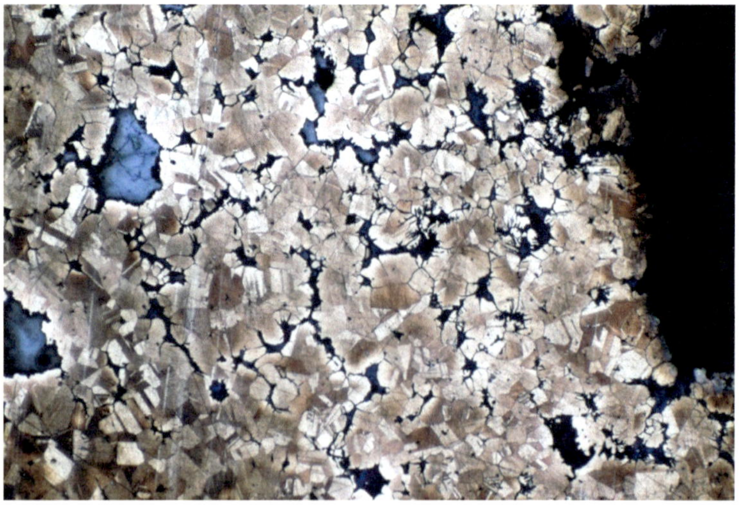

When an annealed object with twinned grains is cold-worked, the twin lines become deformed and will no longer appear straight. This is indicative that the final stage of fabrication was cold-working. Additional features such as strain lines may be visible in the metal after etching, especially if the artifact has been heavily cold-worked. Strain lines, just visible to the edge of Figure 9.16, occur as a result of the slip of crystal planes over one another in the bronze. Very prominent strain lines are seen in Figure 9.17 where this cutting edge of a Bronze Age axe from Austria, shows a bronze deliberately hardened by cold-working, creating strain lines only at the cutting edge and raising the hardness to 270 Hv.

Figure 9.17. Cutting edge of a Bronze Age axe from Austria. Strain lines are the fine striations across the grains in this section. Some twinned grains are also to be seen, showing that the axe was cast, worked and annealed, followed by heavy working of the cutting edge to harden the axe edge as much as possible. Etched in Klemm's no 1 reagent, x300.

Figure 9.18 Slip lines in this copper alloy axe from County Clare, Ireland, towards the cutting edge, shows selective hardening of the bronze alloy in this area. The standard Bronze Age working practice was to attempt to obtain as much tin in solid solution as possible to avoid undue brittleness in working axes and adzes, with cold-working of the cutting edge to increase the hardness up to about 250 HV. The small blue-grey inclusions are copper sulphide from the original smelted material being derived from copper sulphide ores or mixed sulphidic ores. Etched in alcoholic ferric chloride, x 230.

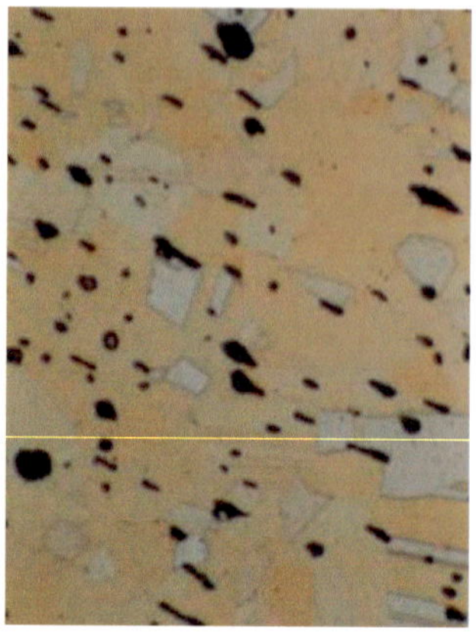

Figure 9.19 Photomicrograph of an Islamic coin from the Ubayid period, showing an extensive scatter of sulphide and lead inclusions in a large grain-size matrix of copper. Magnification x 230.

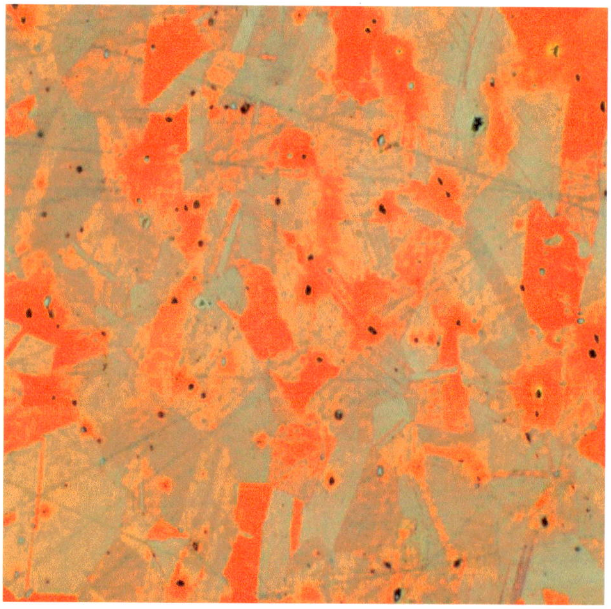

Figure 9.20. Etched section of an Iranian bronze from the Middle Bronze Age, showing twinned and recrystallized grains with colour variation due to coring. Etched in lead thiosulphate etch, x230.

In some alloys such as tin bronze, gold alloys and iron alloys there may be advantages to be gained in quenching the alloy from a high temperature to either retain certain phases which are only present usually at high temperatures, or to prevent phases from forming at room temperature which might lead to embrittlement. Thus, many gold alloys are quenched to prevent ordering reactions occurring, making the alloy softer and less brittle, while many steels are quenched, to form martensite, to make them hard. Martensitic transformations arise from the metastable formation of a single phase from some higher temperature phase, which is important in copper alloys as well as iron and steel. The other common transformation in copper alloys and ferrous materials is the Widmanstätten precipitation.

This precipitation arises from the breakdown of a homogeneous solid solution at higher temperatures to produce two phases in which the orientation of the precipitated phase tends to align with certain crystallographic planes. The precipitate is at some period at

least partially coherent with the matrix, has a preferred shape and bounding surfaces, and a definite lattice orientation. Under some conditions these preferred shapes become visible metallographically and are then known as the
Widmanstätten precipitation, first observed in iron meteorites, discussed later in Volume II of this series. The beta phase brasses include many examples of this, as the illustration shown in Figure 9.21 for 45% zinc shows.

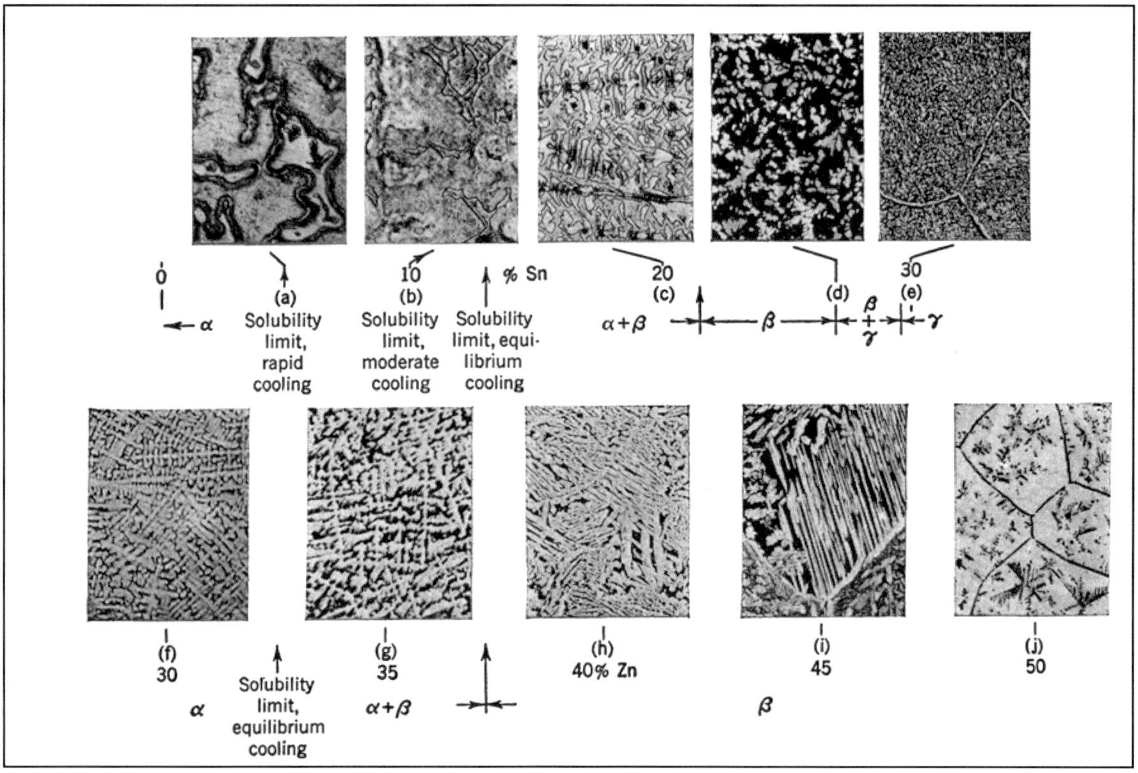

Figure 9.21 The first three drawings show typical microstructures of tin bronze alloys up to 20% tin. Illustrations d and e show the microstructure of beta brasses. f and g show the progression of zinc content from 30 to 35% in the cast state, and h and i, the higher zinc microstructures, with i showing a Widmanstätten structure. At 50% of zinc content, the structure changes to beta grains with precipitates of alpha within the beta grains. After Samans (1963).

The most complete empirical study of the copper-tin system yet published is that by Heycock and Neville in 1904! A series of papers were published on the copper-tin alloys, particularly in the period from 1900 - 1955. Seminal publications include work by: Heycock and Neville (1901a); Heycock and Neville 1901b), Kent (1926), Hanson and Marryat (1927), Matsuda (1928), Chadwick (1939) and Hanson and Pell-Walpole (1951). Further advances were made by Reeve, Bowden and Cuthbertson (1953a and 1953b), and by Ottaway and Wang (2004). Ottaway has been instrumental in several studies related to the different methods of casting a bronze alloy (Eccleston and Ottaway 2002; Ottaway and Seibel 1997; Ottaway 2003.).

Detailed microstructural studies on the cast copper-tin system really began with Heycock and Neville whose work has largely been forgotten in the 21st century. The following nine plates shown in Figures 9.22-9.30 are from their 1904 Bakerian lecture to the Royal Society together with notes which have been summarized here to explain the constitution of each black-and-white photomicrograph. Figure 9.22.1 is 1.85% tin, remainder copper, as cast, x18 showing dark cores in the alpha phase; Figure 9.22.2 is 3.7% tin, as cast, x45, cooled from 1025 °C and has dark primary alpha. Figure 9.22.3 is 3.7% tin, x10, a slow-cooled chill casting, from 957 °C, showing grains of alpha possessing different orientation under oblique illumination. Figure 9.22.3A is a 3.7% tin bronze, a slow-cooled chill casting from 957 °C, x10, showing dark alpha in vertical illumination. Figure 9.22.4 is a 3.7% tin bronze slow-cooled chill casting from 882 °C, x10, showing a homogeneous alpha. Figure 9.22.5 is a 3.7% tin bronze, x18, very slowly cooled chill casting from 770 °C showing homogeneous alpha.

Figure 9.22.5A is a 7.2% tin bronze, x18, a slow-cooled chill casting from 800 °C showing coring in the alpha phase. Figure 9.22.6 is a 7.2% tin bronze, x18, very slow-cooled chill casting from 900 °C showing alpha and matrix. Figure 9.22.7 is a 7.2% tin bronze, x45, very slow-cooled chill casting from 800 °C showing a homogeneous alpha. Figure 9.22.8 is a 7.2% tin bronze, x18, very slow-cooled chill casting from 775 °C, showing a cored alpha phase. Figure 9.22.9 is a 7.2% tin bronze, x18, very slow cooled, not chilled, showing uniform grains of alpha. Figure 9.22.10 is a 10.15% tin bronze, x18, a slow-cooled chill casting from 805 °C showing a light alpha and matrix. Figure 9.22.11 is a 10.15% tin bronze, x18, a slow-cooled chill casting from 805 °C showing light alpha and matrix.

In Figure 9.23.12 is shown a 10.15% tin bronze, x18, very slow cooled, not chilled, showing alpha and a phase called C′ by Heycock and Neville, which is in fact the alpha + delta eutectoid. Figure 9.23.13 is a 10.15% tin bronze, x280, very slow cooled, not chilled, showing patches of the eutectoid. Figure 9.23.14 is a 15.6% tin bronze, x120, cast from 880 °C showing primary and chill primary alpha in a matrix which was liquid at the chill temperature. Figure 9.23.15 is a 15.6% tin bronze, chilled from 777 °C showing a mixture of alpha, light, and beta dark. Figure 9.23.15 is a 15.6% tin bronze, x18, very slowly chill cast from 546 °C, showing a mixture of alpha and beta. Figure 9.23.17, is a 15.6% tin bronze, x18, a very slow-cooled chill casting from 470 °C showing alpha, dark, with eutectoid phase. Figure 9.23.18 is a 15.6% tin bronze, x280, very slow cooled chill casting from 546 °C showing uniform dark beta with a margin of white delta and alpha phase outside. Figure 9.23.19 is a 15.6% tin bronze, x280, very slow-cooled chill cast from 470 °C showing patch of eutectoid, bordered by delta with alpha outside. Figure 9.23.20 is a

20.3% tin bronze, x18, slow-cooled chill cast from 805 °C, showing primary alpha, light, the matrix darker and chill primary beta phase, darkest. Figure 9.23.21 is a 20.3% tin bronze, x18, very slow cooled chill cast from 775 °C showing partially dissolved alpha, light, in a ground of solid striated beta. Figure 9.23.22 is a 20.3% tin bronze, x18, chilled from 559 °C showing a light alpha that has crystallized out of the solid uniform beta. Beta here is dark.

Figure 9.23.23 is a 22.5% tin bronze, x18, very slow-cooled chill cast from 775 °C, showing striated beta free from alpha.

Figure 9.24.24 is a 22.5% tin bronze, x18, cast from 765 °C showing residues of alpha, white, in beta needles, darker, with liquid darkest. Figure 9.24.25 is a 22.5% tin bronze, x45, cast from 775 °C showing a residue of alpha in a darker striated beta. Dark chill primary of beta, liquid light. Figure 9.24.26 is a 22.5% tin bronze, x18, chill cast from 740 °C showing residual alpha, light, with beta dark. Figure 9.26.27 is a 22.5% tin bronze, x18, chill cast from 558 °C, showing new alpha, light, with uniform beta, dark. Figure 9.24.28 is a 23.3% tin bronze, x18, very slow cooled chill cast from 800 °C, showing primary alpha, light, liquid and chill primary of beta, dark. Figure 9.24.29 is a 23.3% tin bronze, x18, very slow cooled chill cast from 675 °C, showing only striated beta. Figure 9.24.30, x18, is a 23.3% tin bronze, x18, very slow cooled from 600 °C, showing new alpha, white, dark and striated beta. Figure 9.24.31 is a 23.3% tin bronze, x18, very slow cooled chill cast from 530 °C showing new alpha, white with uniform beta, dark. Figure 9.24.32 is a 23.3% tin bronze, x18, very slow cooled chill cast from 470 °C, showing alpha, dark, with eutectoid, light.

9.24.33 is a 24.75% tin bronze, x18, very slow cooled from a chill cast at 600 °C showing uniform grains of beta. Figure 9.24.34 is a 25.5% tin bronze, x45, chill cast from 750 °C showing dark primary crystals of beta , chill primary of beta, tin-rich liquid white. Figure 9.24.35 is a 25.5% tin bronze, x45, chill cast from 552 °C showing uniform beta under oblique illumination.

Figure 9.25.36 is a 25.5% tin bronze, x45, chill cast from 452 °C showing eutectoid complex, oblique illumination. Figure 9.25.37 is a 25.5% tin bronze, x550, chill cast from 452 °C, showing eutectoid complex. Figure 9.25.38 is a 26.25% tin bronze, x18, very slow chill cast from 752 °C, showing dark beta with tin-rich liquid, white. Figure 9.25.39 is a 26.25% tin bronze, x18, very slow cooled from 738 °C showing beta filling the ingot. Figure 9.25.40 is a 26.25% tin bronze, x280, not chilled showing crystals of delta in a eutectoid complex. Figure 9.25.41 is a 27.65% tin bronze, x45, very slow cooled from 745 °C showing a maximum of beta in liquid.

Figure 9.25.42 is a 27.65% tin bronze, x18, cast from 731 °C showing dark beta in a tin-rich matrix. Figure 9.25.43 is a 29.05% tin bronze, x18, cast from 734 °C, showing beta decomposing. Figure 9.25.44 is a 30.4% tin bronze, x125, cast from 733 °C showing beta decomposing. Figure 9.25.45 is a 29.05% tin bronze, x45, very slow cooled cast from 728 °C showing a homogeneous gamma matrix. Figure 9.25.46 is a 27.65% tin bronze, x45, cast from 495 °C showing commencement of delta crystals. Figure 9.25.47 is a 27.65% tin bronze, x250, cast from 450 °C showing white delta in a eutectoid complex.

Figure 9.26.48 is a 29.05% tin bronze, x45, chill cast from 550 °C showing a uniform solid solution of beta. Figure 9.26.49 is a 30.4% tin bronze, x45, chill cast from 500 °C showing some white crystallization of delta phase together with beta. Figure 9.26.49A is 30.4% tin bronze, x45, chill cast from 550 °C, showing some crystallization of delta together with beta phase. Figure 9.26.50 is a 29.05% tin bronze, x45, chill cast from 501 °C showing white crystallization of delta. Figure 9.26.51 31.8% tin bronze, chill cast from 726 °C, showing decomposing beta phase in a white tin-rich matrix. Figure 9.26.52 is a 31.8% tin bronze, not chilled, showing white delta in a matrix of alpha + delta eutectoid. Figure 9.26.53 is 30.4% tin bronze, x800, not chilled, showing patches of alpha + delta in a light alpha background.

Figure 9.26.53A is a 31.8% tin bronze, x600, showing patches of alpha + delta eutectoid in a alpha matrix. Figure 9.26.54 is a 33.15% tin bronze, x45, chill cast from 725 °C showing primary gamma and chilled primary phase. Figure 9.26.55 is a 33.15% tin bronze, x45, cast from 650 °C, showing a uniform field of gamma phase. Figure 9.26.56 is a 33.15% tin bronze, chill cast from 590 °C showing a dark eta phase and a light delta phase.

Figure 9.26.57 is a 33.15% tin bronze, x45, cast from 580 °C showing twinned plates of eta phase. Figure 9.26.58 is a 34.45% tin bronze, cast from 733 °C showing primary precipitation of gamma phase. Figure 9.26.59 is 34.45% tin bronze, cast from 690 °C, showing uniform gamma with slender dark bars of eta phase.

Figure 9.27.60 is a 34.45% tin bronze, x18, chill cast from 565 °C showing dark bands of eta phase and a lace-like complex of delta + eta. Figure 9.27.61 is a 34.45% tin bronze, x45, cast from 495 °C showing dark bars of eta phase lace-like bars of delta + eta. Figure 9.27.62 is a 35.8% tin bronze, x 18, chill cast from 620 °C showing crystallization of eta phase in increasing amounts. Figure 9.27.63 is a 37.1% tin bronze, x18, chill cast from 685 °C, showing solid gamma phase, where the crystallization of eta has not been wholly prevented by the chill casting. Figure 9.27.64 is a 37.1% tin bronze, x 18, chill cast from 635 °C, showing dark crystals of eta phase nearly filling the alloy. Figure 9.27.65 is a 38.4% tin bronze, x13, chill cast from 720 °C, showing copper-rich gamma precipitation darker than the tin-rich liquid. Figure 9.27.65A is a 38.4% tin bronze, x45, showing the breaking up of the gamma phase into the complex shown in Figure 9.27.64. Figure 9.27.66 is a 38.4% tin bronze, x18, not chilled, showing bars of differently orientated eta phase. Figure 9.27.67 is a 39.65% tin bronze, x45, cast from 656 °C showing a solid solution of gamma. Figure 9.27.68 is a 39.65% tin bronze, x45, cast from 612 °C showing plates of eta phase. Figure 9.27.69 is a 40.8% tin bronze tin bronze, x18, cast from 647 °C showing gamma with traces of liquid and chill primary of eta bars. Figure 9.27.70 is a 40.8% tin bronze chill cast from 635 ^{o}C showing a gamma solid solution. Figure 9.27.71 is a 40.8% tin bronze, x45, cast from 625 ^{o}C, showing plates of eta phase with inclusions of tin-rich liquid.

Figure 9.28.72 is a 40.8% tin bronze, x18, cast from 565 ^{o}C showing plates of eta with inclusions of liquid. Figure 9.28.73 is a 42.05% tin bronze, x 45, chill cast from 650 oC, showing uniform lobes of gamma, and chill primary of eta with dark tin-rich liquid. Figure 9.28.74 is a 42.05% tin bronze, x45, cast from 625 ^{o}C showing uniform lobes of gamma with primary eta. Figure 9.28.75 is a 47.9% tin bronze, x45, cast from 676 ^{o}C, showing primary gamma with gamma chill matrix. Figure 9.28.76 is a 47.9% tin bronze, x18, cast from 644 ^{o}C showing primary gamma with chilled primary of eta. Figure 9.28.77 is a 47.9% tin bronze, x45, cast from 644 ^{o}C showing primary gamma with chilled primary eta. Figure 9.28.78 is a 47.9% tin bronze, x 45, cast from 628 ^{o}C showing the primary gamma and eta phases. Figure 9.28.79 is a 53.35% tin bronze, x45, cast from 644 ^{o}C, showing primary gamma with chill cast eta.

Figure 9.28.80 is a 53.35% tin bronze, x45, cast from 628 ^{o}C showing incomplete transformation of G. Figure 9.28.81 is a 53.35% tin bronze, x45, showing complete G transformation. Figure 9.28.82 is a 57.5% tin bronze, x18, cast from 600 ^{o}C showing primary and chill cast eta. Figure 9.28.83 is a 53.35% tin bronze, x45, chill cast from 445 ^{o}C, showing primary eta phase.

Figure 9.29.84 is a 65.1% tin bronze, x45, chill cast from 445 ^{o}C, showing primary eta phase. Chill primary is H. Figure 9.29.85 is a 88.2% tin bronze, cast from 433 ^{o}C showing primary plates of eta in a tin-rich liquid. Figure 9.29.86 is a 65.1% tin bronze, x45, chill cast from 420 ^{o}C showing primary plates of eta in a tin-rich liquid. Figure 9.29.87 is a 83.5% tin bronze, x 45, cast from 373 ^{o}C showing plates of darkened eta, margined by white H in liquid. Figure 9.29.88 is a 43.2% tin bronze, x45, chill cast from 380 ^{o}C, showing plates of darkened eta, fringed by H. No liquid. Figure 9.29.89 is a 94.4% tin bronze, x45, not chilled, showing primary H in eutectic of H and tin. Figure 9.29.90 is a 97.25% tin bronze, x45, showing primary of H in a eutectic of H and tin. Figure 9.29.91 is a 65.1% tin bronze, x 45, showing bars of eta with a narrow fringe of H, in H + tin eutectic. Figure 9.29.92 is a 65.1% tin bronze, x18, cast after keeping for ten hours at 380 ^{o}C, showing H reaction incomplete with eta largely transformed.

Figure 9.30.93 is a 65.1% tin bronze, x18, after keeping at 60 hours at 350 ^{o}C, showing traces of eta in bands of H with tin-rich dark liquid. Figure 9.30.94 is a 65.1% tin bronze, x18, after keeping for 21 days at 350 ^{o}C, showing uniform H with holes. No liquid and no eta phase. Figure 9.30.95 is a 57.5% tin bronze, x18. Kept at 350 ^{o}C for 24 days, showing compact H, containing eta. Figure 9.30.96 is a 57.5% tin bronze, x18, kept for 24 days at 350 ^{o}C showing compact H containing eta with many black cavities. Figure 9.30.97 is a

55.4% tin bronze, x45, kept for 21 days at 350 °C showing compact H containing eta. Figure 9.30.98 is a 73.7% tin bronze, x45, kept for 21 days at 350 °C showing light polyhedral of H in the H + tin eutectic. Figure 9.30.99 is a 91.3% tin bronze, x45, kept for 21 days at 350 °C showing two types of H crystals in the eutectic. Figure 9.30.100 is of unknown composition, x45, showing H made by prolonged stirring of a 90% tin alloy at 350 °C by means of a copper stirrer. Figure 9.30.101 is a 65.1% tin bronze, x18, kept for 10 days at 350 °C and the heated to 400 °C, showing eta and liquid produced from H by heating the alloy for a short time to above 400 °C.

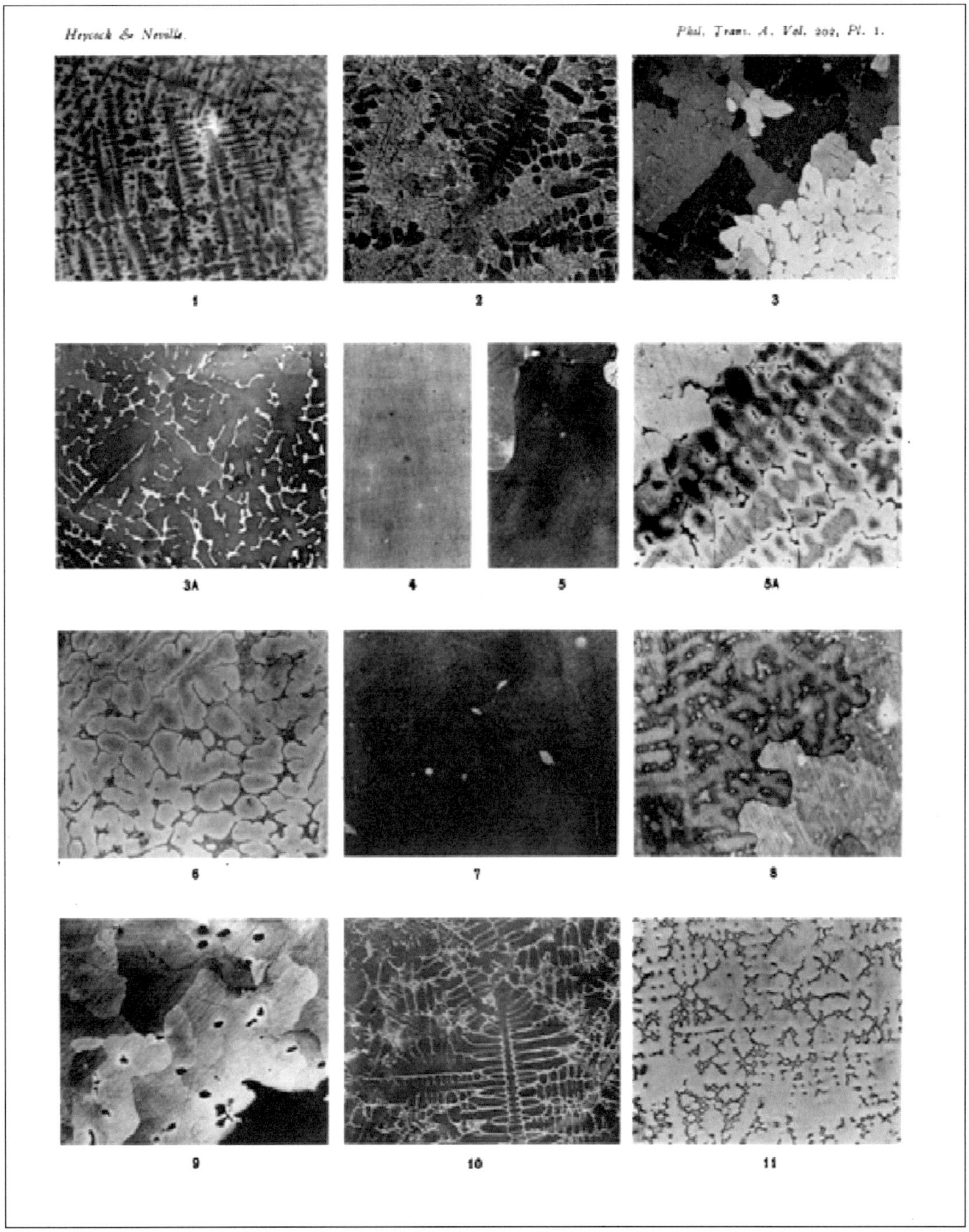

Figure 9.22 After Haycock and Neville 1904 Plate 1.

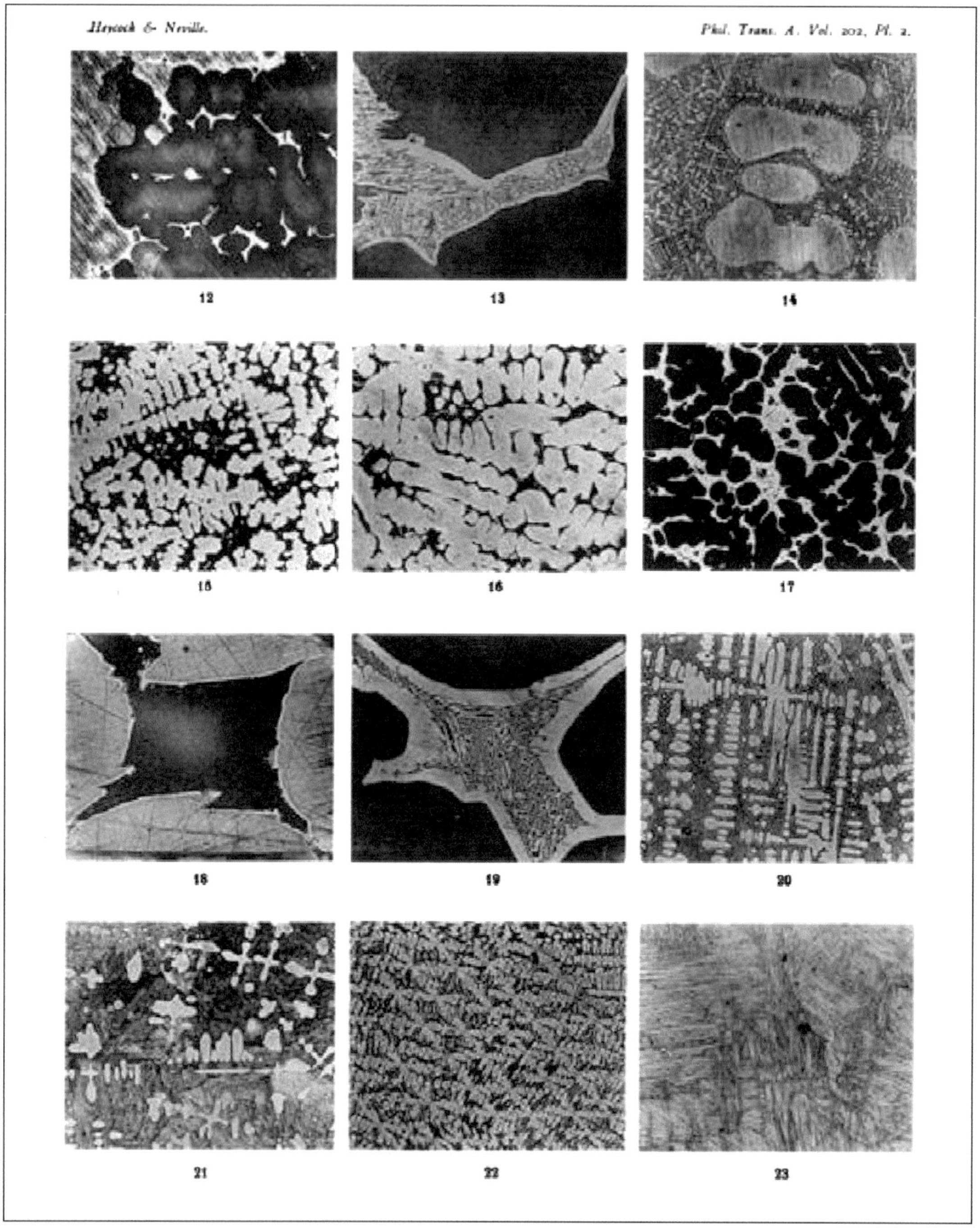

Figure 9.23 After Heycock and Neville 1904 Plate 2.

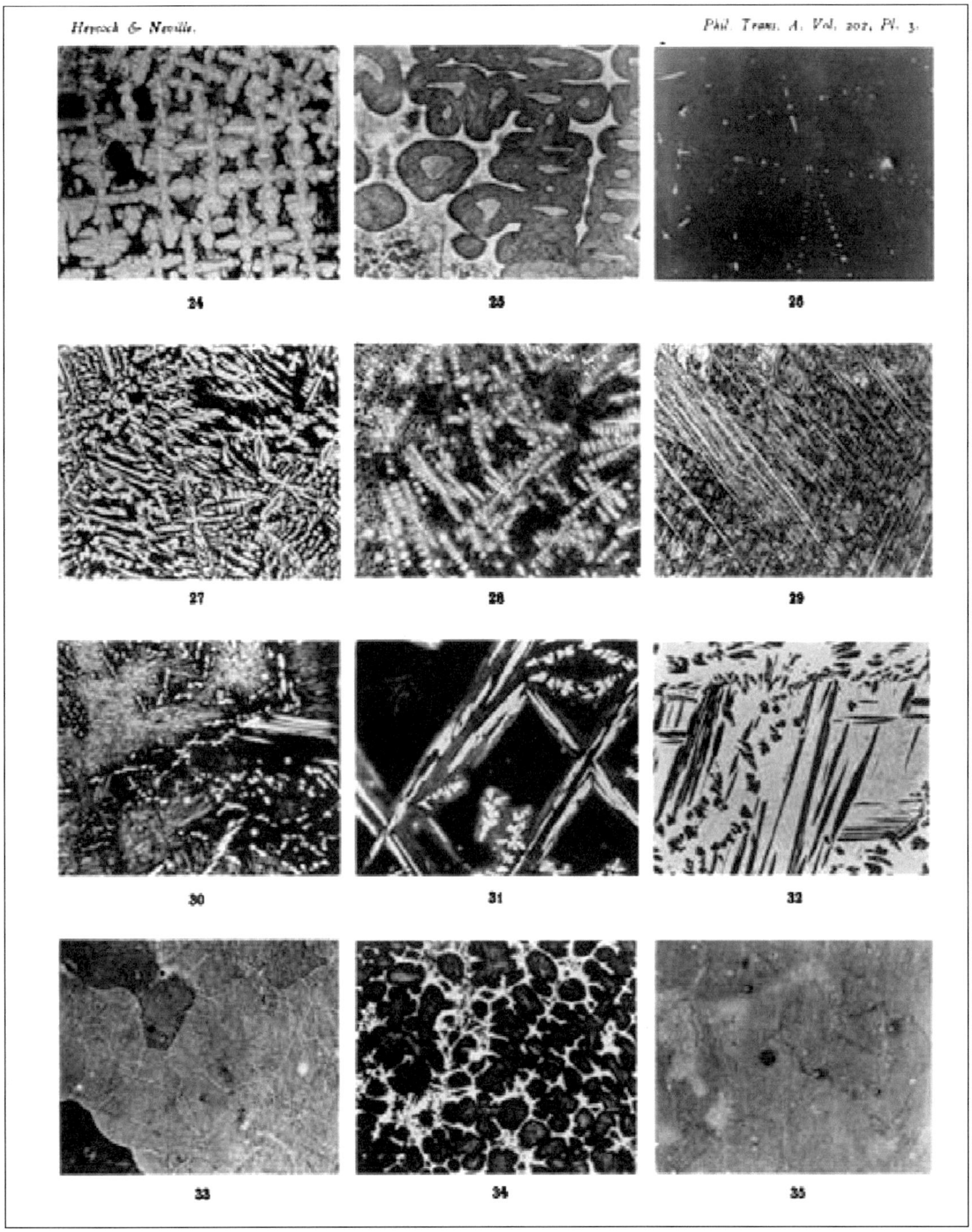

Figure 9.24 After Heycock and Neville 1904 Plate 3.

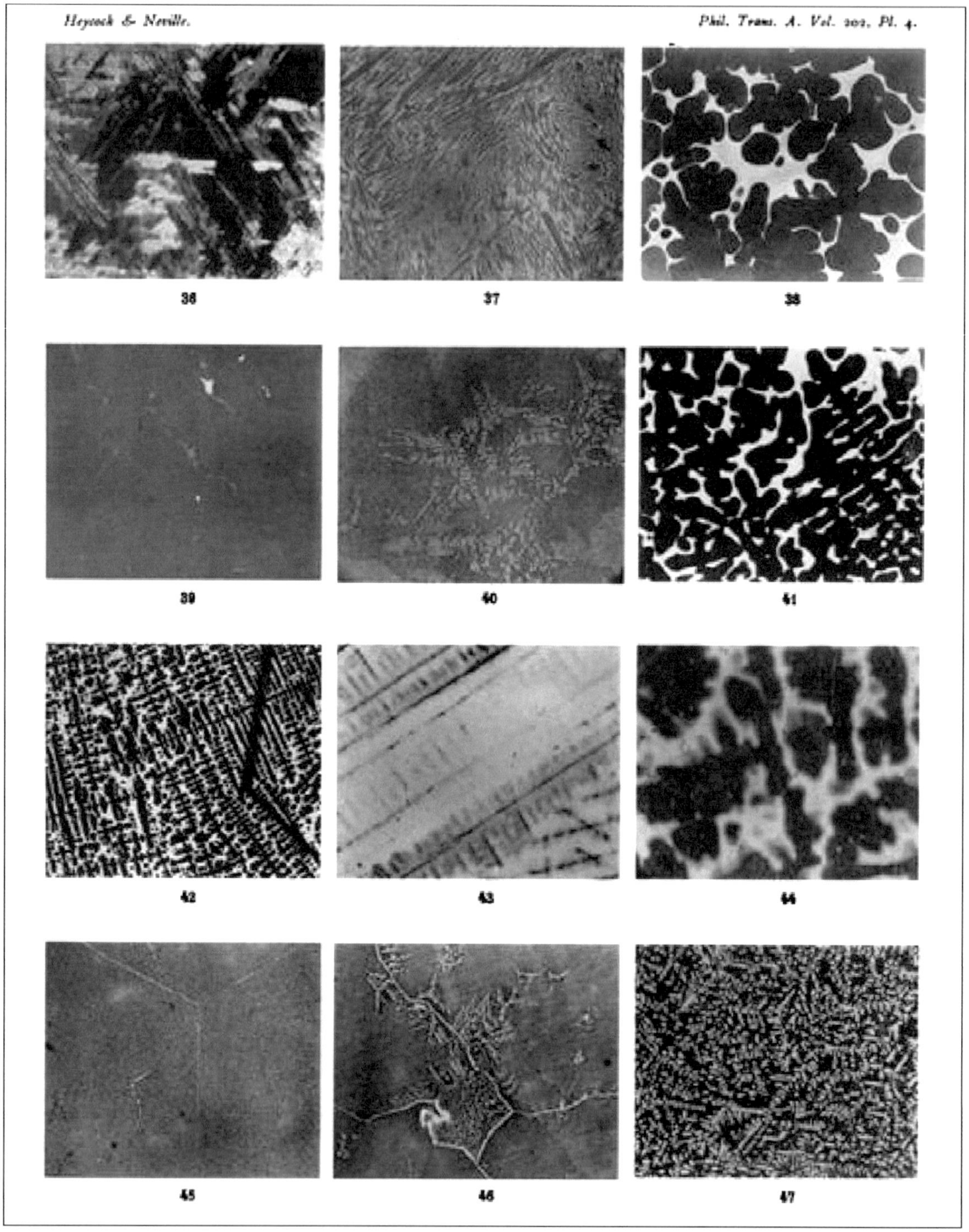

Figure 9.25 After Heycock and Neville 1904 Plate 4.

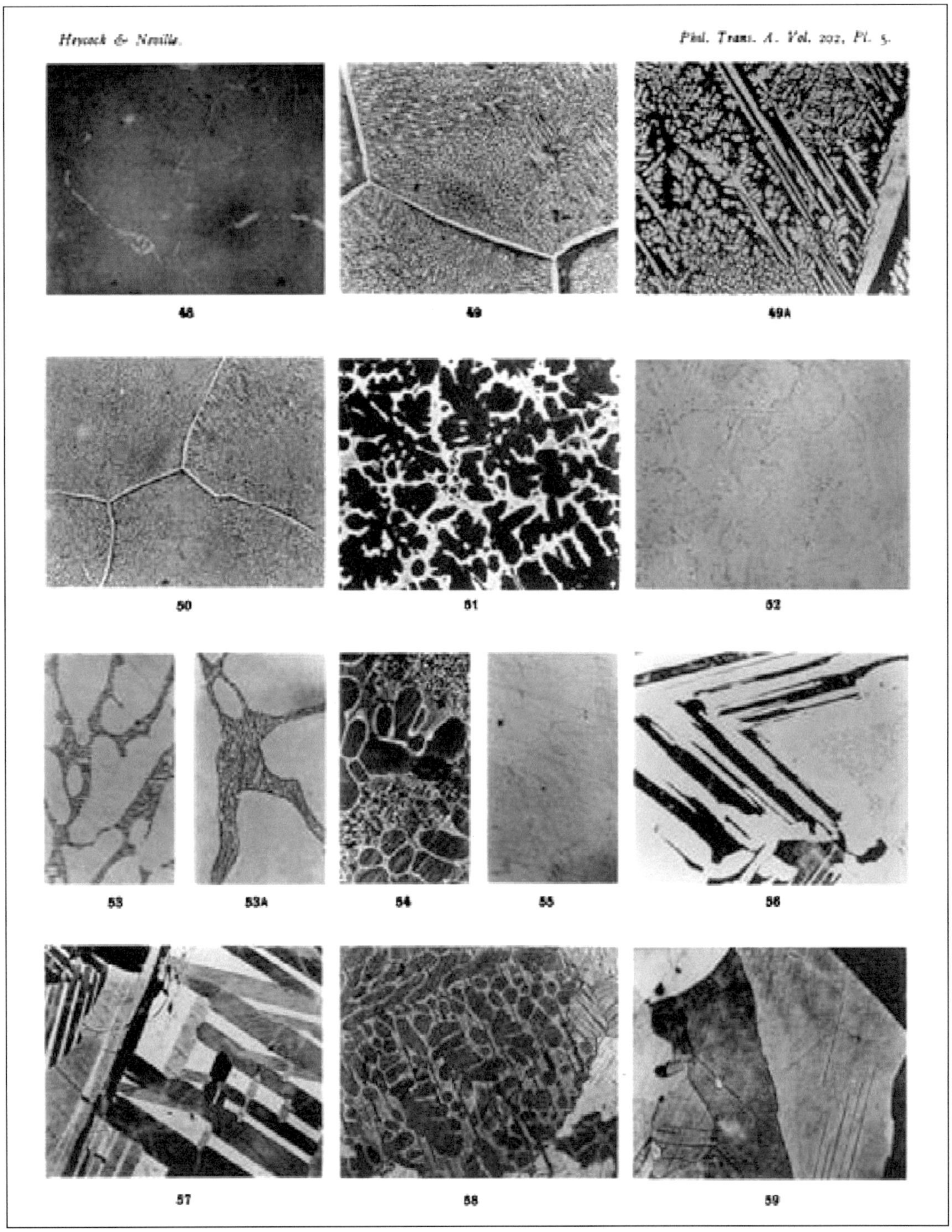

Figure 9.26 After Heycock and Neville 1904 Plate 5.

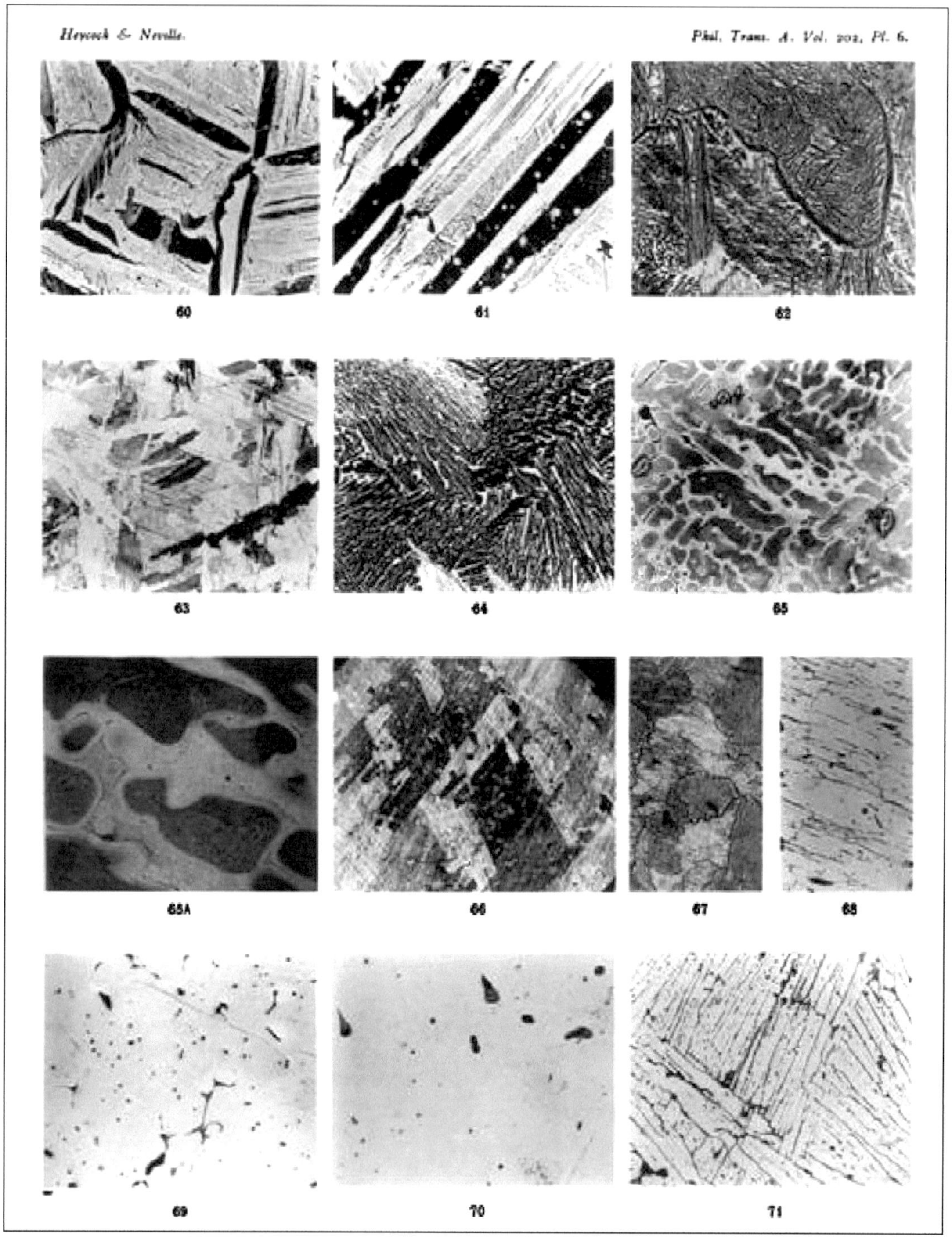

Figure 9.27 After Heycock and Neville 1904 Plate 6.

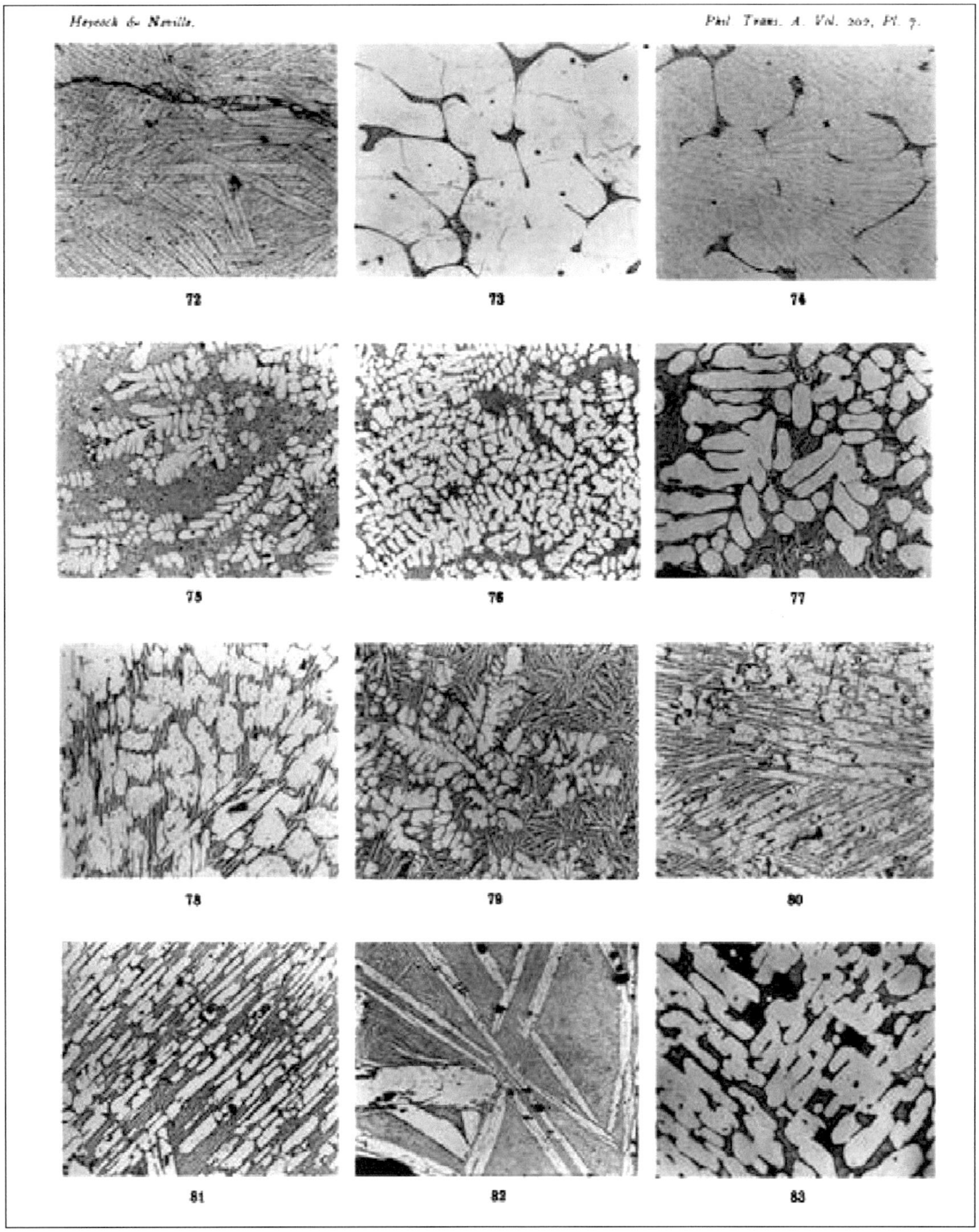

Figure 9.28 After Heycock and Neville 1904 Plate 7.

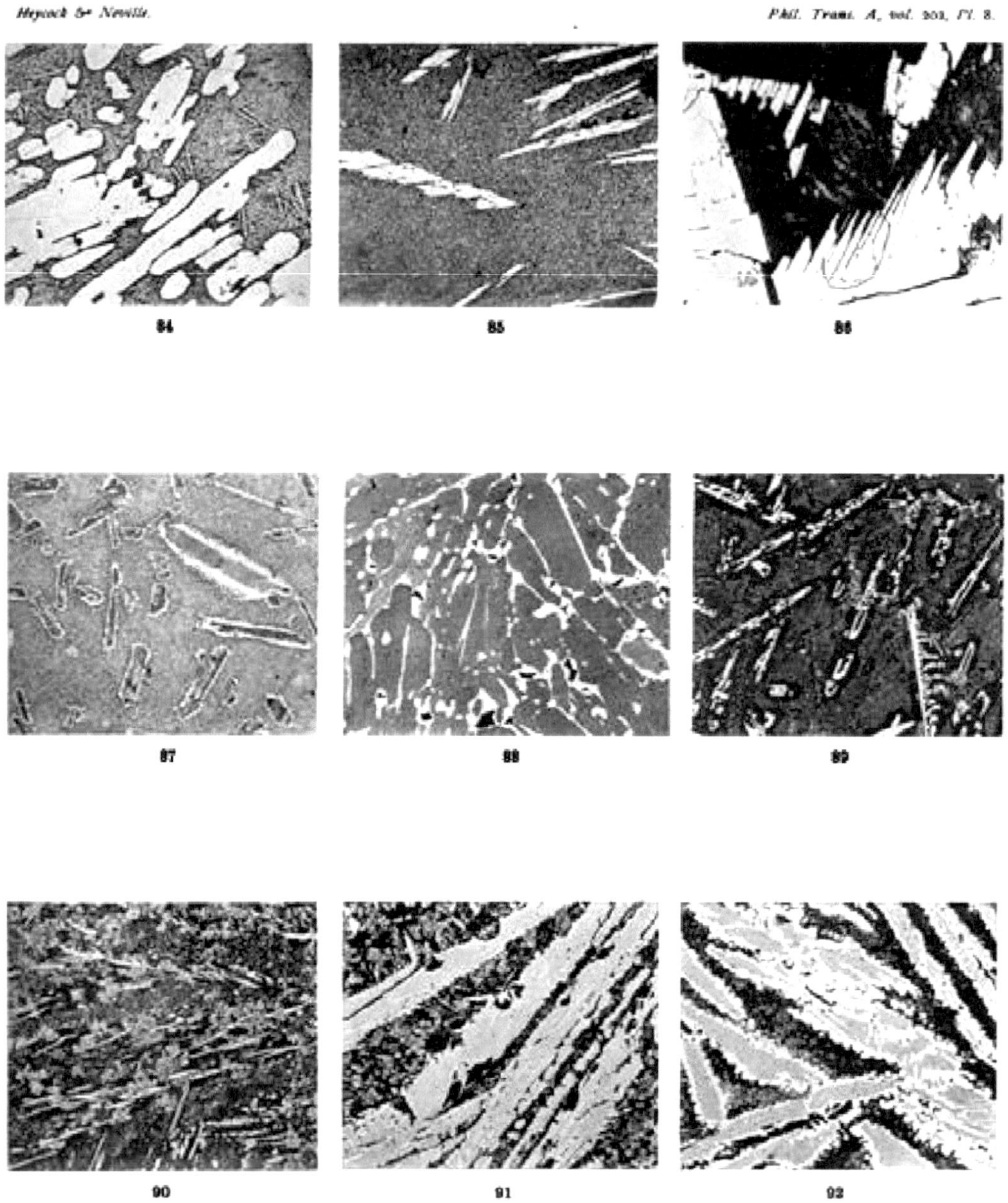

Figure 2.29. After Heycock and Neville 1904 Plate 8

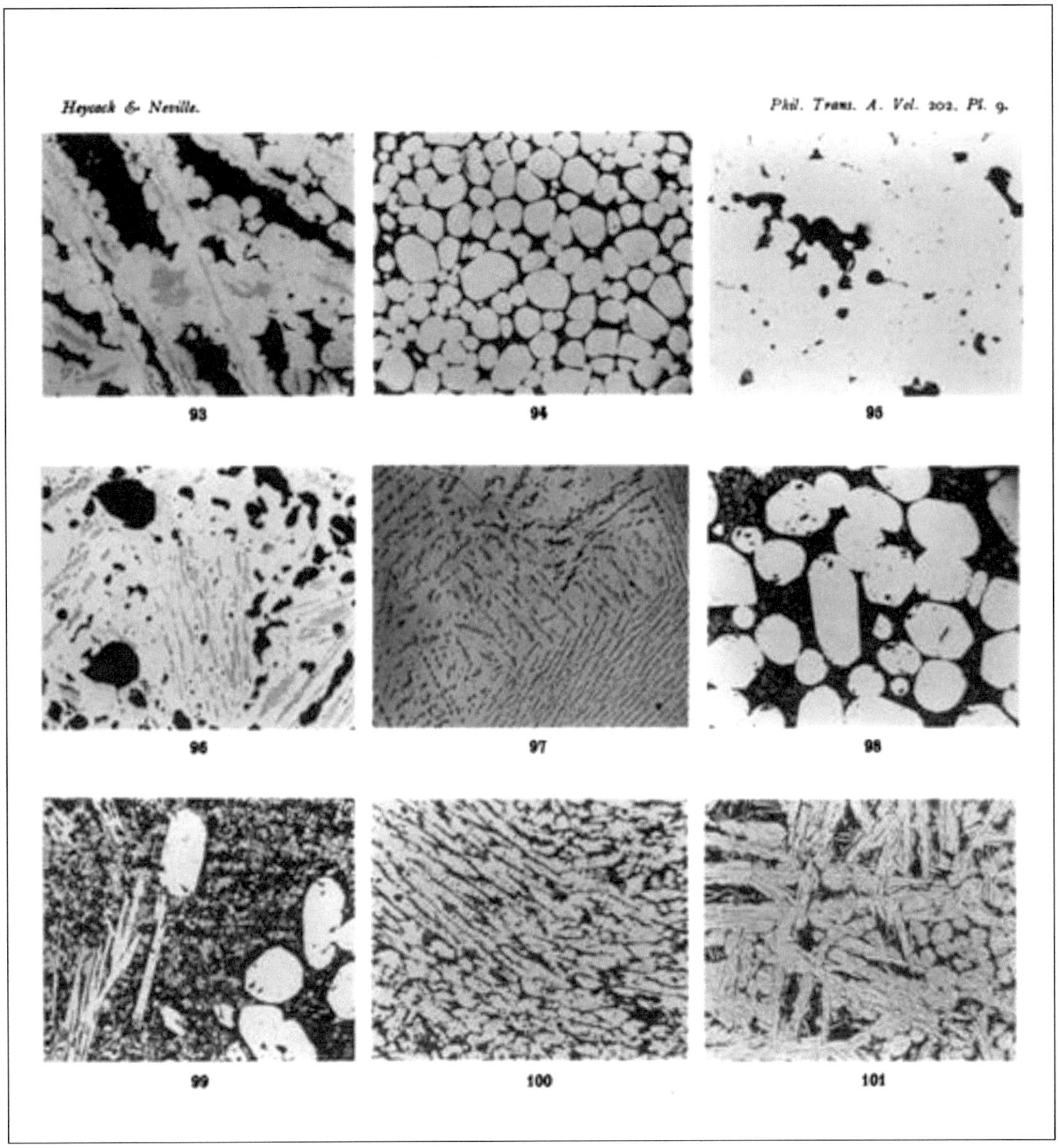

Figure 9.30 After Heycock and Neville 1904 Plate 9.

A Note on Phase relationships

At the time that Heycock and Neville were writing, there was still some confusion concerning the epsilon phase and its stability limits. The phase which Heycock and Neville assign to the complex or phase "H" in the above set of descriptions which have been transcribed from the tables accompanying their plates 1-9 can be read as the epsilon phase. There may also be some problems with their description of the gamma phase in terms of its stability regions, but the phase transitions are essentially correct. For example a 30% tin bronze chill cast from 700 °C will be gamma phase, while we are more familiar with the high-tin bronzes around 21% in which gamma is only stable between 586 °C and 520 °C. As a 30% tin bronze cools from 750 °C it will form gamma, which at 560 °C will break down to a mixture of gamma and delta, which will then decompose to a mixture of alpha + delta and delta. Most of this can be followed on the equilibrium diagram of Figure 9.1. In summary, these plates represent a substantial body of work of use in reference to tin bronzes of a variety of different compositions.

Quenching of Bronze Alloys

For low-tin bronzes, they may be quickly cooled which would affect the extent of dendritic growth, but the overall effects on the microstructure of the examples of bronze shown above would be slight. However, if the tin content is raised to around 20%, then on quenching some of the higher temperature solid phases present in the phase diagram can be retained, producing an alloy with very different working properties than the cast equivalent. For alloys used as mirrors, the differences are not great in terms of working properties but become significant if the cast mirror is finished by turning on a lath or if a cast bowl has to be cut and decorated after casting. Then the advantages of quenching these bronzes become apparent. An example of the microscopic effects of quenching is shown in Figure 9.32, part of a high tin bronze bowl from the site of Ban Don Ta Phet, Thailand, dated to about 100 AD, of composition 22.7% tin and 76.2% copper. The metalwork from this site has been published in summary form by Rajpitak and Seeley (1979). Ban Don Ta Phet is in the district of Phanom Thuan, in Kanchanaburi Province, south-west Thailand. Excavations on behalf of the Thai Fine Arts Department revealed a series of inhumation burials with 163 bronze vessels, 38 bronze bracelets, 7 bronze anklets, 16 bronze rings, 1 bronze ladle, 3 bird finials, 3 bells and 579 iron artefacts and other material.

Etched beads suggest a date in 3rd – 2nd century BC. The bronze bowls range from 5 to 15 cm in diameter (Rajpitak and Seeley (1979), are very thin, with wall thickness of 0.3 – 0.5 mm and with fine incised decoration. These bowls are made in a quenched high-tin bronze, such as that revealed by the sections shown in 9.32 and 9.33. The 21-23% tin bronzes require heating to at least 300 °C before they can be worked (Chadwick 1939), and the microstructure shows that it has been quenched from above 520 °C, which as we have already discussed is in the beta phase field (Reeve, Bowden and Cuthbertson 1953b).

Figure 9.31 An Ifugao Tribesman playing a high-tin bronze gong (tobob). His gong is made of a high-tin bronze. But how can this be if high-tin bronzes are so brittle? How did ancient cultures make this alloy into a workable product? (After Goodway and Conklin 1987). High-tin bronze alloys have interesting musical properties, still used today for cymbals.

Comparable fragments of high-tin bronzes have been reported by Marshall (1951) from Taxila, the ancient Gandharan capital on the North West Frontier of India, mostly from late phases, but Bhir Mound, the Mauryan city, from 3rd – 2nd C. BC contained a pair of bowls one with 21.55% tin, the ideal composition as the microstructures shown in Figures 9.32 and 9.33 are typical and show a preference for some alpha phase together with beta, rather than with slightly higher tin contents which are all beta phase. This shows that the composition of these special alloys was very carefully controlled.

Other important mentions of these alloys are those of: Smith (1973) from Pimai; Phase III at Ban Chiang (Wheeler and Maddin 1976), dating to 500 – 0 BC, a bronze bowl from Coimbatore, South India, and artefacts from Adichanallur in the Nilgiri Hills (Breeks 1873), two Koryu Dynasty bowls from Korea (Voce 1951) and some early Islamic bronze vessels (Melikian-Chirvani 1974). Rajpitak and Seeley (1979) mention the account of Nearchus in his description of India written in the 4th century BC, fragments of which survive in Strabo's geography (Strabo 15.1.67) which states that Indians used vessels which were cast rather than beaten, and which broke like pottery when dropped on the ground. This suggests that the brittle high-tin bronzes may well be described here. Recent work in South Indian by Srinivasan (1998) and Srinivasan and Glover (1998) confirms the evidence for use in this area, as the bronzes are still being made by traditional methods. Metallurgical investigations on bowls and vessels from Nilgiri megaliths and Adichanallur burials showed they consisted of 23-25% tin wrought to between 0.2 mm – 1mm with a Vickers Hardness of 290 – 300 Hv.

A bowl excavated from Taxila dated 1000 BC showed it to be a beta bronze with 24% tin. Three samples excavated from Mohenjodaro thought to be dated to 2500 BC were found to contain 22.1% tin, 22.2% tin with 14% lead and 26.8% tin. Srinivasan and Glover (2007) found that yet another extant tradition of mirror-making in South India utilizes cast alloys with 32% of tin, retaining a cellular delta microstructure such as that shown in Figure 9.11 and Figure 9.26.52, where Neville and Haycock show the same kind of structure, namely delta phase in a matrix of alpha + delta, for their 31.2% tin bronze. The account of the casting of mirrors in the village of Aranmula, Kerala, very close to the composition of the delta phase, $Cu_{31}Sn_8$, whose ideal formula is 32.6% tin, is recounted by Srinivasan and Glover (2007). The technique involves the use of a special closed crucible with attached mould, the account of which is of some interest. The hardness of the delta phase when cast is extraordinary, reaching values of 500 Hv, harder than that of a medium-carbon steel, which is around 400 Hv.

There may be a number of reasons for this choice of alloy. Firstly, the high-tin alloys are not injurious to health for food-keeping as a brass bowl would be; secondly, the quenched alloy can be hot-worked to shape it and can be ground or cut to decorate the surface in a way in which the high-tin alloys cannot if slow-cooled because of the brittle delta; and thirdly for the colour. The 23% tin bronze is a good golden colour which could be visually attractive and simulate gold. The alloy, while not too brittle, is hard, with Vickers numbers between 180 Hv to 260 Hv.

As the alpha phase in these alloys is frequently twinned, we can reconstruct the fabrication methodology from the microstructure.

A. Firstly the alloy is carefully made of exactly the right composition and cast into shape, roughly, probably in a mold.
B. Then the alloy is reheated above 600 °C and hot-worked.
C. The alloy is reheated above 600 °C and quenched. It undergoes a martensitic transformation to the beta phase.

D. The alloy can now be turned on a lathe to smooth the surface or to cut decoration into the surface of the bronze.

E. It can be reshaped if necessary by reheating again, working and quenching, followed by polishing to finish.

We know that operations such as this must have been involved, for otherwise the alpha phase would not be twinned from the hot-working process.

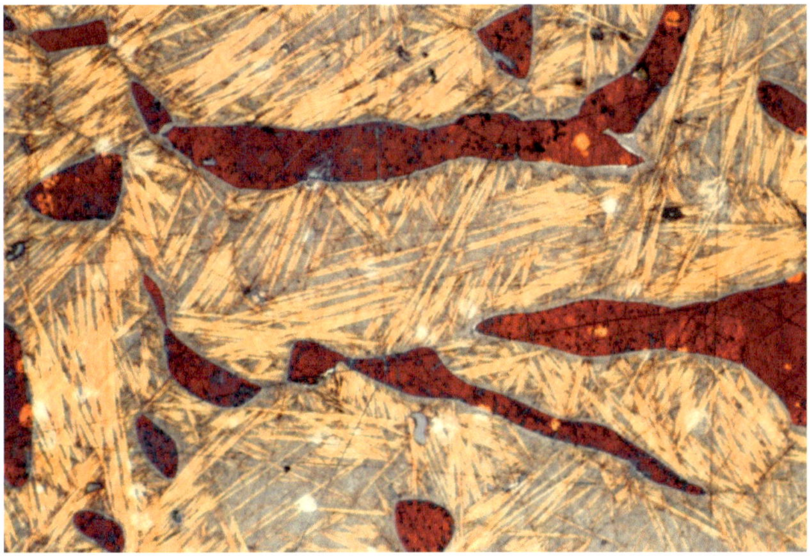

Figure 9.32. Section of a bronze bowl from the site of Ban Don Ta Phet dated to about 100AD of composition 22.7% tin and 76.2% copper, colour etched in Klemm's number 2 reagent. The beta-phase needles appear straw coloured and the copper-rich alpha-phase, red. Twins in the alpha phase can just be seen, showing that the bowl has been hot-worked from the beta-region. X280.

Figure 9.33. Overall view of the microstructure shown in Figure 9.23 of the high-tin bronze bowl from Ban Don Ta Phet, Thailand, showing cast morphology of the α-phase which occurs on boundaries of the former β-grains before there breakdown into martensitic β and islands of α. Etched in Klemm's no. 2 reagent, x87.

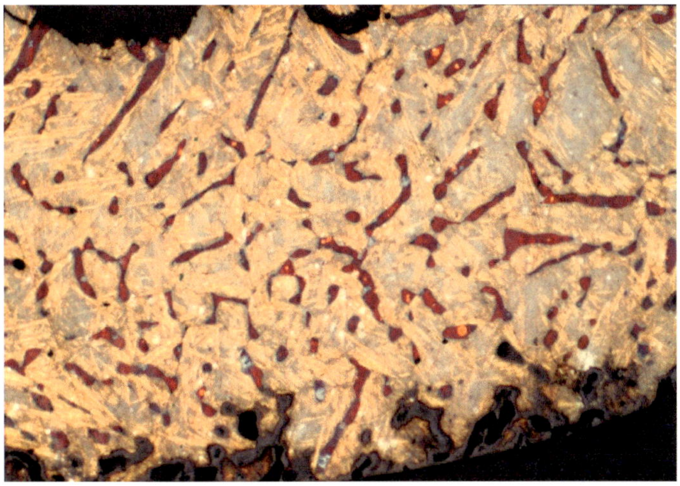

Examination of the etched section shows the presence of some alpha phase copper-rich grains, sometimes acicular and which occur in areas with specific orientation as well as a random scatter. The overall cast structure of these alloys can be seen in Figure 9.33. In the background matrix of this bronze is a banded martensite, a variety of the β phase that proves that the bowl must have been quenched from the temperature region of 520 - 586 °C. This is slightly lower than the usual quenching temperature for these bronzes, which is generally in the 650 - 750 °C range.

Some additional studies of the composition of these alloys confirms the general impression that the metalsmiths were deliberately striving for a composition which includes the alpha and beta phases, without knowledge, of course, of the phase relationships, but purely from their sophisticated observations on the behavior of metals. A study by Park, Park and Lee (2009) showed that, on deformation, the small amounts of alpha present in the 22% tin bronzes, spanning the grain boundaries, act as micro-bridges interlocking the adjacent grains. The alpha phase apparently reinforces the boundary regions, an effect which is very sensitive to the alpha fraction and can be best obtained with tin contents close to 22%. The hardness data obtained by Park is shown in Table 9.2 below in terms of the Rockwell B scale.

Rockwell B-scale hardness numbers measured from the Cu–Sn alloys with varying tin contents and thermal treatments.

#	Sn %	As-cast	Quenched at 750 °C	Quenched at 700 °C	Quenched at 610 °C
1	0	8.3 (8.5, 8.0)[a]			
2	5	24.3 (22.5, 26)			
3	10	52.3 (50.5, 54)		44.5 (45.0, 44.0)	
4	15	69.5 (69.5, 70.5, 68.5)		64.5 (63.5, 65.0)	63.5 (63.5, 63.5)
5	17	80.8 (81, 80.5)	78.0 (77.0, 79.0)	74.0 (73.5, 75.5)	70.0 (73.0, 67.0, 70.0)
6	20	93.8 (93.5, 94)	97.5 (98.0, 97.0)	96.3 (96.5, 96.0)	90.8 (90.5, 91.0)
7	22	99.5 (99.5, 99.5)	103.5 (103.5, 103.5)	105.5 (105.0, 106.0)	92.3 (92.0, 92.5)
8	24	104.0 (103.5, 104.5)		107.5 (108.0, 107.0)	104.5 (106.0, 103.0)
9	26	113.8 (113.5, 114)		112.0 (112.0, 112.0)	
10	28	114.0 (114.0, 114.0)		112.0 (112.0, 112.0)	

Bowls from Korea, from the Unified Silla period (668 – 935 AD) made in this alloy have wall thicknesses of 0.2 mm. Even a minor change in tin content, to 24% where the microstructure is entirely beta would not have allowed this degree of hot-working.

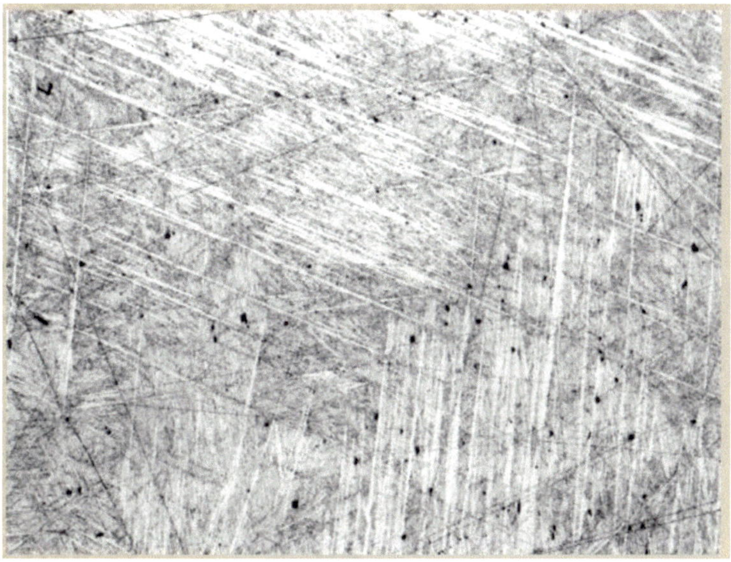

Figure 9.34 A slightly higher tin content laboratory alloy, at 24% tin, quenched in water from 650 °C showing fine, long needles of beta phase predominating. Etched in alcoholic ferric chloride, x250.

Here the phase diagram and its application is an essential component to understanding the microstructure of these quenched alloys. When quenched from high temperatures, the β phase is retained and the alloy undergoes a martensitic transformation.

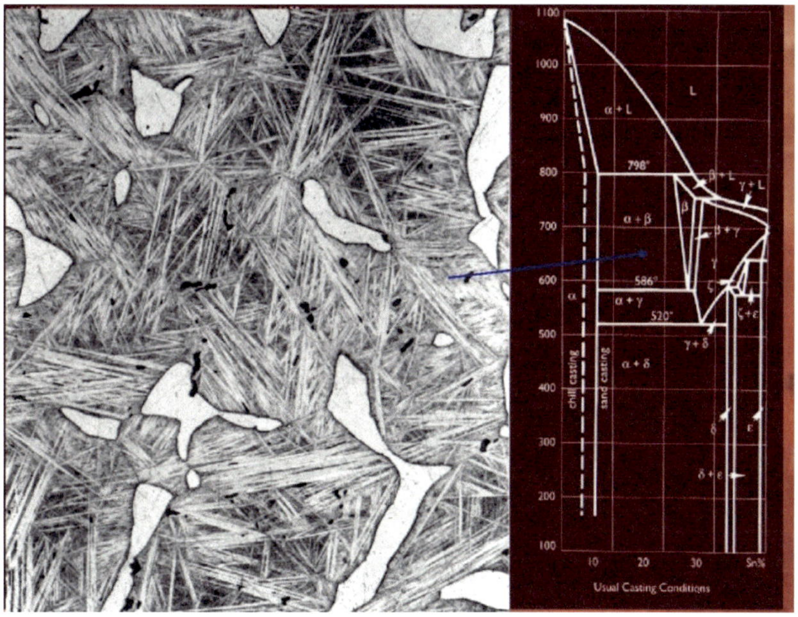

Figure 9.35. Another high-tin bronze alloy from Thailand, from Ban Don Ta Phet. The blue arrow pointing to the phase diagram, shows approximately the quenching temperature of this alloy with about 22% tin content. Magnification x 130, etched in alcoholic ferric chloride.

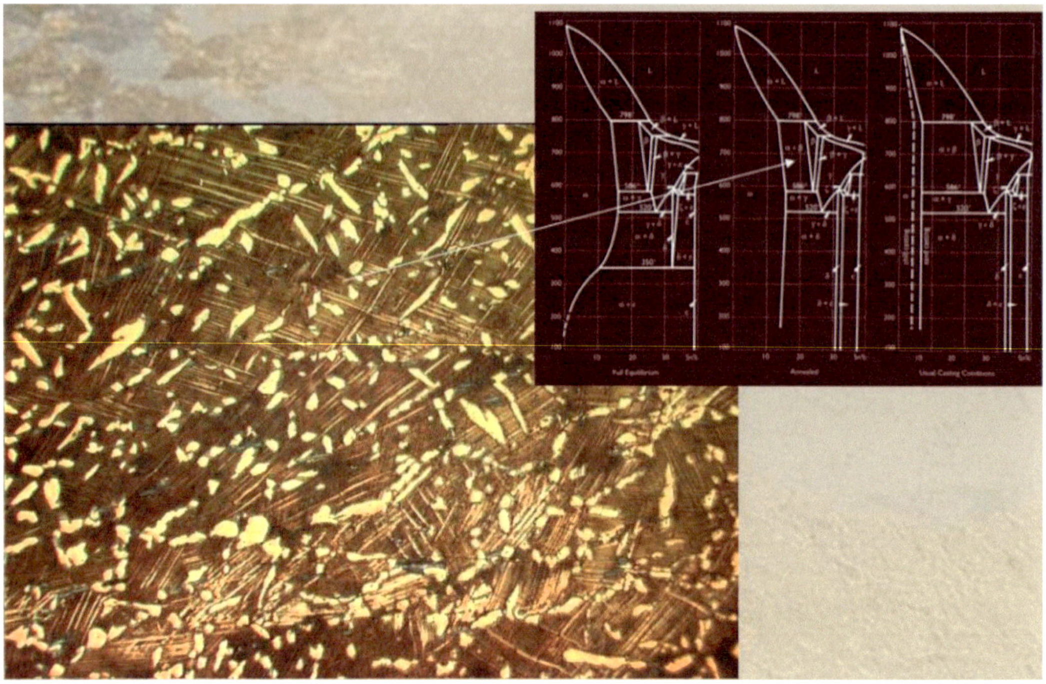

Figure 9.36 A bronze bracelet from Ban Don Ta Phet showing beta needles and twinned alpha grains in a dark etching matrix of gamma phase. This quenched bronze may have been quenched from a temperature not sufficiently high enough to prevent some decomposition to gamma, but is still very serviceable for its function and has been hot-worked, which we know from the twinned alpha grains. Etched in alcoholic ferric chloride, x240

Figure 9.37 Map of some of the regions in which high-tin bronzes, especially quenched high-tin bronzes were important in antiquity. The location of Ban Chiang is shown. Many other sires have been found where high-tin bronzes were made, such as Ban Da Ni and Ban Don Ta Phet, all from the Bronze Age, around 1000 BC.

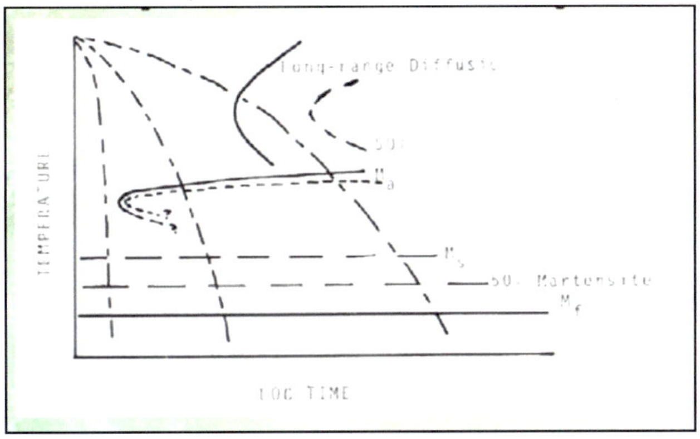

Figure 9.38. Projected onset and finish of martensitic transformations in the high tin bronzes. This is known as a TTT diagram, for time-temperature-transformation diagram, more familiar in the context of iron and steels. After Goodway and Conklin (1987)

Time-Temperature Transformation Diagrams and Continuous Cooling Transformation Diagrams are important in understanding the nature of the changes which may take place on cooling alloys at certain defined rates. The chart shown in Figure 9.38 is the work of Goodway and shows that if very fast cooling takes place in the log time axis closest to the origin, then the whole structure will be martensitic. Less drastic cooling conditions will create a range of different options in which the start of the martensitic transformation will be delayed or result in variations in the type of martensite produced.

Figure 9.39 30% Sn, 70% Cu as cast. This alloy shows a similar microstructure to that shown in Figure 2.29.91. (After Heycock and Neville 1904 Plate 8) in which the eta phase central precipitates are surrounded by whiter epsilon phase set in a fine granular eutectoid ground of alpha + delta. Etched in alcoholic ferric chloride, x300.

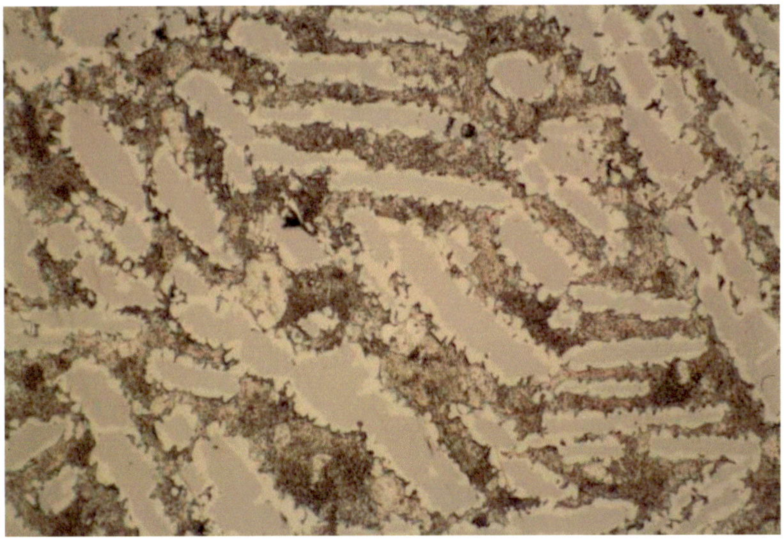

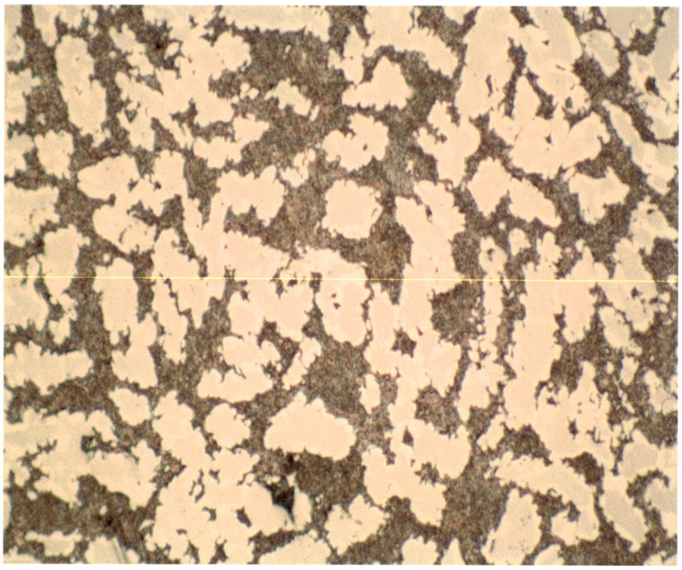

Figure 9.40 Another view in different orientation of the 30% Sn, 70% Cu as cast. This alloy shows a similar microstructure to that shown in Figure 9.29. 91 from the original work of Heycock and Neville (1904 Plate 8) in which the eta phase central precipitates are surrounded by whiter epsilon phase set in a fine granular eutectiod ground of alpha + delta. Etched in alcoholic ferric chloride, x300.

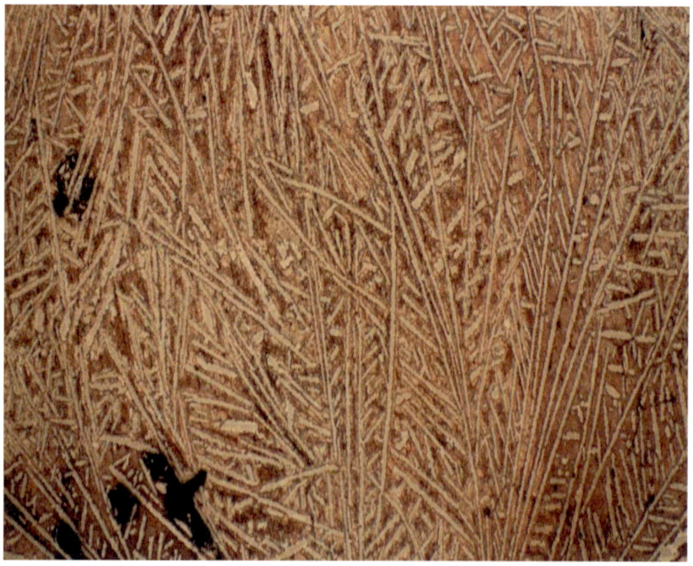

Figure 9.41 35% Sn, 65% Cu cast alloy showing long tendrils of eta phase, surrounded by whiter epsilon, coated with a outer precipitate of what may be epsilon + delta, followed by the eutectoid phase of alpha + delta. Etched in alcoholic ferric chloride, x50

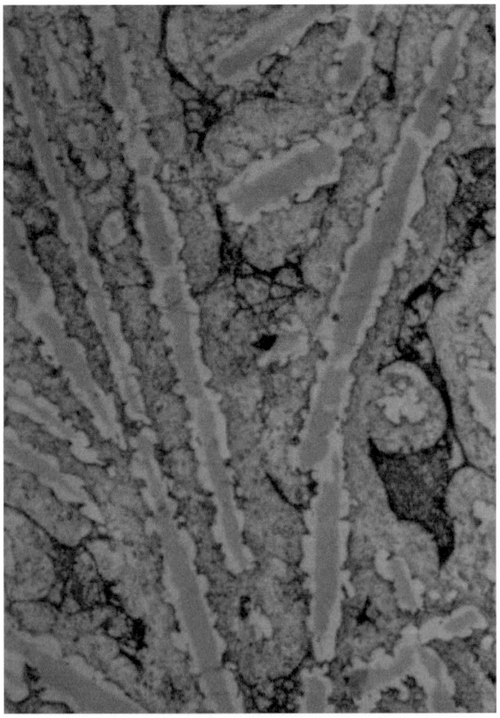

Figure 9.42 35% Sn 65% Cu cast alloy showing long tendrils of eta phase, surrounded by whiter epsilon, coated with a outer precipitate of what may be epsilon + delta, followed by the eutectoid phase of alpha + delta. Etched in alcoholic ferric chloride, x280.

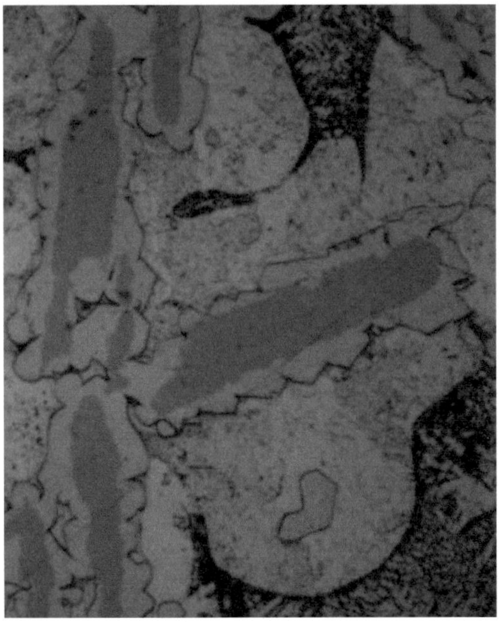

Figure 9.43. An enlarged view of the cast 35% Sn, 65% Cu alloy showing parts of the long tendrils of eta phase, surrounded by epsilon, coated with a outer precipitate of what may be epsilon + delta, followed by the eutectoid phase of alpha + delta. Etched in alcoholic ferric chloride, x400.

Case Study Number 6**:** The Fayum Tablets. These remarkable copper plaques record one of the earliest versions of the Greek script, which may have an eastern Mediterranean connection. One of these plaques was condemned as a forgery in Germany which helped to produce a cloud of suspicion concerning the plaques amongst the linguists of ancient language. Metallographic examination was undertaken to resolve the authenticity of these plaques, which resulted in very strong evidence to support their authenticity.

Figure 9.44 Greek plaques reputedly found in the Fayum region of Egypt, sold at auction in New York. These plaques from the 9th century BC record one of the earliest known versions of the Greek language and are very important for the history of writing and ancient Greek (Woodard 1997, Scott and Woodard 2004; Scott et al 2005).

The three inscribed copper plaques shown in Figure 9.44 through to Figure 9.48 are very important artefacts for the history of writing. The plaque on the right-hand side of Figure 9.44 is housed in the Martin-von-Wagner Museum of the University of Wũrtzburg, Germany. The two plaques shown left and centre of Figure 9.44 are in the Schøyen collection in Norway. These plaques had previously appeared in catalogue 165 (1983) of the New York antiquities dealer, H. P. Krauss, in which they are stated to be of northern Egyptian provenience and labelled as "The Fayum Tablets", being assigned a date of the eighth century BC or earlier. The two from the Norwegian collection measure 215 mm x 135 mm and 212 mm x 137 mm, both about 1.1mm – 1.3mm thick.

Figure 9.45 Another view of the plaques, showing a different view of the middle plaque shown in Figure 9.44. Plaque Schøyen II.

Figure 9.46 The Schøyen plaque side 1, which shows that the plaque was in contact with a surface or covering which has allowed the growth of a concreted corrosion to form across the center of the plaque on this side. Photograph by Courtesy of The Western Semitic Research Centre and Bruce Zuckerman.

Figure 9.47 The Schøyen plaque side two. Note that the more heavily concreted part of the surface is the opposite of that of the reverse side. Photograph by Courtesy of The Western Semitic Research Centre and Bruce Zuckerman.

Figure 9.48 The Wũrtzburg plaque side 1, which shows that the plaque was also in contact with a surface or covering which has created a slightly more extensive corrosion crust towards the two outer reaches of the surface. Photograph by Courtesy of The Western Semitic Research Centre and Bruce Zuckerman.

Figure 9.49 The Schøyen plaque I side two. Under different lighting conditions in which the cuprite appears dark, and the malachite crust pale green with the more heavily encrusted surface across the middle of the plaque. Photograph by Courtesy of The Western Semitic Research Centre and Bruce Zuckerman.

Figure 9.50 The location of the Wűrtzberg plaque, shown is the von Wagner Museum.

The plaques are dispersed to different locations and doubts are raied about the one in Germany, due to the fact that it is made in pure copper and the plaque has a patina of atacamite over cuprite. Scoientists at the Doerner Institute examine the Wűrtzburg plaque and declared that it is a forgery. The reasons are that it is made in pure copper, and since it supposedly originates from the Bronze Age, the plaque shold have been made in pure copper. Another reason was that the symbols suggest some kind of unknown version of the Greek script which cannot be read, and that the patina on the plaque is comprised of basic copper chlorides, and that this is unacceptable in a supposedly buried ancient copper alloy artefact, where a malachite patina should be observed.

Of the four plaques known, the one in the Wũrtzburg collection has been written up in the scholarly literature (Heubeck,1986). The Wũrtzburg plaque, like the two examples in the Norwegian collection, bears an early form of the Greek alphabet, inscribed in sequentially repeating series, beginning at the top right edge of the copper plaque of the obverse face and running left horizontally. At the end of each line, the alphabetic sequence is continued on the line below, from right to left, and so on down the face of the plaque. From the bottom left edge of the plaque, the inscription continues at the top right edge of the reverse side, when the plaque is turned along a vertical axis. In this way, both faces of each plaque are completely filled with alphabetic characters, except for narrow margins at the top and bottom and along each edge. Within these margins, at each corner a small hole has been made through the copper.

The plaques preserve a form of the Greek alphabet which is otherwise unknown. The most unique characteristic is its limitation to 22 characters, like the Phoenician consonantal script on which the Greek alphabet is based, the Fayum alphabet, as the script of the plaques will be referred to here, ends with the consonant tau. All other varieties of the Greek alphabet include at least that vowel character which follows tau, namely upsilon, and most have appended additionally the so-called 'supplemental characters' *phi*, *chi*, and *psi*, with some regional variation, as well as the final vowel symbol, *omega*. The Fayum alphabet also preserves both of the s-symbols known from the local Greek alphabets, *san*, and *sigma*. The various local Greek alphabets choose one and eliminate the other (Woodard 1997:175-188). Also present are the letters *waw* or *digamma* and *qoppa*, both of which were short-lived in the Greek alphabet. While such features place the Fayum alphabet at the very beginning in the development of the Greek alphabet from its Phoenician parent script, as Heubeck pointed out in his analysis of the Wurzburg plaque, certain characters (*alpha, lambda, sigma*) have shapes which have been conventionally identified as more, rather than less, advanced in the evolution of the Greek alphabet. In addition, the Fayum *iota* and *nun* have distinct forms. On the Norwegian plaques, the letter *zeta* alternates between its Z-form and its I-form, even within the space of a few lines (Scott and Woodard 2001).

Provenience

The reported find-place of what is perhaps the earliest known example of Greek alphabetic writing, namely northern Egypt, is somewhat surprising. For Heubeck (1986) the locale is incongruous with his assumption that the alphabet arose in a context of Greek-Phoenician mercantile contact. There is compelling evidence, however, that we should see the origin of the Greek alphabet within an environment of already literate Greek adapters, scribes who wrote with the syllabic Greek script of Cyprus. A find-place in northern Egypt is consistent with a Cypriot epicentre of alphabetic Greek writing, and one could well imagine that the copper plaques were themselves manufactured on Cyprus which was famed for its copper ore.

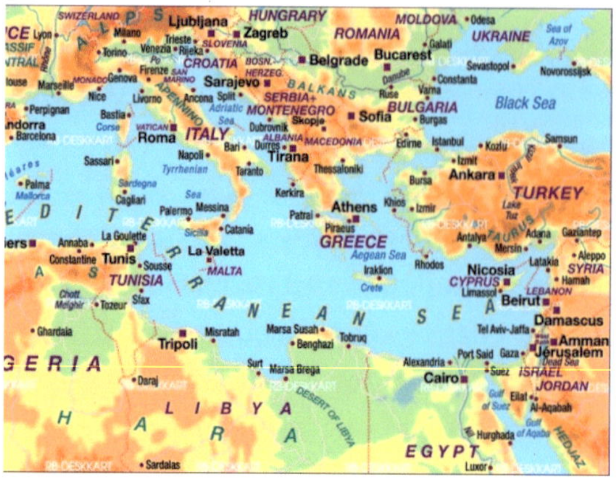

Figure 9.50 The plaques may have a connection to the Eastern Mediterranean and possibly associated with Samos, although they could very well be made of copper from Cyprus and later made their way to the Egyptian Fayum, because in the 9th century Bc, there was nothing much happening in the Fayum region.

Function

The function of the plaques is not known. Plaques of a similar sort, *pinakes*, made of metal or of waxed or whitened wood, were used by the Greeks as writing tablets and sign boards. The possibility that these plaques may have been used for a religious or magical purpose naturally presents itself. Thomas (1992) points out that there is plentiful evidence for the magical use of writing among the Greeks of both the classical and archaic periods. She further draws attention to the practice of intentionally 'deforming the text' of curse tablets, perhaps a type of sympathetic magic. The holes in the plaques may have been used to join them in sequence.

Technical examination and binocular microscopy.

The surfaces of the plaques present a complex mineralogical assemblage in which mineralized organic remains, soil minerals, green corrosion products of copper and dull red cuprite can be seen. There are also many small remnants of wood and mineralised wood which has been converted to a pseudomorphic remnant by preservation in copper corrosion products.

Figure 9.51 One of the Norwegian plaques showing dark malachite over cuprite with a band of different mineralization across the middle of the plaque due to contact with a band of material, probably of textile.

The cuprite layer is here contiguous with the remaining metallic surface, and the inscription is preserved within this cuprite layer, which marks the 'original surface' of the copper plaque. The thickness of the copper sheet is about 1.1 mm. The cuprite has therefore been laid down in an epitactic manner, preserving the shape of the original lettering. In some areas of the surface, the inscription is completely obscured by the overlying corrosion crust, which is hard and compact and therefore not easily removed. During conservation, a little of this corrosion crust was probed with a scalpel under the binocular microscope at x20-x40 magnification. One of the remarkable features of the three plaques is the series of complex differences in aspects of the patina, suggesting that the burial environment was slightly different for each plaque and for each side of the plaques as well. For example, the Wũrtzburg plaque has an unusual patina which contains a high percentage of the basic copper chlorides over a cuprite underlayer, while one of the Schøyen plaques, has a malachite crust over a cuprite underlayer.

Figure 9.52 Photomicrograph of part of the surface of Schøyen plaque 2. The surface shows partial loss of the malachite patina revealing part of a letter preserved in a cuprite underlayer. Magnification x 25.

Figure 9.52 shows a common feature in the natural corrosion of copper alloys (Scott 2002). Layering is present, and the topmost green coating of malachite is hard and compact, but partially delaminated from the granular, crystalline cuprite underneath. The texture of this cuprite is hard and varies in colour from a deep dark red to a lighter bright red, but never assumes a yellowish colour often associated with cuprite which has grown fast due to artificial corrosion processes. There are real difficulties in creating this kind of cuprite crust over copper artefacts and it has long been a standard visual clue for conservation scientists regarding the authenticity or age of the artefact in question. Creating a real depth of cuprite corrosion and producing a convincing metallographic replica of a real patina was proved to be extremely difficult to create, one of the criteria which we employ in assessing authenticity, dating back to before World War II (Plenderleith 1938). I should stress here that this is one of the criteria which we can employ: it is not the only one that we rely on.

The corrosion crust from the Wũrtzburg plaque is shown in Figure 9.53 where the outlines of the inscribed lettering is clearly preserved. The letters are preserved within the cuprite patina, partially overlaid with atacamite. Probing of this cuprite layer close to the letters reveals a considerable depth to the cuprite layer. It is not superficial as is sometimes found in recent reproductions. Cleaning and surface investigation was limited to the removal of a few microsamples for x-ray diffraction study and for scanning electron microscopy; the integral nature of the corrosion on those surfaces of the plaques which have not been cleaned should be kept as evidence for burial and association with the environment which may well prove useful in further studies of the plaques.

Figure 9.53 Photomicrograph of the Wũrtzburg plaque, showing two letters clearly outlined in corrosion. The lettering is so well preserved that the order of the strokes of the chisel can even be made out. The light green-blue corrosion overlying the cuprite is principally atacamite. Magnification x 30.

Figure 9.54 Schøyen plaque number 1. On this side the morphology of the malachite assumes a raised series of rounded deposits which incorporates soil minerals, often of small quartz grains, rounded rather than angular which are cemented to the malachite crust. In the photomicrograph some cuprite can be seen under the malachite layer. Magnification x 25.

The incorporation of soil minerals within the patina, cemented to the surface and within the malachite crust is illustrated in Figure 9.54. The mineral grains trapped within the crust are predominantly of quartz, rounded by marine or river transport, some being practically spherical.

Figure 9.55 View of part of Schoyen plaque 1 in which the inscribed lettering is partially obscured by a compact cuprite layer with another cuprite layer beneath in which the letters are preserved.

Figure 9.55 shows another feature often associated with ancient surfaces of copper alloys. A hard and compact crust of cemented cuprite obscures part of the original surface beneath, in which the letters are well-preserved, but in some parts of the surface totally obscured by the corrosion crust.

The examination by binocular microscopy showed the presence of a number of mineralized and carbonized wood fragments preserved within the corrosion crust, an occasional small flake of gold, as well as firmly embedded soil minerals, an extensive depth to the cuprite patina and the presence of complex and heterogeneous phases within the corrosion layers, as seen in Figure 9.56- 9.59.

9.56 View of the plaque 2, showing that corrosion has grown in contact with another surface which has restricted growth of the blue-green corrosion products here. x 78.

9.57 The patterning revealed in this photomicrograph from plaque 1 suggests that the malachite growth has been influenced by a textile surface. This observation is consistent with the banding across the surface of the plaque shown in Figure 9.51

Figure 9.58 Lettering preserved in an outline of cuprite with the inscribed lines filled in with malachite and with cuprite below. x34.

Figure 9.59 Plaque 2, showing entrapped wood fibres, the lower part turned green with copper corrosion products and partially mineralized. A small piece of wood and a charcoal fragment are also seen on this malachite corrosion crust. x56.

Metallography

The extent of penetration of the corrosion within the metallic substrate was assessed using metallographic examination. A small "V" shaped section about 1 mm across was removed from each of the plaques, from an area where no damage could be caused to any inscribed text. Each sample was removed using a fine jeweller's saw and then mounted in Buehler epoxide resin mounting compound. The mounted samples were prepared and polished in the manner described in Chapter 3. Two of the resulting photomicrographs are shown in Figures 9.60 and 9.61. The most significant observation is that both photomicrographs reveal that a considerable amount of corrosion has occurred with extensive conversion to cuprite, copper (I) oxide.

Figure 9.60 Detail of the corrosion crust of the copper plaque, number two, showing copper to the left hand side, with cuprite inclusions, followed by crystalline cuprite corrosion, then a complex green corrosion crust with malachite and some copper chlorides and occasional tiny flakes of charcoal. x145. Figure 9.61 (right) Strong metallographic evidence for authenticity, showing a crystalline cuprite crust under the outer green malachite and copper chloride crust. The grains are twinned and worked and annealed to shape the plaque. Etched in alcoholic ferric chloride, x230.

Corrosion has therefore attacked the copper plaques fairly extensively and this cuprite layer can be seen passing into sound metal to a depth which varies from 30 to 80 microns. This cuprite layer is overlaid with a thick malachite deposit of variable thickness, well illustrated in Figure 9.60. The depth and type of this cuprite crust is sufficient, in general, to already establish that the plaques are authentic. Etching in alcoholic ferric chloride revealed that both of the Norwegian plaques have well-developed twinned crystal structures with straight twin lines, showing that annealing was the final working stage after cycles of working and annealing to shape the plaques.

The inclusions present within the copper itself derived from the smelting stage of extraction were shown to be of cuprite. Cuprite inclusions are typically associated with the earlier stages of metallurgy and the extraction from oxidized ores, while the presence of copper sulphide inclusions are more commonly associated with the developed bronze age material from many different cultural contexts.

X-ray fluorescence spectrometry

X-ray fluorescence analysis of the plaques was carried out, non-destructively, employing a Kevex 0750A x-ray fluorescence spectrometry and a barium/strontium secondary target at 55 kV, 3.3 mA for 200 seconds acquisition time. The analytical data showed that the plaques are made essentially of unalloyed copper, containing a trace of arsenic, with no other alloying elements detected. This is not unexpected; the copper would be much softer than bronze for the purposes of inscribing the text, and every effort would have been taken by the metalsmith to employ clean, unalloyed copper for such an important task. Scanning x-ray fluorescence analysis was employed to look at elemental variations over parts of the uncleaned surface of the Norwegian 1 plaque. There is a significant variation in the distribution of some elements within the malachite patina, including copper, iron, gold, sulphur, and silver. There is a filamentous and particulate character to the iron distribution, derived from the precipitation of iron salts from the burial environment within the malachite surface. Besides the numerous small quartz grains, which occur in regions of the patina, there are occasional calcite inclusions, as revealed by the shape of the particles and the presence of calcium, where isolated small inclusions and one larger inclusion occur. The burial environment of the plaques has been both varied in chemical type and has produced a complex series of corrosion crusts.

Figure 9.62 Metallographic section of the second Norwegian plaque. The depth of penetration of well-crystalline cuprite seen in this polarized light view of the polished section within the metal itself confirms its authenticity. Magnification x80.

The scanning x-ray images show that the sulphur which is present precisely matches that of gold and silver areas, suggesting that some tarnished patina is present on the gold flake in the form of silver sulphides or mixed gold-silver sulphides. The complexity of the corrosion crust was also revealed by scanning elemental distribution maps in the electron microprobe, as shown in Scott and Woodard (2001, 143, Figure 8), where the maps for chlorine, calcium, and potassium reveal the depth of penetration of these elements within the corrosion crust below the level of the original surface of the plaque.

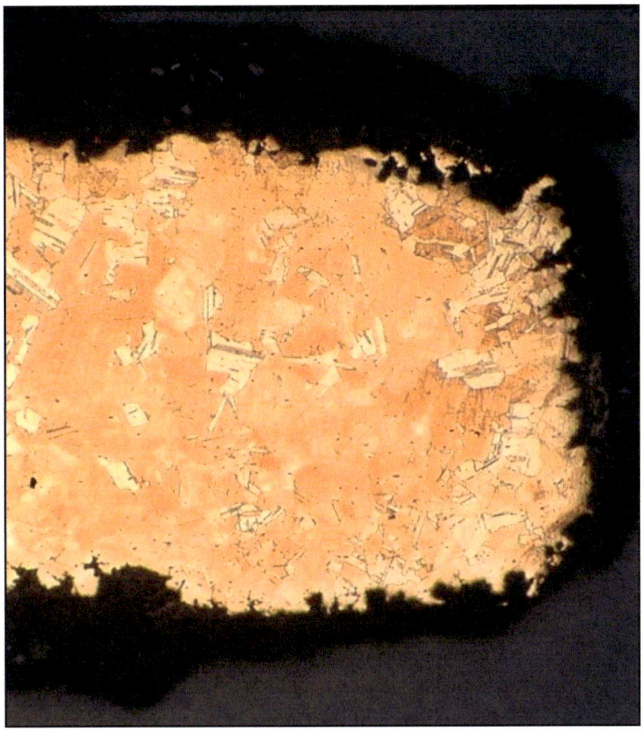

Figure 9.63 etched view of the plaque 1 showing worked and recrystallized copper grains. Etched in alcoholic ferric chloride, x120.

X-ray diffraction

X-ray diffraction was carried out, in situ, using a Siemens D5005 x-ray diffractometer and a Gobels mirror assembly at 40 kV, 40 mA. This examination confirmed that the corrosion crust on the Norwegian plaques consist of malachite overlying the cuprite layer. Some evidence for the presence of atacamite was found but this was present only as a minor component, as malachite was predominant as the green-coloured copper corrosion product in the surface samples examined and from the in situ x-ray diffraction studies.

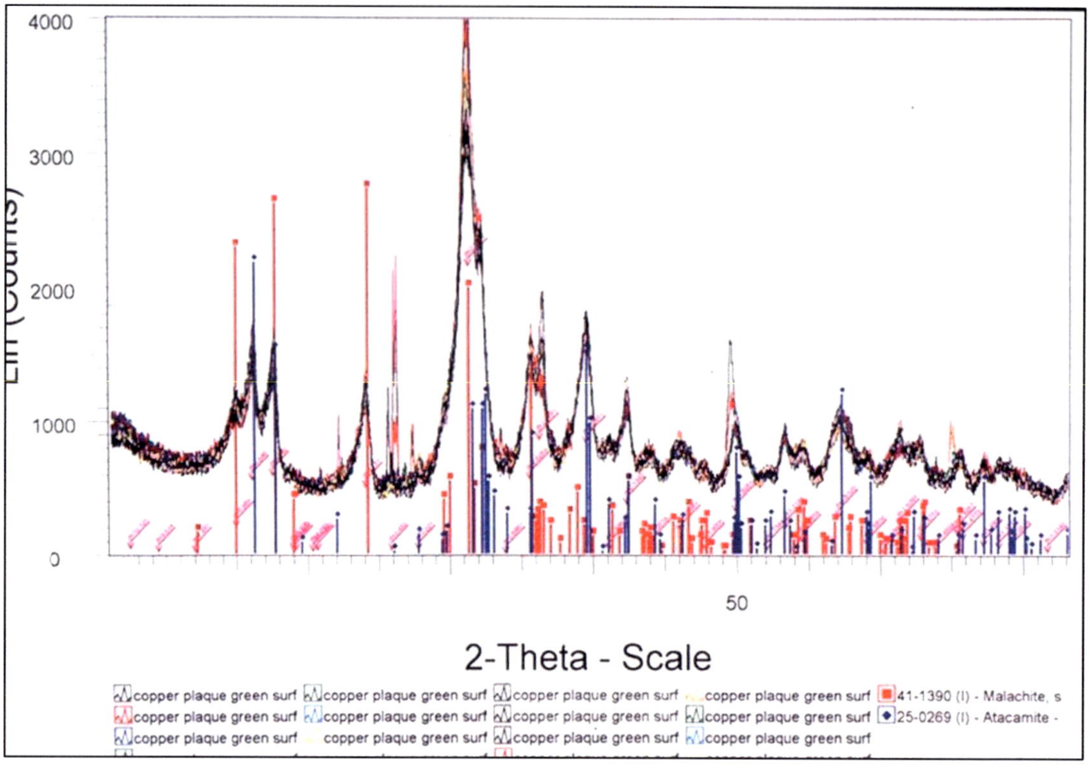

Figure 9.64 Scans in situ showing the presence of malachite in the corrosion crust. These scans are considerably rougher in quality than those taken from small samples but have the advantage that they can be taken from the object without the need for any sampling.

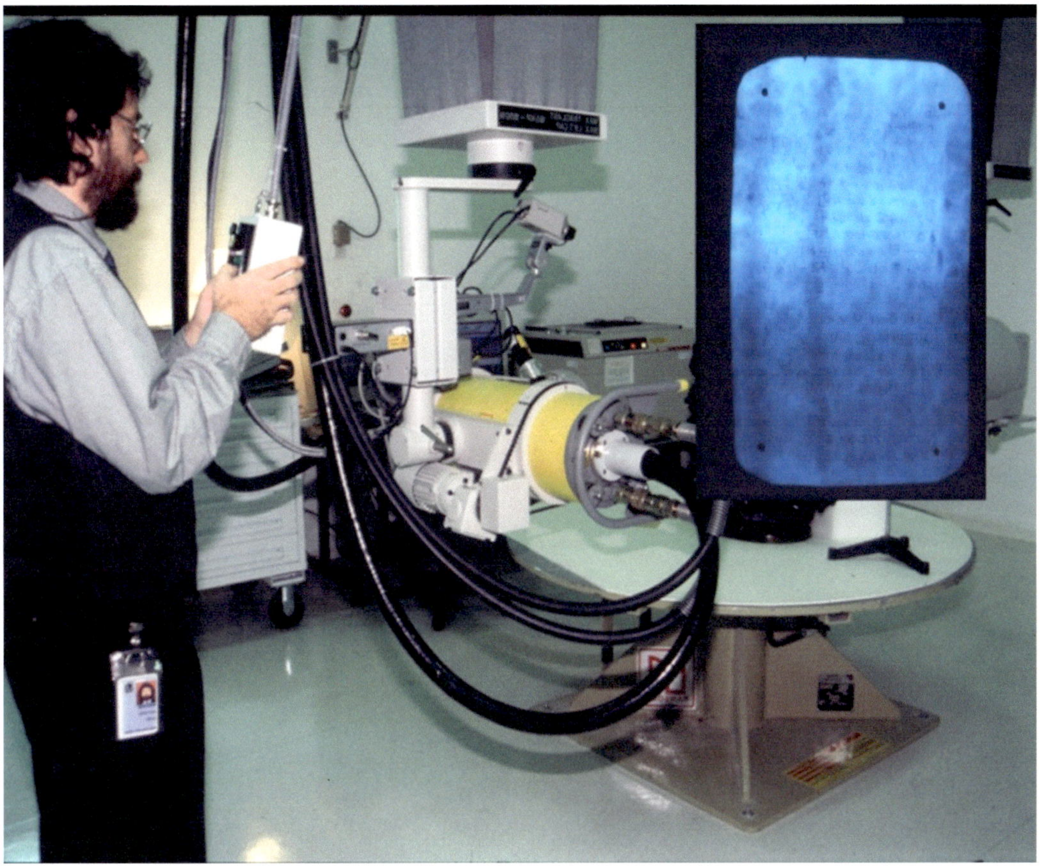

Figure 9.65. Taking an x-radiographic image of the copper plaques at the Getty Centre Museum Services Laboratory in Brentwood in 1999. This machine is equipped to reach 400 kV but for copper only 100 kV at a few mA of current is needed to create the images to see the letters.

X-ray radiography was employed to attempt to characterize whether the copper plaque had been cast, rolled, or hammered to shape, and to determine whether the inscription could be read more clearly from an x-ray radiograph than by eye, especially for those areas obscured by the hard malachite corrosion crust. For this examination, a Phillips x-ray tube was used, operating at 100 kV, 5 mA for 120 seconds and recording the x-ray image using Kodax Industrex M x-ray film, now practically obsolete! A typical x-ray photograph resulting from this study is shown in Figure 9.66 for the plaque illustrated in Figure 9.67.

Figure 9.66 X-radiograph of the Norwegian plaque II clearly reveals the hammering marks, which all the plaques have in common! This means that the plaques were all cut from the same sheet of copper. A remarkable discovery.

The fact that the plaques all bear the same pattern of hammering marks, not only shows that the copper plaques were worked and annealed to shape from a cast ingot, but that the hammering produced a distinctive and individualistic pattern which passes down along of the plaques. The cross striations at the top of Figure 9.66 may well indicate that this plaque formed one end of the length of the hammered copper before it was cut into a number of sections for each of the plaques. The slight curvature of the hammered copper is typical for the hand-working of such a long length of copper. The ability to keep such a length of hammered copper even relatively straight along the entire length is a very skillful example of copper working. Copper is tough and tenacious, and is not easy to hammer out over large surfaces as used for the plaques here.

Figure 9.67 The plaque from which the x-ray image was taken, and which is shown in Figure 9.62.

A series of x-ray radiographs were taken of the plaques to determine the optimum conditions for revealing the inscribed text. The useful range was between 90 kV – 115 kV at 5 mA, and the best examples were those selected for the textual analysis. As the plaques are inscribed on both the obverse and reverse, it is difficult to follow the pattern of the lettering revealed on the x-ray radiograph, although they are not exactly inscribed into the copper on the same relative position on both sides of the plaque, which enables the letters to be deciphered. In fact, the x-ray radiographs were very successful in revealing the letters previously hidden under the malachite crust. The periodic mottling pattern revealed on the plaques show that the copper was cold-worked and annealed to shape rather than being rolled. This hammering technique is what would be expected for the presumed period of manufacture of the plaques. An interesting fact emerged from the x-radiographic study, that all of the plaques reveal the same pattern of hammering marks, resulting in two pronounced deformation bands running down the length of the plaques with another one towards the other side.

The plaques reveal the same spacing and pattern of deformation in the radiographs which means that all of them were cut from the same length of copper. We can therefore reconstruct the size of the hammered blanks used to make these plaques: a long length of hammered and annealed copper sheet was ordered whose width was about 135 mm. This long strip was then cut into the required length of each plaque, about 215 mm, with the result that the same pattern of hammering marks are preserved on each.

A detailed analysis was carried out by electron microprobe analysis with the results shown for the three plaques in Figures 9.68, 9.69 and 9.70, in which relatively pure copper was confirmed, the EPMA results also showing a trace of nickel and very occasional inclusions containing higher levels of arsenic and antimony. The general picture revealed by this set of analytical data is that there is no reason to suppose that it represents modern manufacture. The tiny inclusions which are richer in arsenic and antimony are quite in keeping with an ancient origin for the copper. The presence of some nickel is of interest as one or two elevated levels of nickel are shown in analysis of copper from Cyprus by Fasnacht and Northover (2002). Small amounts of arsenic, antimony and nickel are often found in early copper alloy analyses, along with iron and sometimes cobalt and lead. From the analytical data we can conclude that the three plaques were made from the same batch of copper. Until further EPMA analyses for Greek copper is available, we do not have comparable data for the tiny inclusions that were found in this study. Analysis by ICP-MS or other bulk techniques will only give an average composition of the sampled area or drilling of the alloy.

Plaque 1 "Norway 1"

	S	O	Pb	Sb	Sn	Cu	As	Ni	Fe	Total
#1	0.007	0.410	0.003	0.000	0.000	99.967	0.000	0.129	0.018	100.534
#2	0.000	0.562	0.000	0.000	0.000	100.113	0.000	0.241	0.013	100.929
#3	0.020	6.544	1.571	7.205	0.000	80.105	0.000	6.361	0.020	101.826
#4	0.010	0.525	0.003	0.000	0.000	99.386	0.000	0.173	0.015	100.112
#5	0.000	0.369	0.031	0.000	0.000	100.105	0.000	0.106	0.016	100.627
#6	0.009	0.409	0.110	0.053	0.000	98.744	0.000	0.276	0.016	99.617
#7	0.004	0.414	0.000	0.000	0.000	99.500	0.000	0.176	0.016	100.110
#8	0.010	0.350	0.016	0.000	0.000	99.708	0.000	0.170	0.005	100.259
#9	0.000	0.403	0.027	0.000	0.000	99.718	0.000	0.165	0.011	100.324
#10	0.009	0.365	0.000	0.000	0.000	100.260	0.000	0.148	0.013	100.795
#11	0.011	0.339	0.000	0.000	0.000	101.519	0.000	0.135	0.011	102.015
#12	0.009	0.318	0.000	0.000	0.000	99.843	0.000	0.149	0.004	100.323
#13	0.012	0.364	0.000	0.000	0.000	100.961	0.000	0.138	0.018	101.493
#14	0.014	0.346	0.029	0.000	0.000	100.000	0.000	0.292	0.011	100.692
#15	0.006	0.348	0.006	0.000	0.000	100.532	0.000	0.152	0.016	101.060
#16	0.001	0.395	0.074	0.030	0.000	99.884	0.000	0.327	0.016	100.727
#17	0.006	0.341	0.027	0.000	0.000	101.059	0.000	0.128	0.013	101.574
#18	0.000	0.368	0.000	0.000	0.000	101.361	0.000	0.143	0.014	101.886
#19	0.000	0.352	0.020	0.000	0.000	101.279	0.000	0.155	0.011	101.817
#20	0.000	0.489	0.042	0.000	0.000	100.082	0.000	0.488	0.009	101.110

Figure 9.68 EPMA results for the 1st Norwegian plaque, showing relatively pure copper with occasional tiny Inclusions with elevated contents of arsenic and antimony, Traces of nickel appear to be present throughout.

Plaque 2 "Norway 2"

	S	O	Pb	Sb	Sn	Cu	As	Ni	Fe	Total
#21	0.004	0.540	0.063	0.094	0.000	99.028	0.000	0.529	0.008	100.266
#22	0.000	0.349	0.000	0.000	0.000	99.343	0.000	0.089	0.012	99.793
#23	0.002	0.350	0.028	0.000	0.000	99.134	0.000	0.119	0.014	99.647
#24	0.004	0.454	0.008	0.093	0.000	98.556	0.000	0.308	0.012	99.435
#25	0.002	0.443	0.002	0.000	0.000	99.045	0.000	0.326	0.012	99.830
#26	0.005	0.369	0.006	0.000	0.000	99.901	0.000	0.201	0.011	100.493
#27	0.000	0.328	0.006	0.000	0.000	99.778	0.000	0.171	0.017	100.300
#28	0.071	1.326	7.016	0.000	0.000	88.168	0.890	0.112	0.008	97.591
#29	0.006	0.321	0.004	0.000	0.000	98.917	0.000	0.124	0.015	99.387
#30	0.007	0.419	0.069	0.000	0.000	98.637	0.000	0.171	0.017	99.320
#31	0.000	0.354	0.039	0.000	0.000	100.265	0.000	0.077	0.013	100.748
#32	0.010	0.357	0.000	0.000	0.000	99.981	0.000	0.163	0.013	100.524
#33	0.010	0.509	0.036	0.000	0.000	98.153	0.000	1.282	0.013	100.003
#34	0.007	1.559	0.179	0.721	0.000	94.903	0.000	2.519	0.005	99.893
#35	0.003	0.334	0.013	0.000	0.000	99.446	0.000	0.163	0.011	99.970
#36	0.005	0.357	0.000	0.013	0.000	99.229	0.086	0.418	0.008	100.116
#37	0.006	0.356	0.014	0.000	0.000	99.919	0.000	0.266	0.015	100.576
#38	0.000	0.376	0.014	0.000	0.000	101.216	0.035	0.143	0.015	101.799
#39	0.008	0.468	0.008	0.071	0.000	99.938	0.000	0.289	0.014	100.796
#40	0.000	0.337	0.025	0.000	0.000	100.107	0.000	0.162	0.011	100.642

Figure 9.69 EPMA Data for Norwegian plaque 2

Plaque 3 "Wurtzberg"

	S	O	Pb	Sb	Sn	Cu	As	Ni	Fe	Total
#41	0.015	0.323	0.000	0.000	0.000	97.892	0.000	0.152	0.000	98.382
#42	0.009	0.293	0.000	0.000	0.000	98.107	0.000	0.153	0.003	98.565
#43	0.024	0.329	0.006	0.000	0.000	98.349	0.000	0.154	0.003	98.865
#44	0.009	0.342	0.000	0.000	0.000	98.212	0.000	0.152	0.007	98.722
#45	0.011	0.351	0.023	0.000	0.000	98.713	0.000	0.180	0.000	99.278
#46	0.012	0.365	0.011	0.000	0.000	97.476	0.014	0.157	0.009	98.044
#47	0.012	0.311	0.019	0.030	0.000	98.683	0.000	0.354	0.000	99.409
#48	0.012	0.321	0.032	0.000	0.000	97.780	0.000	0.240	0.000	98.385
#49	0.011	0.324	0.028	0.000	0.000	98.391	0.162	0.133	0.000	99.049
#50	0.017	0.339	0.023	0.000	0.000	99.193	0.000	0.133	0.007	99.712
#51	0.008	0.337	0.000	0.000	0.000	98.444	0.000	0.114	0.004	98.907
#52	0.020	0.366	0.532	0.000	0.000	97.835	0.000	0.112	0.003	98.868
#53	0.011	0.337	0.009	0.000	0.000	98.618	0.141	0.145	0.003	99.264
#54	0.022	0.346	0.000	0.000	0.000	98.501	0.000	0.146	0.000	99.015
#55	0.019	0.360	0.013	0.000	0.000	98.459	0.000	0.147	0.000	98.998
#56	0.024	0.342	0.014	0.000	0.000	98.329	0.000	0.183	0.009	98.901
#57	0.006	0.284	0.040	0.000	0.000	98.501	0.139	0.138	0.001	99.109
#58	0.012	0.725	0.230	0.383	0.000	95.637	0.000	0.637	0.002	97.626
#59	0.020	0.333	0.024	0.000	0.000	99.152	0.000	0.162	0.000	99.691
#60	0.031	0.320	0.000	0.000	0.000	98.271	0.000	0.191	0.007	98.820

Figure 9.70 EPMA data for the Würtzberg Plaque

Environmental scanning electron microscopy

Environmental scanning electron microscopy (ESEM) was employed to examine in greater detail some of the organic remains on the surface of the plaques. The samples were examined at 15 kV, 0.3 mA at a pressure of 10 torr. Wood fragments embedded in the corrosion could already be determined by binocular microscopy, and an example of a large fragment was shown by Scott and Woodard (2001, 144, Figure 10). ESEM examination was used to record the grain structure of the wood fragments; it has not yet proved possible to determine the genus or species of wood preserved on these plaques. The extraordinary detail of the microstructure is well preserved. The original wood cell constituents have been replaced with copper corrosion products, principally malachite, as a positive cast of the original cellular material.

Figure 9.71 Preserved spiral thickening of a tracheid cell on the surface of a plaque. Pseudomorphic preservation within malachite corrosion products. Environmental scanning electron micrograph x 670.

Conclusions

The evidence derived from the technical and analytical studies described here leave no doubt as to the authenticity of the plaques. The depth of corrosion, the method of fabrication, the preservation of the letters within the cuprite corrosion crust and the presence of completely mineralized wood remnants, all attest to the considerable age of the plaques; there is no reason to doubt that they originate from the 8th-9th century BC. The plaques are made in quite pure copper and have not been intentionally alloyed. Epigraphic examination based on the x-radiographs, has revealed that each plaque is related to the other and that a particular order exists between them. The research has revealed the importance that these plaques have for the development of the Greek language and has rehabilitated them, in view of doubts which have been expressed concerning their authenticity as a result of previous studies.

Figure 9.72 A Luristan bronze from the collections of the Iran Bastian Museum, courtesy of Rasoul Vatandoost-Haghayi, showing heavily segregated cast bronze with about 7% tin content and twinned redeposited copper from preferential corrosion of the δ phase of the copper-tin system. Used as the frontispiece to this chapter.

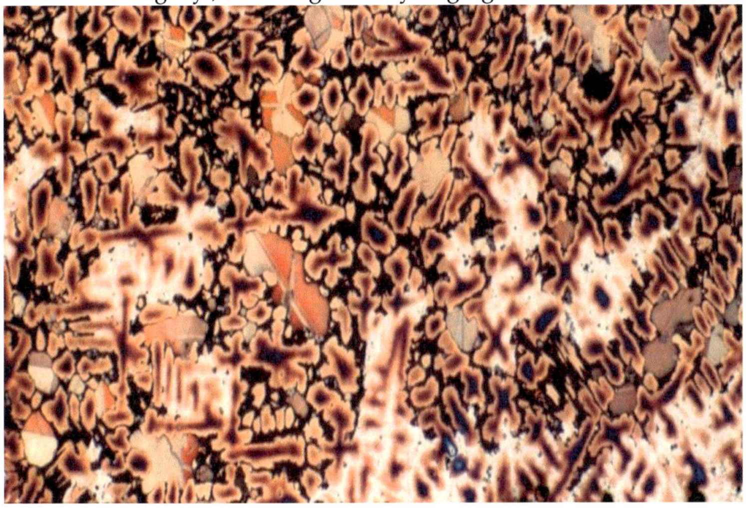

Figure 9.72 illustrates the presence of twinned crystals of copper, deposited at some depth in a 14% tin bronze handle from a Luristan bronze. In the artefact in question, the bronze handle with about 14% of tin had been cast-on to a lower tin bronze dagger blade, in the knowledge that the higher tin bronze casting would not destroy the blade. Indeed the handle and blade are still perfectly joined together after nearly 2000 years. Corrosion during burial has resulted in the dissolution of some of the eutectoid phase within the bronze alloy, resulting in its partial replacement with pure copper, which is naturally twinned from chemical deposition.

This results in the interesting microstructure shown in Figure 9.72, where a dendritic structure contrasts with a twinned grain structure of the pure copper. The presence of redeposited copper at some depth in this Luristan bronze shows the extent of penetration of corrosion solutions within the matrix of corroding bronzes during burial. The presence of redeposited copper is a common event in the corrosion of buried bronzes, usually resulting from a corrosion-redeposition phenomena during burial in which cupric solutions may be reduced to cuprous, or cuprous to metallic copper.

Case Study Number 7. The Greek Herm in the Getty Museum. Metallographic examination was undertaken to try to obtain further information concerning the casting technology of the partially hollow bronze rods which pass up the base of the sculpture on all four sides.

Beginning around 500 BC., Greeks placed herms, pillars surmounted by a head of the god Hermes, at physical boundaries, such as crossroads or even doorways. Such places were sites of ritual and worship, where the herms served a magical, protective function. By the Hellenistic period, the repertoire of heads found on herms had expanded to include other gods and even famous mortals, and people began to use the herms for non-religious, decorative purposes. This herm carries a head of Dionysos wearing a turbanlike headdress of loosely wound ribbons. Both the hair and the beard of Dionysos have been carved in a style that would have looked old-fashioned to the contemporary viewer. Greek bronze statues typically had eyes made of contrasting materials. In most instances the eyes have been lost, but this herm still retains the ivory inlay of the white of its left eye, giving a better idea of the work's original appearance (Getty Handbook 2002). Boethos worked as a bronze caster and silversmith in the second century B.C. As with many ancient sculptors, we know of his work primarily from passing references by Roman writers such as Pliny and Pausanias and from later Roman copies of his sculptures.

One of the interesting technical discoveries made about the Getty Herm is the existence of bronze tubes of the same alloy composition passing up the centre of each of the long box-like sides. These tubes are not completely solid and may represent a means of supporting the wax slabs with reeds or another material that formed part of the original model for the base of this impressive bronze. The casting of the long base demands technical mastery of the bronze-casting process, and a number of small areas of porosity in the base were cut into little rectangular shapes which were then filled in with small bronze plugs, hammered into place. These can clearly be seen on the bronze in display at the Getty Villa Museum in Malibu, California.

Figure 9.73 Bronze, leaded, with ivory inserts. Herm of Dionysos. Workshop of Boethos of Kalchedon or more likely, a copy made from the Boethos original, 100 – 50 BC. H. 103.5cm, W 23.5cm (base). The J. Paul Getty Museum, Villa Collection, Malibu, California. 79.AB.138.

One signed original statue by Boethos survives: a bronze herm from a shipwreck off the coast of Tunisia near the city of Mahdia, now in the Bardo Museum in Tunis. The shipwreck was found by fisherman in 1907 and had sunk around 80 BC, probably on its way to Italy from Greece. A black and white photograph of the two herms together is shown in Figure 9.74.

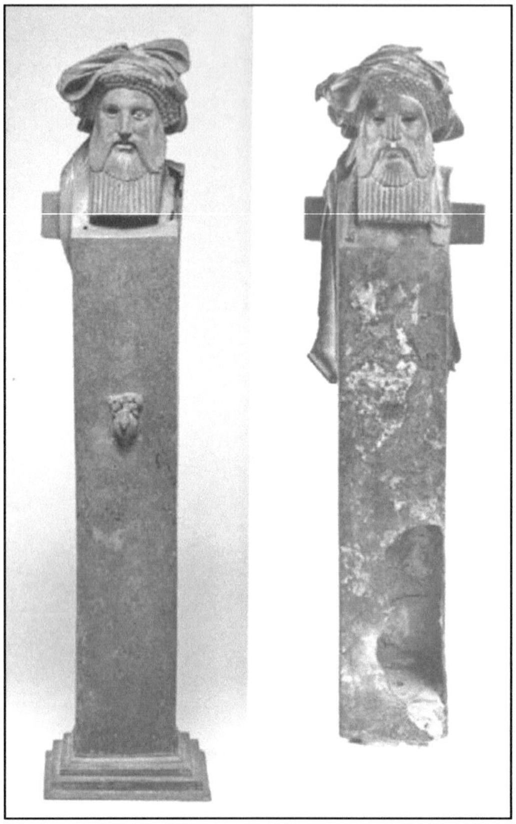

Figure 9.74. The Getty Herm, on the left and the Mahdia herm on the right compared. After Schwab et al 2008:Figure 2.

Schwab et al (2008) carried out a lead 210 determination on the Getty Herm, as if there was a problem with the authenticity of the piece, which is clearly impossible on even a cursory examination of the surface of the Getty Herm. Lead 210 is an isotopic decay product of naturally occurring uranium 238. Schwab et al (2008) explain this: "uranium 238 is collected in the smelting of the metal as the uranium itself fully disappears in the slag. Without the mother nuclide of the lead isotope, the lead 210 decays with a half-life of 22,3 years so that it is usually measurable if the metal was produced in less than 100 years..."[trans. by Joseph Lehner, 2011].

There is a lead socket and leaded bronze used for the Mahdia Herm which have a common lead isotope signature compatible with the Laurion deposits of Attica, which suggests that the production occurred in the same workshop as the installation of the Mahdia Herm on its base as that is what the lead was used for.

Interestingly, there is neither an isotopic nor chemical correlation between the Getty Herm and the Mahdia. Both pieces were also fabricated differently. The Getty Herm has four vertically mounted interior sprues, which have not been found on other published ancient bronzes. These rods are contiguous with the centre of each rectangular base of the Herm and extend to the attachment of the head. Vegetal structures on the surface of these sprues suggest that they were originally formed from plant stems, possibly from reed stalks impregnated with wax which were then burnt out in the casting. These sprues strengthened the body and its connection to the head. They are cut and concealed at the transition between the shoulder and the head.

Schwab et al (2008) write that: *"external channels are typical of ancient castings remnant of the gating system, which were removed after casting. Even the construction of the head, with its turban-like head decoration is very different. So it seems that the elaboration in this area of the Getty Herm were carried out differently than solely by the indirect lost wax casting on the Mahdia Herm, whereby a simpler, less detailed surface design was possible.*

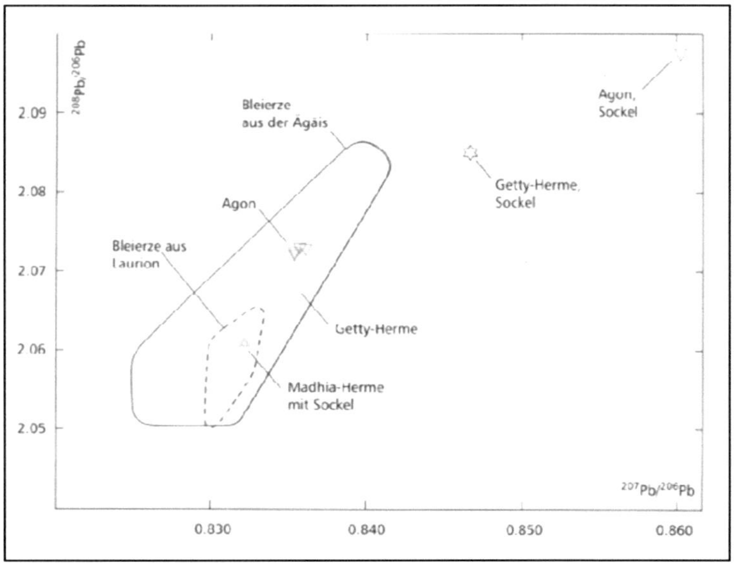

Figure 9.75 Lead isotope data for the Getty Herm and the socket area of the Getty Herm, compared with the data for the Mahdia Herm. The differences in the lead isotope data for the two suggests different regions for their manufacture. After Schwab et al (2008)

The sharp decline in the quality of manufacturing of the Getty Herm is particularly noticeable on the back of the head. The head-dress seems designed rather arbitrarily and without consideration of the natural characteristics and material quality, as if one could base the design only to cast an imperfect form, and as if the artists had no quality template at hand. This would be explained by the fact that the four strands of hair on the crown were not worked out....especially at the back of the head were relatively coarse revision marks (smoothing) made by a spatula can be seen, which may originate from the concealing of a joint assembly of the wax model. The head of the Mahdia

Herm is constructed of eight segments, corresponding to the ancient manufacturing technique. This does not allow for strong undercuts....consequently the elaborately designed areas of each individual section were divided. On the other hand the head of the Herm in the Getty only has a partial single seam visible, and this supports the use of a flexible molding material. Such is not used in the ancient world and it is hardly conceivable. Although the use of bitumen is mentioned in ancient literature concerning the production of copies, this material served only as a release agent between the model and the form. The Mahdia Herm displays in relation to the Getty Herm a much more detailed, naturalistic sculptural elaboration with strong undercuts, openings and additional applications, such as vine leaves, which were achieved by direct modeling with wax. ...the Herm in the Getty has been created as a copy of the Mahdia Herm..". (Trans. By Joseph Lehner 2011).

The Getty herm does not have a provenance, unfortunately. The measurements of the two bronzes are very close, which may suggest that the Getty Herm is a copy from the signed original.

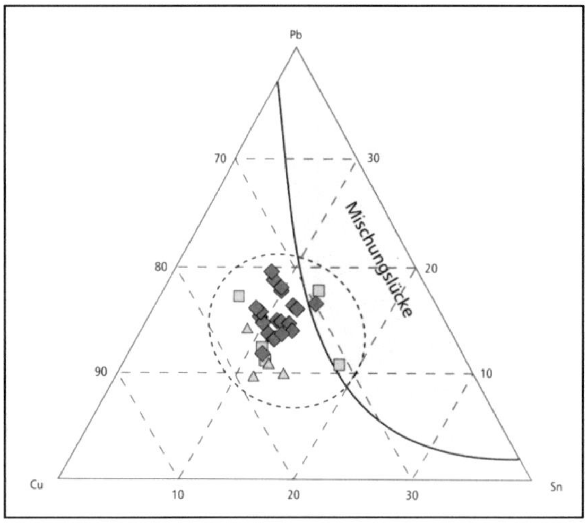

Figure 9.76 Mahdia shipwreck bronze data plotted on a ternary copper-tin-lead alloy. After Schwab et al (2008).

The shafts are only 0.04 m difference in height for example, because they must be indirect casts taken from the same original model. Piece moulds must have been made from the original and the cast wax parts had added wax in the case of the Mahdia herm to produce a higher class copy than that in the Getty Museum. The Mahdia herm had been installed, but was ripped from its base where it had been attached with lead solder, a common fate for sculpture taken away on board ship, as the Getty Bronze Youth is an example of the same event: a sculpture torn from its base, with broken feet as a result. A marble head of a Herm, the same as the Getty Herm and the Mahdia Herm, was found at Pompeii which was destroyed in 79 AD, so the style at least continued for 160 years. (Mattusch 1995).

The two similar herms, one signed and one unsigned, raise questions about the role of the artist and the organization of workshops in ancient Greece. In bronze casting, numerous versions of a work can be produced from the same mould, making the precise role of the artist signing the work unclear. Does the signature of Boethos mean that he created the original mold from which these two slightly different castings were made? Or does it mean that he was the artist who customized this mould with the more elaborate detailing found on the Mahdia herm? The herm has been cast in a heavily leaded tin bronze.

Duplicate drillings from the base gave the following results in weight percent (Scott and Podany 1991 : First result: Cu 68.5%, Sn 11.4%, Pb 18.4% Zn nd, Fe 0.07%, As 0.4%, Sb 0.04%, Ni 0.04%, Bi 0.04%, Ag 0.02%, Au 0.02%. Second result: Cu 69.7%, Sn 13.9%, Pb 16.1%, Zn nd, Fe 0.09%, As 0.47%, Sb 0.24%, Ni 0.04%, Bi 0.02%, Ag 0.06%, Au 0.02%,. Cobalt was found in both at 417 and 586 parts per million, which is significant. The data obtained by Schwab et al (2008) gave results which found for the Getty Herm:
Cu 74%, Sn 8.8%, Pb 17%, Zn 0.1%, Fe 0.26%, As 0.22%, Sb 0.164%, Ni 0.04%, Bi 0.02%, Ag 0.039% Au nd, Co 0.16%, Se 0.01%, and lead isotope rations of :Pb208/Pb 206: 2.070; Pb 207/Pb206: 0.836 Pb206/Pb204 :18.76. Lead-tin solder from the region of the base of the Getty Herm gave: Sn 39%, Pb 60%, Te 0.02%, Sb 0.02%, Ag 0.03%, Zn 0.7%, Fe 0.3 with Pb208/Pb206 2.091, Pb207/Pb206 0.847 and Pb206/Pb204 18.47. The Mahdia Herm had Cu 76%, Sn 6.6%, Pb 16%, Fe 0.23%, Co 0.23%, Ni 0.02%, Zn 0.2%, As 0.39%, Se 0.02%, Ag 0.029%, Sb 0.030% and lead ratios of Pb208/Pb206: 2.061 Pb207/Pb206 :0.832 and Pb206/Pb204 :18.82.

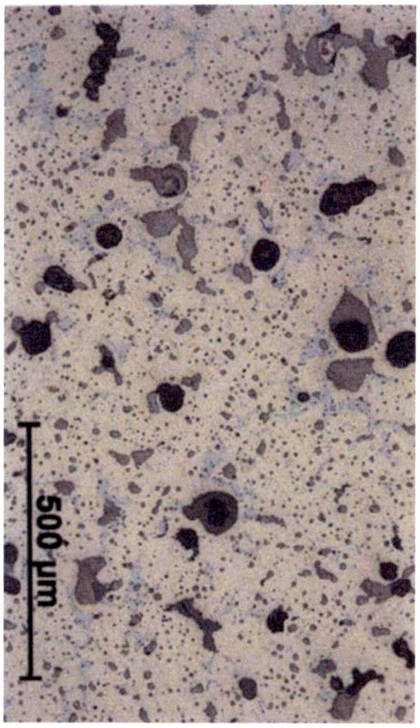

Figure 9.77 The microstructure of the Mahdia Herm showing a clear dendritic morphology with some lead globules in interdendritic regions. The space bar is the scale for 500 microns. After Schwab et al (2008).

This unusual feature of the Getty herm was the focus of a metallurgical study to see if the rod-like structure might provide any clues to the fabrication process. X-radiographs revealed some interesting structural features of the herm, the most striking one, apparent on the inside of the square boxlike base, being the presence of the four bronze rods that traverse most of the length of the interior at the midpoint of each section. Each rod stops before the head of the herm, but each passes over a wax-to-wax join close to the head. There is no sign of any join between the head of the herm and the base adjacent to it, and this, together with evidence of the wax-to-wax join appear to show that the tube is not completely solid, suggesting that it might have been made from an organic supporting material, such as reeds, to support the box-like base when it was still sheets of wax.

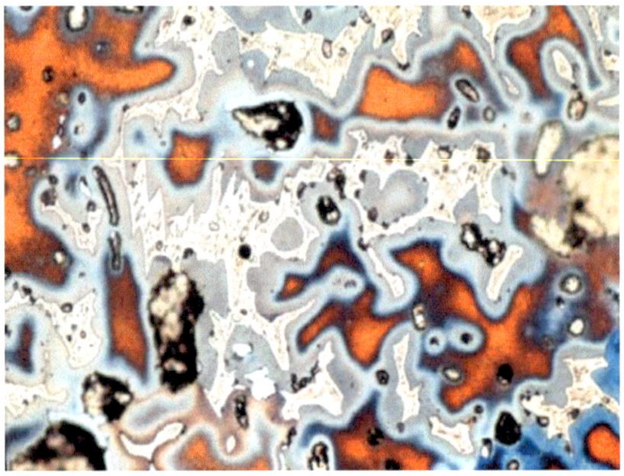

Figure 9.78 Heavily segregated cast microstructure contiguous with the wall of the Herm showing that the rod passing down the inside had been cast at the same time as the rest of the Herm itself. Etched in lead thiosulphate complex etch, x160.

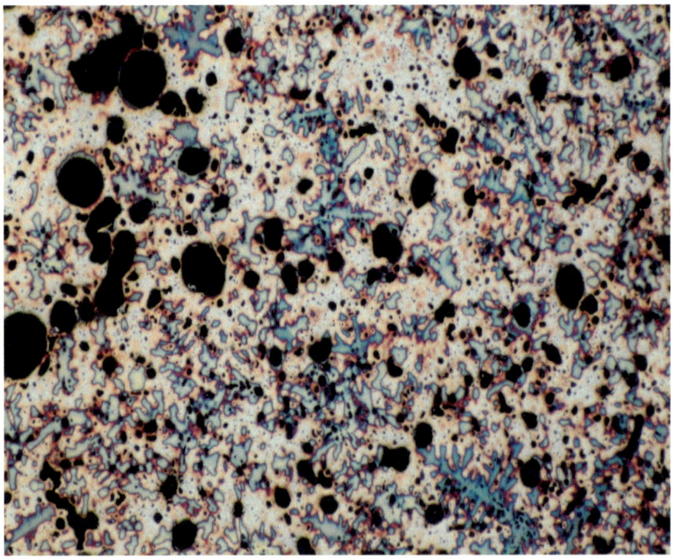

Figure 9.79 Another view of the Herm microstructure etched in Klemm's number II reagent, x 140 showing remnant cast dendritic structure interrupted by lead globules and large regions of porosity.

Case Study Number 8. Adriaen de Vries: Bronze casting of the Juggling Man, or Allegory of Alchemy in the collections of the J. Paul Getty Museum. Metallographic examination was carried out to understand the corroded patina of this bronze, which had been mistaken for a piece of garden sculpture.

This is a sculpture with an interesting history illustrated in Figure 9.80. The Times of London newspaper for 1989 described the auction at Sotheby's at which the sculpture was purchased by the Getty Museum with the headline "100 pound sterling garden ornament fetches 6.82 million pounds in sale" This is the reason this "indoor" bronze was found to have an "outdoor" patina, consisting principally of brochantite with small amounts of posnjakite over cuprite (Scott 2002: 375). The bronze was originally cast and installed in Prague, but must have been stolen by Swedish troops when they sacked many of the treasures of Prague, removing them to Sweden.

It was sold at Sotheby's in 1952 from an unidentified private English collection and remained in private hands until 1989 when it was purchased by the percipient Cyril Humphris. The composition of the alloy is 87.5% copper, 11.0% tin, 0.8% antimony and 0.2% lead with traces of nickel. (Scott in Scholten et al 2000: 201). The subject matter of this bronze is also hard to define and may be an allegory of alchemical mystery. Figure 9.80 shows that this bronze must have been mistakenly placed outdoors for a considerable period of time, judging by the considerable thickness of the corrosion crust on the small pour of excess metal removed from the base for metallographic examination. The bronze still retains the original casting core material and armature. The two hands and plates are in solid bronze. X-radiographs show that the armature is very similar in construction to the wire support used in the group in Washington, the Hercules, Nessus and Deianeira. which is dated to between 1603-1608 and originally from Prague. The modeling of the Juggling Man gave the figure a waxy look, suggesting that not much cold-working or chiseling of the casting took place after the initial cast.

The bronze is based on a classical sculpture of a dancing faun with two cymbals attached to each hand who is pumping a scabellum, a kind of foot-organ. Various motifs indicate that the meaning of the sculpture is related to alchemy, with the cymbals being replaced with small plates and the foot-organ with a bellows. The plate in the figures right hand bears a mark suggestive of the moon, while the circle in the left suggests the sun. the sun and moon stand for the two precious metals, gold and silver which are commonly associated with the two planets in many different cultural settings. For further fascinating details concerning this sculpture the reader is referred to the Catalogue by Scholten (2000).

Figure 9.80 Adriaen de Vries. Juggling Man, circa 1615 AD. 90.SB.44. Height 76.8cm, Width 51.8 cm, Depth 21.9cm.. Photograph by courtesy of the J. Paul Getty Museum, Brentwood, California.

The section shows the cored dendritic structure of the cast bronze, a composition that De Vries was very fond of, rather than the brass alloy which is usually erroneously described as "bronze" in catalogues of Renaissance art. Note the corrosion along strain lines showing that the cast surface has been finished by cold-working. The existence of cuprite under the outer green corrosion here, at some depth, argues for this bronze to have been outdoors for hundreds of years rather than a few decades.

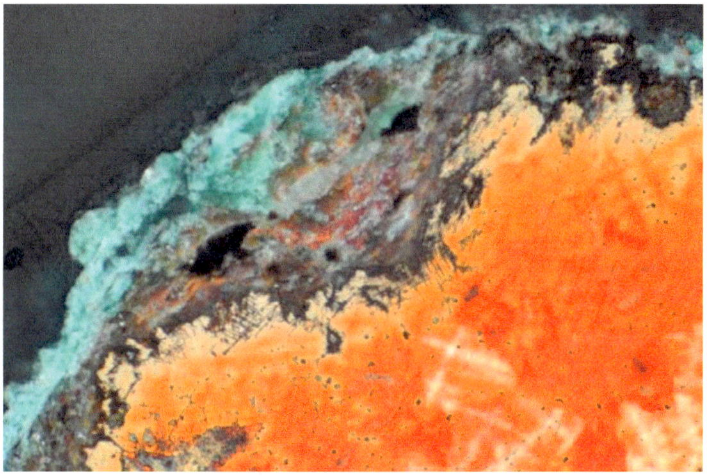

Figure 9.81 Etched in 10% aq. potassium peroxydisulphate, followed by saturated sodium thiosulphate and 1 gram of potassium metabisulphite in 100ml of solution. X200.

Figure 9.82 Etched in 10% aq. potassium peroxydisulphate, followed by saturated sodium thiosulphate and 1 gram of potassium metabisulphite in 100ml of solution. X400. Showing corrosion along slip lines in the outer regions of the casting and extensive corrosion crust.

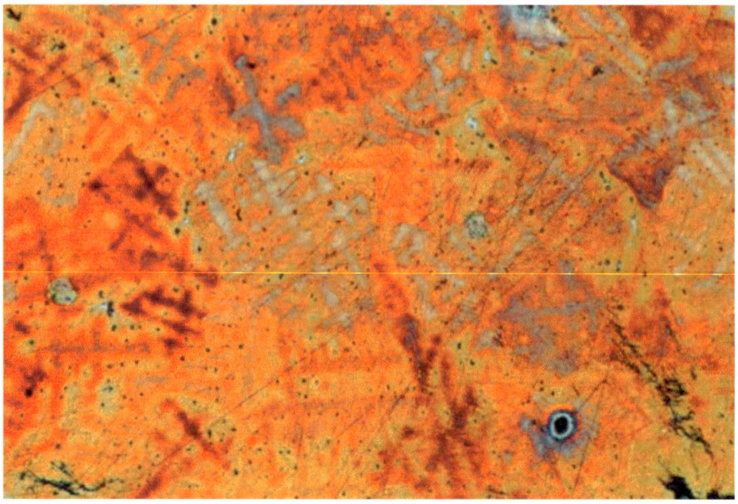

Figure 9.83 Etched in 10% aq. potassium peroxydisulphate, followed by saturated sodium thiosulphate and 1 gram of potassium metabisulphite in 100ml of solution. X100. Here the dendritic nature of the alloy has been brought out by the etchant very clearly.

Case Study Number 9: Sardinian Archer in the collections of the J. Paul Getty Museum. Metallographic examination was carried out to assess authenticity of the bronze casting. The elongated and simplified form of this bronze is characteristic of the bronze statuettes produced on the island of Sardinia between 1000 and 500 B.C. The bow and quiver of the figure clearly identify him as an archer, one of the various warrior types that were favorite subjects for Sardinian bronze casters (Getty 2002). Located off the western coast of Italy, Sardinia had a long established native culture, which came into contact with the Phoenicians when they began expanding their trade routes in the western Mediterranean during this period. This statuette was once thought to be a forgery because the composition of the bronze alloy differed from that of most Sardinian bronzes known at the time. Subsequent study and testing has shown that, in fact, the alloy is well within the range of other Sardinian bronzes, and the statuette is genuine. Analysis was conducted due to doubts about the authenticity of this piece. Microscopic examination showed an outer surface with various oxide, carbonate and chloride corrosion products. There are signs of mechanical cleaning having been used to reduce these corrosion products. Probing showed that the copper chloride areas are active and deeply pitted, with a degree of undercutting of the cuprite layer. There are numerous casting "bubbles" visible at the surface. A thick coating of what appears to be an acryloid or Incralac covers the entire surface.

These observations are not contrary to ancient bronze surfaces and support the antiquity of the archer. Tylecote et al (1976), discuss copper and bronze metallurgy in Sardinia, a good starting point for a study of the metallurgy of the island. The authors undertook a survey of the mineral and metal resources of the island in a collaborative effort to define and treat the problems of the significance of metallic technology and trade for the Bronze Age nuragic populations. The first verifiable smelting slag was found and a hoard of ancient tin confirmed that local smelting as well as casting was taking place. Evidence for early interaction with Mycenaeans and Phoenicians were found, supporting a concept of Sardinia as an emporium after 1200 BC.

Figure 9.84 Statuette of an Archer. Sardinia, about 600 BC, 75.AB.110. Gift of Gustav Dalla Valle. Height: 24.5cm. Photograph by courtesy of the J. Paul Getty Museum, Villa Collection, Malibu, California. Analysis gave: Cu 79.4, Sn 3.7, Zn 0.3, Pb 7.5, Ni 0.02, Fe 0.9, Sb nd, Ag nd.

Figure 9.85 Metallographic evidence confirming the binocular observations that the green corrosion is sometimes under the cuprite surface. Unetched, cross-polar illumination, x 130.

Figure 9.86 The same view as Figure 9.63 this time in plane polarized light to allow the corrosion to be seen very clearly, but not the pleochroic colours of the cuprite and malachite which stand out well in Figure 9.63 under crossed polars.

Figure 9.87 Etched in 10% aq. potassium peroxydisulphate, followed by saturated sodium thiosulphate and 1 gram of potassium metabisulphite in 100ml of solution. X100. The dendritic segregation can clearly be seen.

Figure 9.88 Etched in 10% aq. potassium peroxydisulphate, followed by saturated sodium thiosulphate and 1 gram of potassium metabisulphite in 100ml of solution. X200. The same view as 9.65 at higher magnification showing alignment of sulphide inclusions with the alpha + delta eutectoid.

Figure 9.89 Etched in 10% aq. potassium peroxydisulphate, followed by saturated sodium thiosulphate and 1 gram of potassium metabisulphite in 100ml of solution. X400. An enlarged view of Figure 9.81.

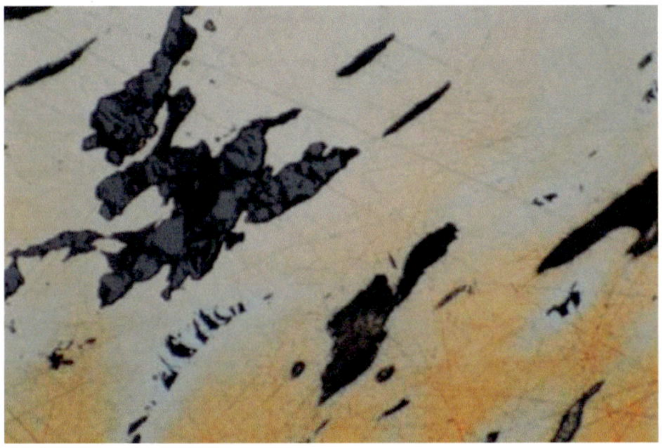

Case Study Number 10. A Roman Intaglio seal in the collections of the J. Paul Getty Museum. Metallographic examination was carried out to see if the seal had been made in a high-tin bronze or if had been tinned on the surface.

Figure 9.90 Roman Oval Intaglio Seal, 3rd-2nd C. BC. 81.AN.76.207. Height: 2.4cm, Width: 2cm, Depth: 0.3cm. Photograph by courtesy of the J. Paul Getty Museum, Villa Collection, Malibu, California.

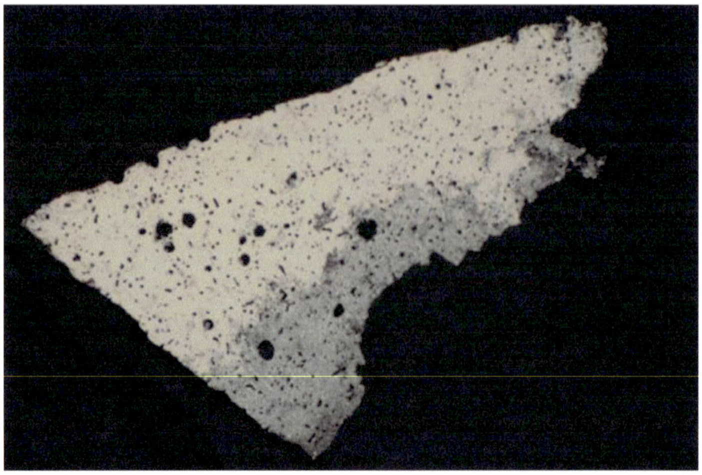

Figure 9.91 Photomicrograph of Roman Intaglio Seal at x 100 magnification. Unetched. Natural corrosion has etched the lower part of the section for us, by selective removal of the alpha phase. This alloy is high-tin and as-cast.

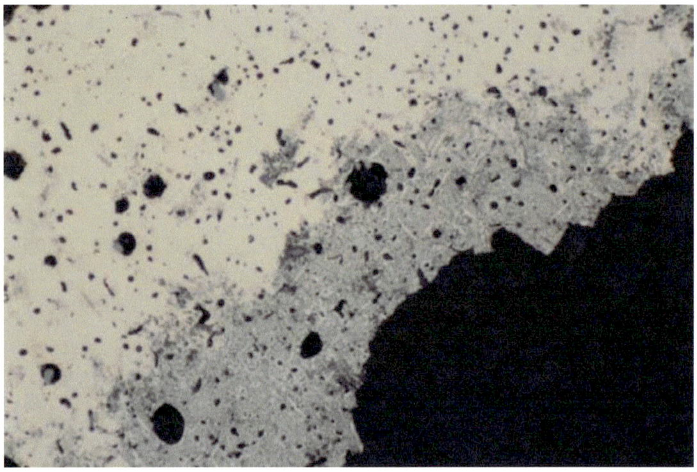

Figure 9.92 At this magnification, x370, the structure can clearly be seen to be a leaded tin bronze, the dark regions are the lead areas between the eutectoid phase. The composition of the seal is about 6% lead, 22% tin, 78% copper, a typical Roman high-tin alloy of the kind that the Romans used for making mirrors.

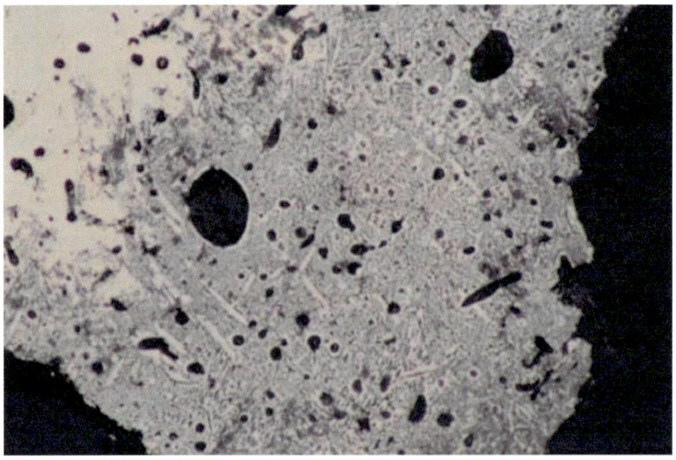

Figure 9.93 Roman Intaglio seal, at x490 magnification. The fine structure of the naturally etched surface is a little clearer and consists principally of delta phase.

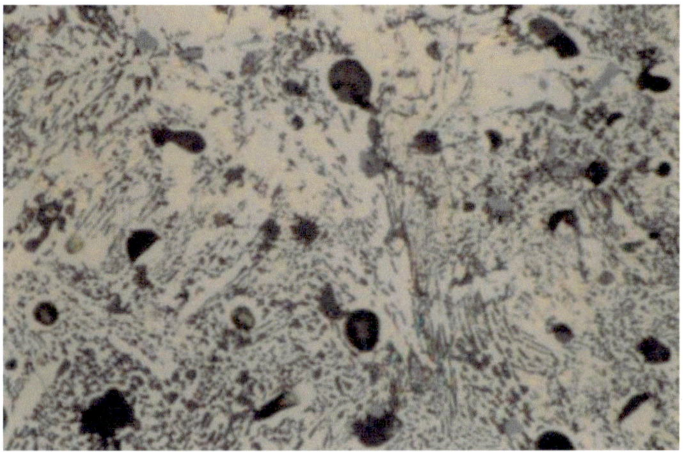

Figure 9.94 The same view at higher magnification showing the remnant alpha grains as= as well as eutectoid phase and lead globules. X510.

Case Study Number 11. Etruscan Kouros in the collections of the J. Paul Getty Museum. Metallographic examination was carried out because of doubts about the authenticity of the bronze.

Figure 9.95 Statuette of an Etruscan Kouros, about 490 BC. 85.AB.104. Height: 22.5cm. Photograph by courtesy of the J. Paul Getty Museum, Villa Collection, Malibu, California.

The Getty Handbook (2002) states that: *"This Etruscan Kouros is based on a type which was widely known in 6th century BC Greece. This statuette was originally intended to serve as a votive offering in a sanctuary. The God worshipped was probably Apollo. In imitation of its Greek model, the Etruscan statuette stands upright, with left leg projecting slightly forward. Unlike Greek kouroi, the Etruscan one is clothed in a tebenna, a precursor of the Roman toga..."* The kouros was originally called into question by an art historian who thought that the way the tabenna was depicted was not consistent with ancient examples. The kouros was analyzed using a drilled sample by atomic absorption spectrometry which gave the following results in weight percentage (Scott and Podany 1991):
Cu 77.13 Sn 6.23 Pb 14.41 Zn tr Fe 0.22 As 9.04 Sb 0.05 Ni 0.04 Bi 0.11 Ag 0.07. Greek bronzes before the Archaic period seldom contain lead as an alloying element. For example the "Dead Youth" in the Getty collections, thought to be Greek of about 480 BC has the lowest lead content of any of the bronzes analysed in the Getty collections. Craddock (1976) suggests that bronze alloys used for statues or statuettes in Greece were unleaded until the 4th century BC, although many contemporary bronzes from different regions contained lead at that time.
The Etruscan kouros might be considered anomalous due to its being cast in an alloy of 14.5% lead and 6% tin. There are therefore three avenues of investigation which leave unanswered questions about this Etruscan bronze: firstly the art historical doubts, secondly, the fact that the bronze alloy is heavily leaded at an early date, and thirdly, the metallographic investigation reveals that the patina may be unsatisfactory for age. In this case, it is the metallographic evidence which provides the most compelling evidence, due to the lack of any cuprite in the corrosion crust, upon which doubts rest.

Figure 9.96 Etched in 10% aq. potassium peroxydisulphate, followed by saturated sodium thiosulphate and 1 gram of potassium metabisulphite in 100ml of solution. X100.

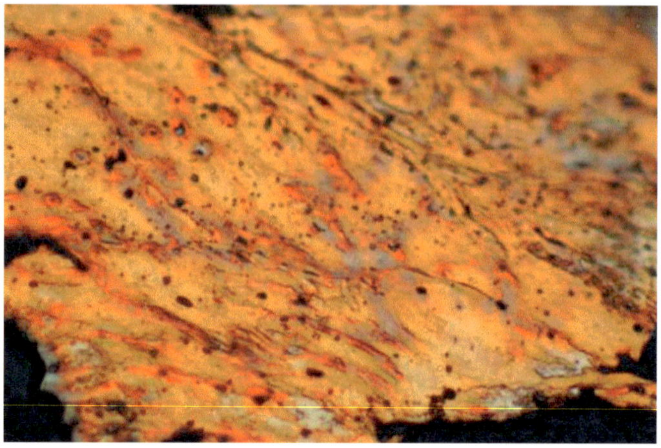

Figure 9.97 Etched in 10% aq. potassium peroxydisulphate, followed by saturated sodium thiosulphate and 1 gram of potassium metabisulphite in 100ml of solution. X200.

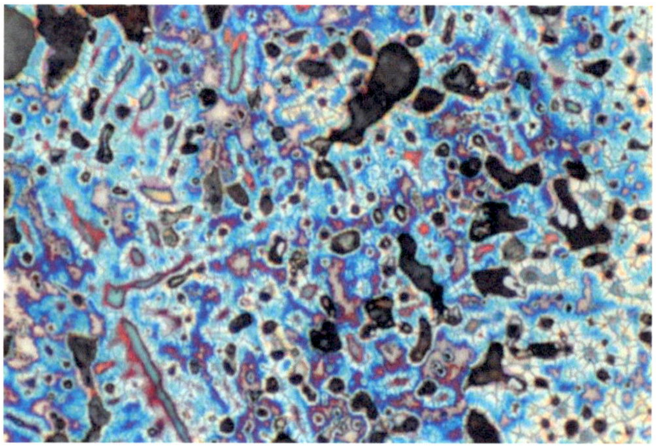

Figure 9.98 Etched in 10% aq. potassium peroxydisulphate, followed by saturated sodium thiosulphate and 1 gram of potassium metabisulphite in 100ml of solution. X100. A colourful result, showing casting segregation. The fine lines in the photomicrograph are an artifact of the colour etching process and not a metallographic feature.

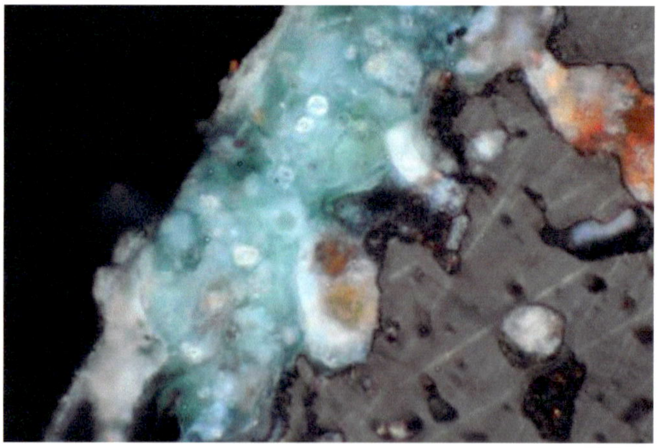

Figure 9.99 Photomicrograph, unetched of the corrosion crust shown under polarized light at x100 magnification. The problem with this corrosion crust is that it lacks penetration, and lacks a well-developed cuprite layer at all. In fact, the predominantly green corrosion crust down to the existing metal is not a good sign for authenticity and is more commonly associated with modern reproductions.

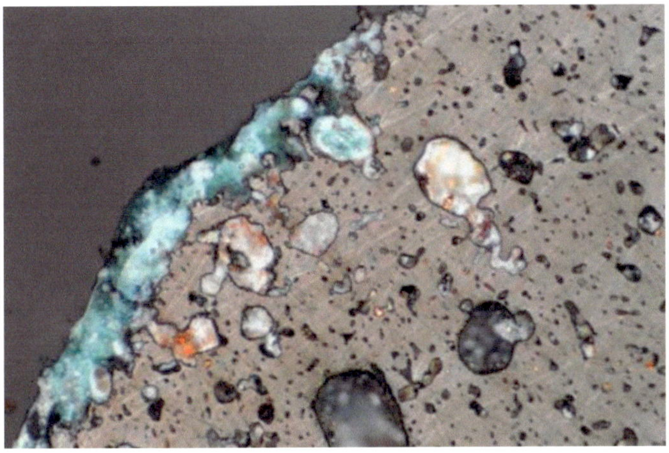

Figure 9.100 Photomicrograph, unetched of the corrosion crust shown under polarized light at x400 magnification. The problem with this corrosion crust is that it lacks penetration, and lacks a well-developed cuprite layer at all. In fact, the predominantly green corrosion crust down to the existing metal is not a good sign for authenticity and is more commonly associated with modern reproductions.

The conclusion from the present investigation is that more work needs to be carried out to investigate the kouros, as a result of this re-evaluation of the patination which is present on this piece. Metallographic sections can occasionally be found in which the original metallic alloy passes directly to a malachite crust, but such cases are unusual and would not be expected here.

Case Study Number 12. Fragment of a Greek shield in the collections of the J. Paul Getty Museum. Metallographic examination was carried out to assess the degree of working and alloy type.

Figure 9.101. Bronze shield fragment. Aristodamos of Argos. About 575 BC.
Villa Collection, Malibu, California. 84.AC.11. H.16.2cm, W. 8cm. Photograph by courtesy of the J. Paul Getty Museum. Aristodamos was a Greek bronze metalsmith, known from a signature on this bronze shield strap which says "Aristodamos of Argos made it". This has helped in the identification of other works by the metalsmith as the name has also been found at the Sanctuary of Zeus at Olympia.
The city of Argos in southern Greece was one of the regions in which bronze was being used for a variety of purposes, including the usual array of Greek sculpture and armor. By 500 BC the Bronze Age was well-developed.

This shield fragment has been made in a low-tin bronze with about 5% tin content. The alloy has been chosen for its ability to be worked and annealed to shape without undue brittleness which would have been created by having, for example, 12% tin. The alloy is also unleaded, since lead is well-known to create planes of weakness in heavily hammered sheet above a level of 0.5% lead (Archbutt and Prytherch 1937: 46). The worked and annealed alloy should also have a small grain size on working and annealing to shape. Grain growth must be avoided during annealing which would make the shield component weak. These conditions have been met, as seen in Figures 9.80-9.83. Some planes of weakness have been introduced by the type of copper used for the shield. This is a smelted copper derived from the smelting of copper sulphide ores.

Figure 9.102 Greek Bronze Shield Fragment: Etched in 10% aq. potassium peroxydisulphate, followed by saturated sodium thiosulphate and 1 gram of potassium metabisulphite in 100ml of solution. X100.

Figure 9.103 Greek Bronze Shield Fragment: Etched in 10% aq. potassium peroxydisulphate, followed by saturated sodium thiosulphate and 1 gram of potassium metabisulphite in 100ml of solution. X200.

Figure 9.104 Greek Bronze Shield Fragment: Etched in 10% aq. potassium peroxydisulphate, followed by saturated sodium thiosulphate and 1 gram of potassium metabisulphite in 100ml of solution. X400.

Case Study number 13. Large left foot of a Roman sculpture in the collections of the J. Paul Getty Museum, Metallographic examination was carried out to assess casting and patination.

Figure 9.105. Fragment of a leaded bronze foot from a large sculpture. Left Foot, late 2nd-3rd century AD. L. 18.5cm. The J. Paul Getty Museum, Villa Collection, Malibu, California. 72.AB.103

In hoards of Greek or Roman bronzes there are often found collections of parts of bodies, destined to be re-used or molten down as scrap. This foot is a typical example of the survival of a solitary foot from what may have once been an important Roman masterpiece. Certainly the modeling of the details of the foot are carried out with skilled craftsmanship and the patina present on the alloy is fairly typical for leaded bronze castings of the Bronze Age. The lead content tends to darken the patina as shown in the illustration in Figure 9.98.

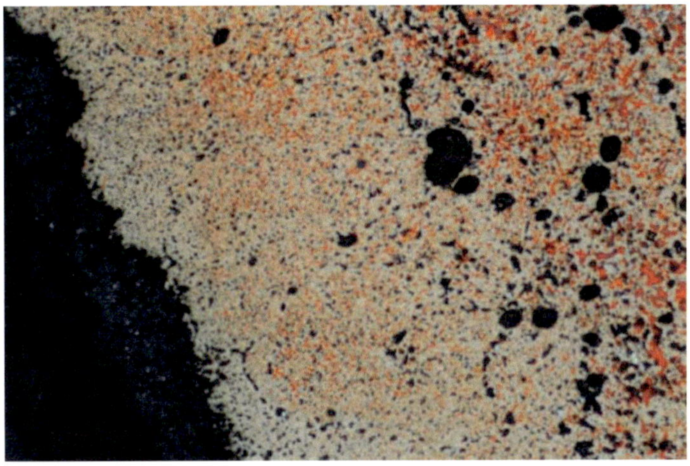

Figure 9.106 Large Left Foot: Etched in 10% aq. potassium peroxydisulphate, followed by saturated sodium thiosulphate and 1 gram of potassium metabisulphite in 100ml of solution. X100.

There may be some normal segregation taking place in the casting of this large foot. Notice that the outer regions of the casting are more compact and contain less lead, while the central region of the casting displays larger droplets of lead. This may be because the outer surface of the mold was sufficiently cool to solidify the alloy quickly in this region, while the central part had a longer period of solidification, allowing larger globules of lead to form. The casting is of high quality if we compare this leaded cast to that of the two Sphinxes with volute scrolls shown in Figure 2.2.

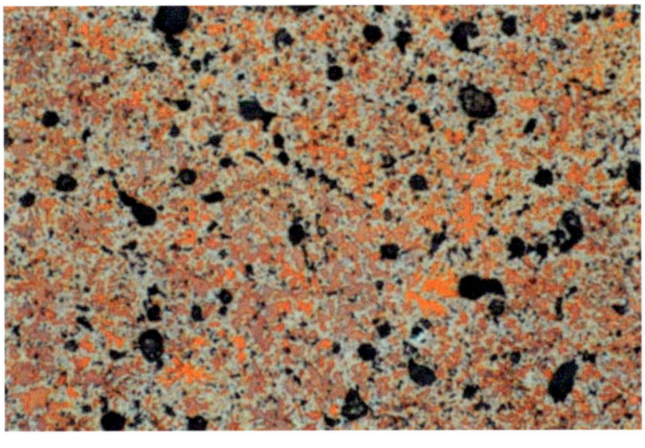

Figure 9.107 shows that very little porosity was created in this casting and consequently the finished foot has taken a fine polish and an even patina has built up over the alloy.

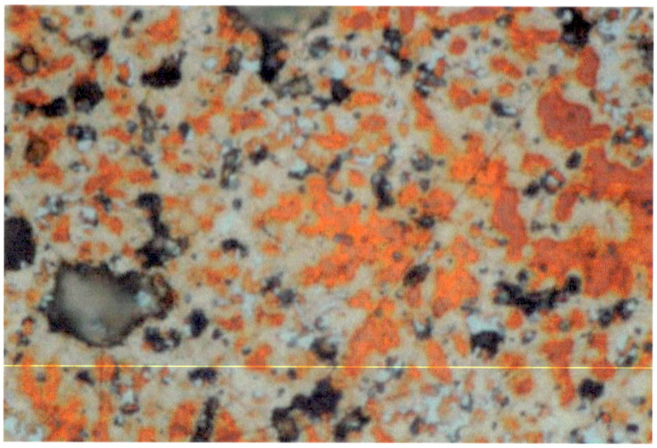

Figure 9.108 Large Left Foot: Etched in 10% aq. potassium peroxydisulphate, followed by saturated sodium thiosulphate and 1 gram of potassium metabisulphite in 100ml of solution. x200.

Figure 9.109 Solder fragment from another large foot from a sculpture in the collections of the J. Paul Getty Museum. No 71.AB.229. Etched in 10% aq. potassium peroxydisulphate, followed by saturated sodium thiosulphate and 1 gram of potassium metabisulphite in 100ml of solution. x100.

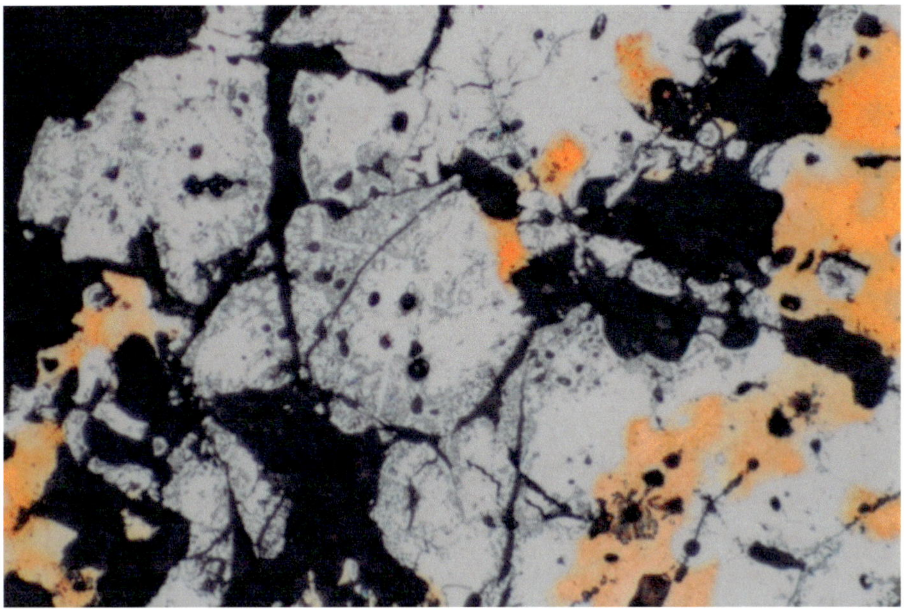

Figure 9.110 Solder fragment from another large foot from a sculpture in the collections of the J. Paul Getty Museum. No 71.AB.229. Etched in 10% aq. potassium peroxydisulphate, followed by saturated sodium thiosulphate and 1 gram of potassium metabisulphite in 100ml of solution. x400. At this magnification the two-phase nature of the high tin bronze microstructure is evident with the delta phase naturally etched by corrosion.

Figure 9.111 Roman Situla from Gaul, Europe. Courtesy of the J. Paul Getty Museum, Villa Collection, Malibu, California. Accession number 96.AC.55. The Getty Handbook (2002) states that: "Dionysos, the god of wine, and his retinue decorate this bronze vessel: the shape referred to as a situla. Across the middle of the vessel, Dionysos reclines in a chariot drawn by two panthers, one ridden by Eros, the young god of love. Various followers accompany Dionysos: a maenad, a goat-legged Pan playing the pipes, and satyrs. This bucket-shaped

vessel, now missing its original handle, imitates a particular basket shape that the Greeks called a kalathos. Since the kalathos was used to gather the grape harvest, this vessel's form is cleverly connected to its Dionysiac subject matter "

The technique used to decorate this situla is unusual. Strips of tin, either plain or with figural cut-outs, were laid over the bronze for contrasting colour effects. X-ray diffraction studies carried out, *in situ*, suggested that the tinned surface was now mostly eta phase, the Cu_6Sn_5 commonly associated with tinned surfaces.

The bronze alloy itself is a low-tin bronze and another situla gave the microstructures shown in Figures 9.112 and 9.113. The colour etching illuminates the segregation which is still present from the cast state, underpinning the worked and annealed grains which can be seen here. The segregation appears surprisingly intense here, given the fact that the situla must have been worked and annealed to shape it from the original cast ingot.

Figure 9.112. Roman bronze situla. Etched in 10% aq. potassium peroxydisulphate, followed by saturated sodium thiosulphate and 1 gram of potassium metabisulphite in 100ml of solution. x200.

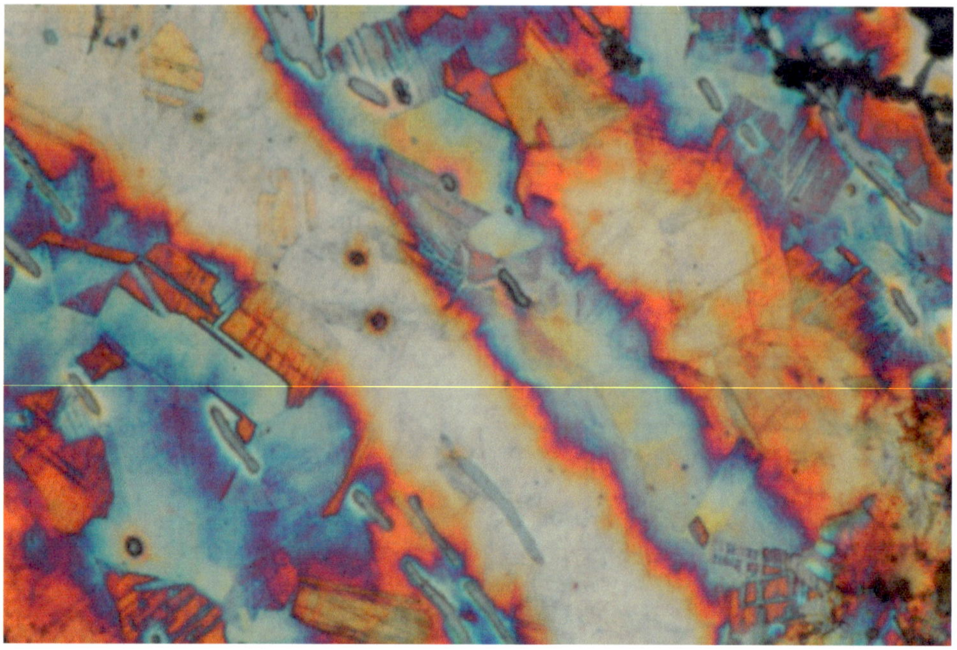

Figure 9.113. Roman bronze situla. Etched in 10% aq. potassium peroxydisulphate, followed by saturated sodium thiosulphate and 1 gram of potassium metabisulphite in 100ml of solution. x400.

Some additional microstructures

Figure 9.114. Coin of Gordius III. Etched in 10% aq. potassium peroxydisulphate, followed by saturated sodium thiosulphate and 1 gram of potassium metabisulphite in 100ml of solution. x200. The bright blue inclusions are of copper sulphide and the alloy is a low-tin bronze with about 4%

tin. The grains are worked and recrystallized, probably from hot-working the cast coin blank to shape.

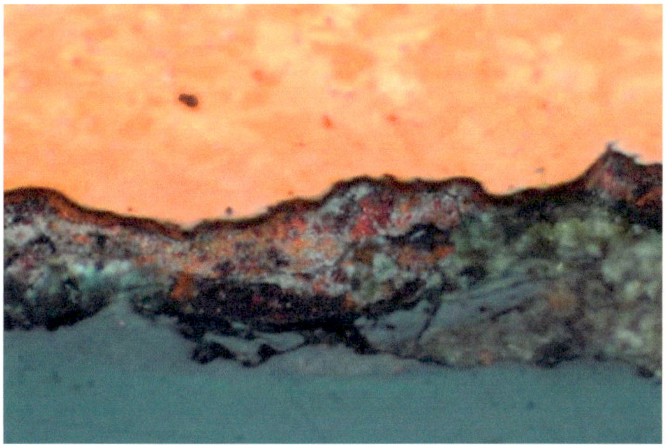

Figure 9.115. Corrosion crust is in focus here to show the extent of corrosion on this coin of Gordius III. Etched in 10% aq. potassium peroxydisulphate, followed by saturated sodium thiosulphate and 1 gram of potassium metabisulphite in 100ml of solution. x200. The depth and complexity of this crust would normally be sufficient to suggest that the coin is authentic.

Figure 9.116 Grain structure of a late Bronze Age sword from Palestine made in a tin bronze with about 8.2% tin (Ghandour 1981). Etched in 10% aq. potassium peroxydisulphate, followed by saturated sodium thiosulphate and 1 gram of potassium metabisulphite in 100ml of solution. x100.

Figure 9.117 Etched in 10% aq. potassium peroxydisulphate, followed by saturated sodium thiosulphate and 1 gram of potassium metabisulphite in 100ml of solution. x400. The twinned grains, strain lines through some, and intergranular corrosion, together with large grain size can all be seen clearly in this photomicrograph of the Palestine sword blade (Ghandour 1981).

The blade illustrated in Figures 9.116 and 9.117, cast in a 5% tin bronze, would have been rather weak in practice, as the grain size is too large, there are only 8-10 grains across the thickness of the blade, which is insufficient to create enough grain boundaries to produce a strong alloy. This is why modern copper alloys have very small grain sizes. The small size creates many grains moving against each other which result in a better performance under stress or working. The grains of the blade are annealed with straight twin lines showing that annealing has been the final operation. Perhaps the worked alloy was over-annealed resulting in grain growth.

Tinned surfaces

Tinning is an easy operation to carry out, as tin flows readily at low temperatures, and is able to flow evenly across the surface, creating a good quality coating. Many iron and copper alloy artefacts are tinned for decorative purposes. In the case of copper alloys, tinning can be carried out in a number of different ways, either by directly applying the molten tin, preferably with a flux, to the surface to be tinned, by dipping the object to be tinned into a bath or tray of molten tin, or by the use of tin amalgams, which are illustrated in the following Figures. There tends to a progression in the phases present from metallic tin, to the eta phase, Cu_6Sn_5, to the epsilon phase, Cu_3Sn and then to the delta phase, $Cu_{31}Sn_8$.

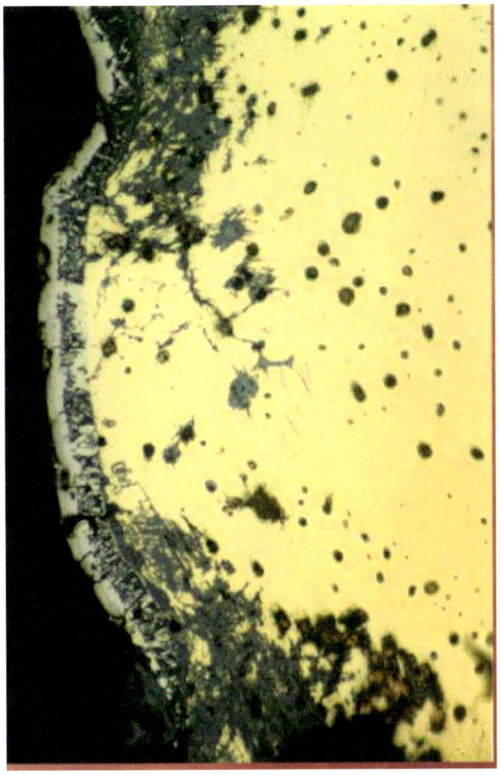

Figure 9.118. Tinned bronze surface characteristics for a Chinese bronze belt plaque, amalgam tinned, from the Spring and Autumn period, 6th-5th century BC (After Ma and Scott 2002), unetched, x130. Has a few residual areas of tin at the surface, followed by eta phase (η) Cu_6Sn_5 with 61%Sn 39%Cu, then epsilon phase (ε), Cu_3Sn, with 38.2%Sn 61.8% Cu, then the eutectoid phase of $\alpha + \delta$ where delta is $Cu_{31}Sn_8$ with 32.6% Sn 67.4% Cu.

The most commonly determined tinning phase present on the surface is the eta phase, Cu_6Sn_5. Because these phases are ordered, with specific compositions, they can readily be determined, often non-destructively, by x-ray powder diffraction. We have carried out this procedure on the Rigaku, R-axis Spider X-ray diffractometer operating in Weissenberg geometry on small objects very successfully.

When the tinned surface is examined in section, it is often apparent that there is a successive progression in the tin phases from the outer surface of the artefact to the inner regions. Notice in Figure 9.118 that the eta phase has a slightly bluish colour when compared with the epsilon phase, which is whiter, as it the remnant of metallic tin at the surface of this artefact which is a similar white colour to the epsilon phase.

Figure 9.119 Tinned bronze surface characteristics. Amalgam tinned belt plaque from the Spring and Autumn period showing the typical progression of these phases in tinned surfaces (After Ma and Scott 2002). Unetched, x 340. Same phase transitions as Figure 9.117.

Tin-sweat

In some castings, inverse segregation forces tin, the last constituent to solidify, to the surface of the casting, creating a silvery surface as a direct result of the casting process, rather than deliberate tinning. The hardest microstructures to resolve in this case, are those which have more delta phase and possibly tin at the surface of the castings. A microstructure such as that shown in Figure 9.119 cannot be formed as a result of tin sweat, but with simpler surface microstructures, this may not be true. Tin sweat has been associated with Bronze Age castings of axes or palstaves in which the substantial amount of metal cast into the mould may allow tin to be extruded, as it were, to the outer surfaces. In such cases, further research into the culture, the use of tin, and whether tinned surfaces are only found on cast axes or other implements, may help to answer the question. Metallography is essential in these cases in any event.

Joining of bronze alloys

The joining of bronze statuary has traditionally employed three different methods. The most sophisticated method is to create a cast-on join in which a separate mold is made and essentially the same bronze alloy is then cast-in to join the bronze parts together. When the cast join is then smoothed and polished, the join is practically invisible, often essential in cast bronzes. We usually distinguish between soft soldering and hard soldering, also called brazing. Neither of these will be metallurgically the same alloy as the bronze components to be joined. Soft solders are essentially tin-lead or tin-rich alloys which have a melting point far below that of bronze. Hard soldering often refers to the use of a silver-copper or gold-silver-copper alloy and in more recent metalwork, to a copper-zinc brazing alloy. Brazing alloys usually have a melting point closer to that of a bronze, creating a stronger join than that possible with a tin-rich alloy, although the alloy may have to be coloured during patination, otherwise the join will be evident. Since copper alloys cannot be welded together, the other methods used for component parts is to join them mechanically by using sleeve-joins, rivets, or other mechanical methods, such as tab-and-slot fabrication, often used in South America in the cultures of Peru, Chile and Ecuador. It is only in iron and steel alloys, and occasionally with gold, that joining by hot-welding of the component parts together is practical. The oxide film which develops on copper alloys during heating prevents this.

Chapter 10: Leaded Bronzes and Other Ternary Alloys

Alloys with more than two components, ternary alloys, are quite common in many ancient alloying systems. The most pervasive is the copper-tin-lead ternary alloys. Although lead does not alloy with copper and tin as the alloy cools down, and the lead simply occurs as a series of globules or interdendritic solid particles, the alloys at high temperatures have very good casting characteristics which allow them to be molten between 800 - 900 °C. The fluidity of the cast ternary alloy is very good, and it cools taking a good replica of the surface of the mould with the high-tin varieties.

The leaded bronzes perform poorly if they are worked and annealed to shape them as sheet: this is because the lead creates planes of weakness in the hammered metal, resulting in a tendency to delaminate if the lead content is much over 0.5%. Below that value, the lead may even be beneficial for producing cold-worked and annealed ternary copper alloy, but higher lead contents are generally avoided, although with skillful working, one still finds a few examples of leaded bronzes which have been formed into worked sheet. The lead content of many cast bronzes varies from around 5% to 15%. Very high percentages of lead, say 50%, are avoided because of dull colour and possible pooling of lead during solidification.

Lead is kept in solution at high temperatures in the form of an emulsion and when the alloy cools, the lead is scattered through the solid metal. If cooling in the mould is too prolonged, then the lead may begin to sink through the bronze and collect in larger globules, as seen in Figure 10.21 in this chapter. We have already examined the microstructure of a few leaded tin bronzes in Chapter 8, such as the Greek Herm, and here a series of examples of higher-tin leaded bronzes are illustrated drawn partially from a technical examination of 100 Chinese bronze mirrors in the Cotsen collection in Beverly Hills, Los Angeles which is currently in the course of publication as a multi-authored monograph. In these alloys, the casting conditions have been carefully controlled to minimize porosity, and in the classic compositional range, to create a finely dispersed Widmanstätten microstructure, producing a surface which takes a very high polish. Dendritic solidification is suppressed in some of these examples, while in others, the cast morphology is obvious. Beta-quenched microstructures are also possible, adding to the variety seen in this ternary system.

A fine dispersion of lead globules is the aim of the casting process in the case of the Chinese bronze mirrors, in other cases where the addition of lead is simply to cheapen the casting and to bulk out the alloy, the lead may be present as sizeable inclusions in the bronze, and may be mixed with sulphide inclusions or other impurities which have solidified last in the casting of the bronze alloys of this type.

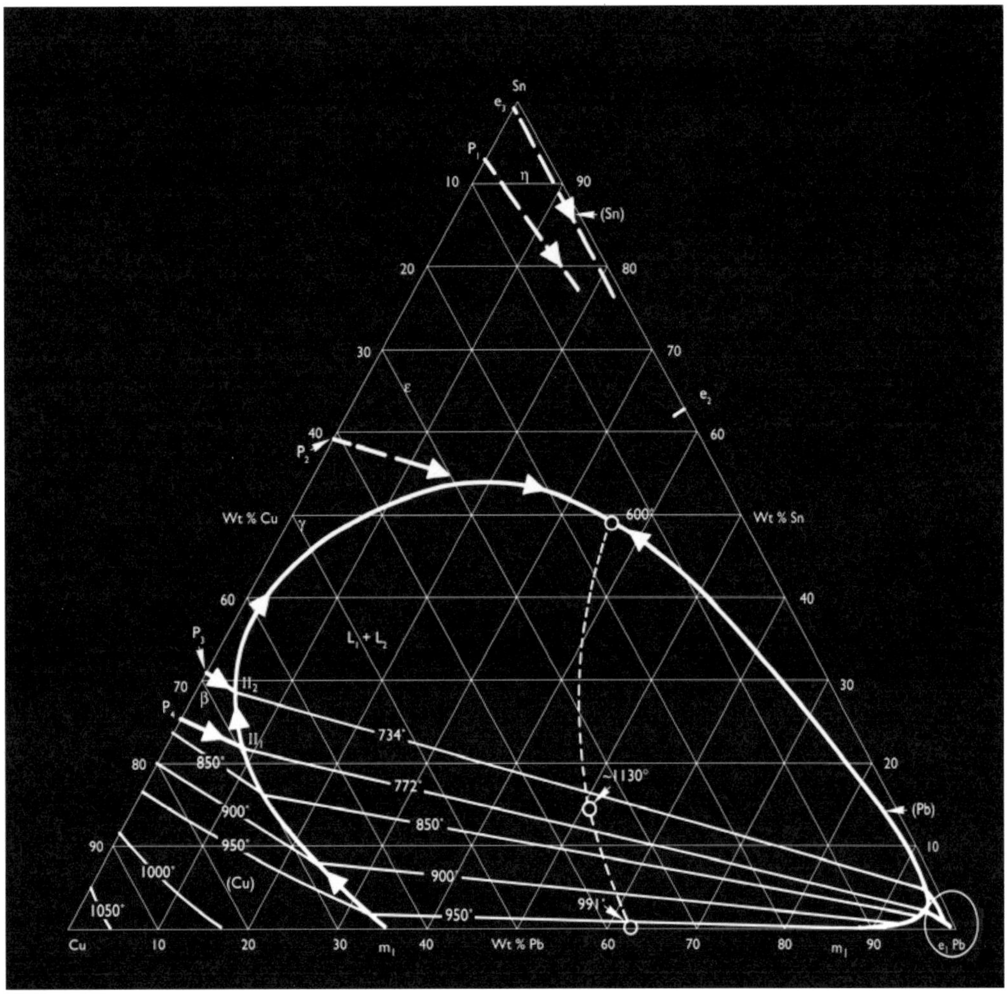

Figure 10.1 The ternary liquidus for the copper-tin-lead alloys. After Scott (1991) showing the melting points of the alloys of different compositions. The ternary alloys have been the backbone of cast bronzes for thousands of years. The leaded bronzes result in microstructures in which the lead is segregated in the alpha phase as the alloy cools from the liquid state, which results in a fine dispersion of lead in most cast bronzes. If the leaded bronze is allowed to cool down slowly in the mold, then some lead segregation due to gravity segregation may occur, as the heavier lead separates out and floats down through the casting. Heavily leaded bronzes usually have 15-20% lead, further additions much above this level are not desirable.

Chinese mirrors of the high-tin variety have liquidus values close to the arrow on the P4 tie-line shown on the diagram. The ternary liquidus is a slice through the triangular prism of the entire system, and often these slices are taken at different temperatures to show the variation of phases which occur. This triangular phase diagram shows the tie lines across the two-phase filed showing show the liquidus region moves downwards in temperatures in the two-phase field. The alloys with high lead and high tin contents with low copper are of limited use. Normally, alloys with up to 30% of tin with between 10% - 20% lead and the remainder copper are the upper range of the ternary alloys used. Chinese bronze mirrors usually have about 5% lead, 24% tin and 71% copper, with a variation of plus or minus 3% in each constituent.

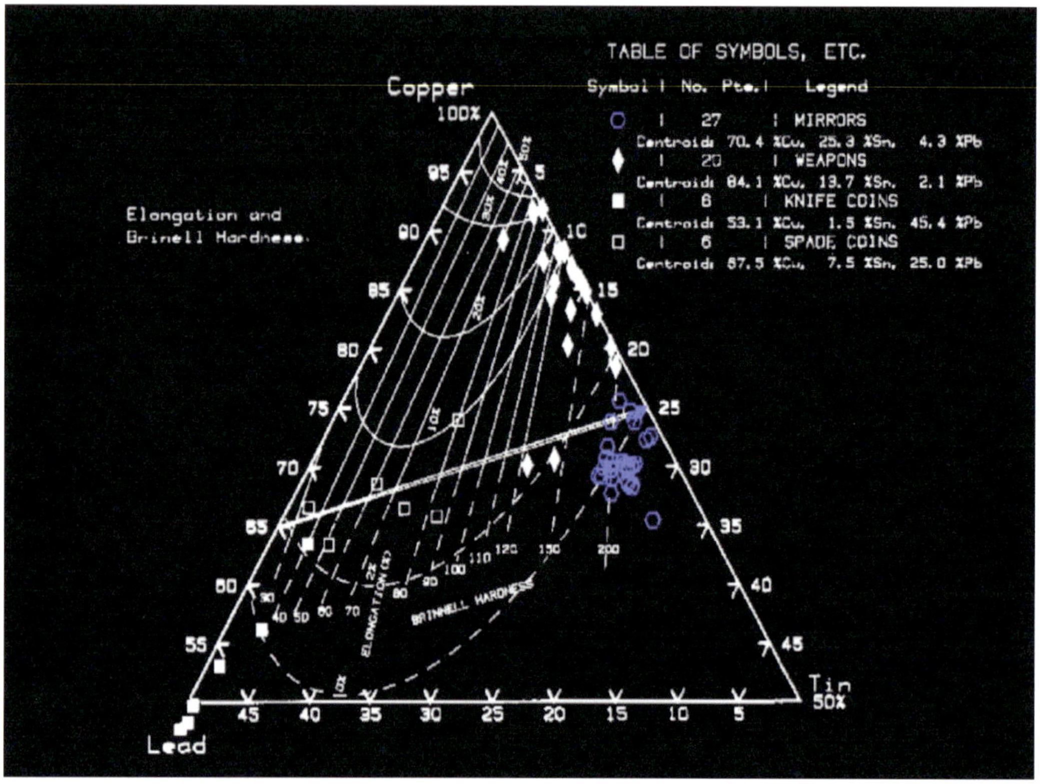

Figure 10.2. The hardness figures given in Brinell figures together with the elongation for the ternary Copper-Tin-Lead system (after Chase 1993) Blue circles show the typical composition of some of the ternary Chinese mirror alloys. Diamonds show the composition of several different weapons, squares are for Chinese knife coins and open squares for spade coins

A good example of the microstructures and problems of authenticity of some of these alloys is afforded by the Chinese cast copper-tin-lead alloys. The inscription on the mirror shown in Figure 10.3 boasts of its excellent workmanship and it is not an idle boast: it is indeed of lustrous material and excellent workmanship, cast in a ternary copper-tin-lead alloy of exactly the right composition, its reflection would have been bright and undiminished. By the time this mirror was cast around the 7th century AD, there was already two thousand seven hundred years of bronze mirror development, which continued into the 18th century AD, representing four thousand years of bronze mirrors across many different cultures and geographical regions.

The earliest known mirror was excavated in Guinan county, Qinghai province in 1977, dated to about 2000 BC (Shoukang and Tangkun 1993: 51). These early examples tend to be of modest size and from 1-3 mm in thickness. A number of them from excavated Bronze Age contexts from Xinjiang are discussed by Mei ((2000). For example, a mirror from the Tianshanbeilu cemetery, Hami, Xinjiang, 5.6 cm in diameter (Mei 2000: Fig 2.10), was cast in bronze, quite thin, and dated between 1900 - 1100 BC, while another mirror, from the Wupu cemetery, dated to 1400 – 1000 BC, also from Hami was 5 cm. in diameter (Mei 2000: Fig 2.11).

Figure 10.3 Cotsen mirror 0-0363, Sui or early Tang. Cast in the classic mirror composition, the high quality alloy of 70% copper, 25% tin and 4% lead. Photograph by courtesy of the Lloyd Cotsen Corporation

The earliest mirror in the Cotsen collection, 0-0803, 4.6mm in diameter, 2mm thick, is thought to be from the Qijia culture, dated to 2300-1800 BC, at the inception of the Chinese Bronze Age. The Qijia has been suggested by Fitzgerald-Huber (1995), to be connected to cultural influences from the Eurasian steppes, such as the Andronovo culture. The use of star-shaped designs as seen on the Cotsen example is comparable to design elements from the Andronovo culture. The discovery of several excavated mirrors with handles from Xinjiang also suggests these cultural connections, since handled mirrors became increasingly common in Eurasia from the first millennium BC and are associated with several nomadic peoples (Jacobson 1995: 182). Handles on Chinese mirrors are therefore seen in these very early examples, later to return to mirrors from the Song Dynasty onwards.

The Andronovo culture, sometimes called the Sintashta-Petrovka complex, represents a collection of western Siberian and west Asiatic steppe peoples from 2300–1000 BC which had connections with the Chinese Tianshanbeilu culture, (Mei 2000:72), suggesting that Chinese bronze technology may have been influenced by the eastward extension of Andronovo influence. Stone mould casting, and the use of arsenical copper, a tradition widely seen over the Eurasian steppes during the second millennium BC, is also seen in early metallurgy in Xinjiang province.

Despite the early use of arsenical copper and ternary alloys of copper-arsenic-lead in Xinjiang, the evidence suggests that Xinjiang mirrors were already made in copper-tin-lead alloys, much like the early Tang example illustrated above. Two mirrors from the Hami region were cast in tin bronze with 22-23% tin, (Mei 2000: 35), implying that mirror craftsman were already intentionally selecting specific alloys for mirror production which were out of the ordinary range of copper alloys used for other artefacts. There is a significant similarity between the Qijia Cotsen mirror and those from Xinjiang, and that is the very low lead content in Xinjiang alloys. The Cotsen mirror has 75.2% copper, 18.6% tin and 0.3% lead, showing that higher tin content alloys without lead additions were already in use. The Cotsen mirror also has two small holes towards one edge, perhaps for attachment, a feature often seen in early mirrors, such as those from Aidinghu, Turfan, Banfanggou, and Urumchi (illustrated in Mei 2000: Figure 3.14). A single or double hole near the perimeter, seen in these examples, is a feature which rapidly disappears in later Chinese cultures. Mirrors with central bosses and a hole for suspension from a cord, or with a handle were common in early periods, but only the central boss remained as an enduring motif.

Figure 10.4. Bronze mirror from the Qijia Culture. Cotsen 0-0803. Photograph by courtesy of the Lloyd Cotsen Corporation.

The possible use of stone mould technology is suggested by the difficulty the craftsman had in producing the circular bands around the outer rim, which are hard to carve in stone in a smooth circle, but much easier to produce in clay. This would also conform to what is known about some of the casting technology from this region, before the more general use of ceramic piece moulds in the centuries to come. The ideal range of alloy composition for Chinese mirrors is about 26% tin, 3% lead, 71% copper, with a variation of plus or minus 10% for each element. In both very early and later periods there may be much deviation from this, as the Qijia mirror illustrates. Another mirror 0-0427, ascribed to the Shang Dynasty, 1600-1100 BC is made in a low-tin leaded bronze with some arsenic content. Analysis gave: 75.3% copper, 11.2% lead, 1.4% silver, 3.7% tin, 1.2% arsenic. Alloys with more lead than tin, and with some arsenic content are anomalous as with this amount of tin, the alloy colour would be yellow-grey, and not white at all. Analysis of Shang dynasty bronzes from Hanzhong, Shaanxi province (Mei et al 2009), shows that a circular disc was made in an alloy with 18% tin, but many bronzes of different types contained high amounts of lead. The mirror examined here would fit well into this general range of Shang bronzes: lead is usually kept low in examples from the Warring States.

Figure 10.5. Ascribed to the Shang Dynasty. Cotsen 0-0427. Low tin leaded bronze. This mirror has more lead than tin and would not be white in colour. Photograph by courtesy of the Lloyd Cotsen Corporation.

An early Warring States example in the Cotsen, 0-0201, also has more lead content than tin, and the composition, of roughly 60% copper, 14% tin and 19% lead, places the alloy in the low-tin, heavily leaded category. Because of the extensive contacts with nomadic tribes, Xinjiang and Qinghai province are areas where foreign influences may be most apparent. The mirror shown in Figure 10.5, from Qinghai province and dated by association with other material found with it to the 6th century AD, is entirely non-Chinese in terms of its metallurgy. Microstructural investigation showed that it is a β-quenched bronze, that is a bronze, typically of 21-22% tin with no lead, which is quenched from between 700-600 °C, from the β phase region, as we have already seen above. As far as we know from present data, this is the only β-quenched bronze found from Chinese territory, which must have been an exotic import.

Figure 10.6. Mirror from Qinghai Province, dated to 6th century AD. Made in an alloy of 21% tin 79% copper, with no lead. Photograph by courtesy of Melanie Roy and TK Asian Antiquities.

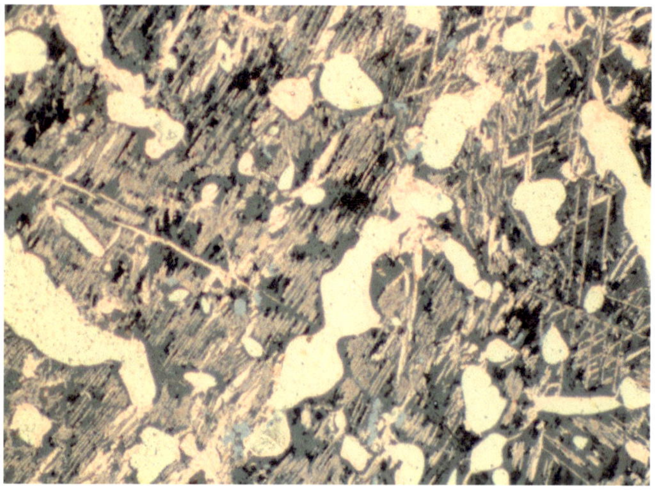

Figure 10.7. Photomicrograph of the Qinghai mirror from 6th century AD, revealing α -islands in a groundmass of slightly corroded β-phase, showing that the mirror has been quenched from about 650 °C. Chinese mirrors were not quenched, therefore this is an exotic import. Etched in alcoholic ferric chloride. Magnification x 450.

We can contrast this type of micromorphology with that of the classic high-tin bronze of the Tang dynasty mirror illustrated in Figure 10.7, thought to be from the 8th century AD, which is made in a copper-tin-lead alloy, which has cooled below the 520 °C isothermal of the copper-tin system. The microstructure of this mirror represents the archetypical Chinese high-tin leaded bronze mirror morphology, consisting of a finely interlocked Widmanstätten precipitation of the (α + δ) eutectoid, arising from the decomposition of the higher temperature β solid. It is quite possible that Chinese mirrors could have cooled quite quickly in the moulds, but we know from their structures that they were never quenched products. The Widmanstätten precipitation ensures that the lead globules are finely dispersed and that the surface will take a highly reflective finish on polishing.

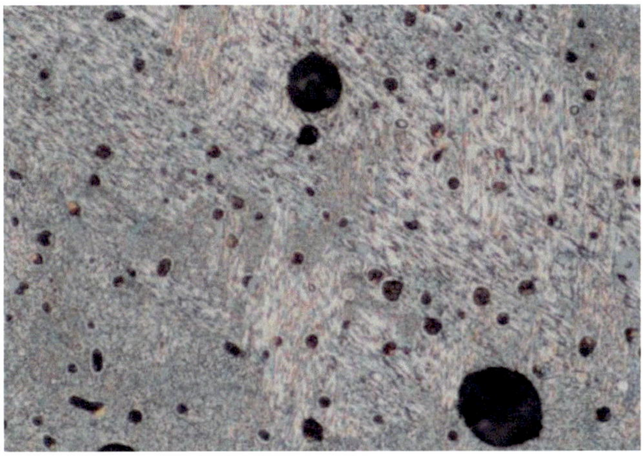

Figure 10.8. Etched photomicrograph of the microstructure of the Tang mirror, illustrated in Figure 10.3, revealing fine Widmanstätten precipitation of the (α + δ) eutectoid phase with a scatter of small lead globules and two larger areas of porosity. Magnification x 450. Etched in alcoholic ferric chloride.

Chinese mirrors are usually made in alloys of copper-tin-lead. For castings, lead is often added as it lowers the melting point of the mixture and aids in the fluidity of the casting and ease of polishing. Mirrors made in copper-lead alloys or binary copper-tin alloys with no lead were also made, as we have seen. In later periods, when zinc becomes available, tin is no longer used, and the mirrors are made of copper-zinc, or copper-zinc-lead alloys. The constraints on casting a relatively thin flan of metal meant that efforts had to be made to purify the starting materials as much as possible to avoid surface imperfections or defective castings, which would ruin the reflective surface on polishing. The most common minor element detected by X-ray fluorescence spectroscopy (XRF) in the Cotsen collection is arsenic, which appears as a trace impurity in several. The prevalence of arsenic as a significant minor component of bronze alloys is revealed by the analyses given by Chase et al (1995), for Eastern Zhou ritual bronzes in the Freer Collection, in which most contain arsenic. In fact, arsenic is very beneficial, assisting with the whiteness of the colour of the mirrors and acting as a deoxidizer, improving the soundness of castings. Arsenic is not necessarily present at all in Bronze Age alloys from other regions of the world which suggests that the Chinese metalworkers had access to copper ores which contained some arsenic.

In the literature of the Western Han (206 BC – 25 AD), three different terms are used to indicate the purity of the copper used: *Sanlian*, *Jiulian* and *Bailian*. These can be translated as the three, nine, or hundred-refinings (Soukang and Tangkun 1993: 52). By fire-refining the molten copper, impurities such as sulphur, iron, arsenic, and antimony, would be sequentially reduced, resulting in a very pure copper which was then used for the highest quality mirrors. The copper was then alloyed with lead and tin, easy to cast and which took a good polish of silver-white colour. On Mirror 0-0133, from the Three Kingdoms Period, 3rd century AD, in the Cotsen collection, part of the inscription reads:

"I made a bright mirror, in seclusion refining the three metals"

The copper, lead and tin in this case were all carefully selected and refined in order to produce a final casting, free from defects formed from dross or inclusions. The only additional elements found by non-destructive XRF analysis of this mirror were traces of arsenic, nickel and silver, none of these impeding the quality of the casting.

During the earlier periods, from the Xia to Shang dynasties, about 2000 BC - 770 BC, the bronze alloys used might be variable in composition and include not only the high tin bronze alloys with 22-27% tin, but also those technically defined as low-tin bronzes, with tin contents from about 5% to 14%, examples of which have already been discussed. This kind of bronze composition was essentially the same as that used for bronze vessels and ornaments. The colour of these mirrors are golden rather than white, as the addition of tin to copper changes the colour from the salmon-red of pure copper to the golden hue of bronze, and only with higher tin additions, usually from 20 – 28% tin, does the colour of the alloy turn to a pale silver-white, or grey-white colour.

By the time of the later Zhou dynasty, and especially from the Warring States period onwards, from around 400 BC, the high-tin bronzes with lead additions became prevalent and lasted into the Song Dynasty (960 AD – 1279 AD), until α-brass mirrors with about 20% zinc content appeared. Some Qing Dynasty mirrors (1644 – 1911 AD) can have zinc contents even higher than this, above 40% zinc, which places them in the β-brass category of the copper-zinc alloy system, meaning that the alloy is a two-phase mixture of grains of β in a matrix of α. One such mirror made of β-brass is in the collections of the Victoria and Albert Museum ((Kerr 1990: plate 89, p103). The presence of zinc in ancient mirror alloys is potentially useful in establishing a *terminus post quem*, the earliest point in time at which zinc may have been used in the manufacture of Chinese bronze mirrors, which according to the analyses carried out on the Carter collection by Christman (1995:358), would have been during the later Tang Period, around 900 AD, although zinc does not become common until sometime around 1100-1200 AD. Ming Dynasty (1368 -1644 AD) examples from the Carter collection were found with over 10% of zinc (Carter Collection 1995: 370), while several from the Jin (1115 -1234 AD) and Song (960 -1279 AD) contained over 20% of lead (Carter Collection 1995: 381, 382, 383, 384). The increase in the percentage of lead would have cut the cost of production while impairing the quality of the polished surface of the mirror.

The presence of substantial amounts of zinc during the Tang Dynasty does not accord with the chronology of zinc use proposed by Zhou (2007), and these Tang Dynasty mirrors with high zinc contents may well be later reproductions. With rising tin contents, the alloy becomes harder, sonorous, but increasingly brittle, which is why so many mirrors have been broken and subsequently restored: the eutectoid phase has practically no ductility. Over the period from the Warring States to the Song, the composition for the high-quality mirrors changes relatively little, usually from 69 – 72 % copper, 22 – 25% tin and 3- 6% lead. By referring to the ternary phase diagram for the copper-tin-lead alloys we can see that the liquidus temperatures of these alloys is in the range from 800- 850 °C, which greatly facilitates the casting process.

Remarkably, very similar compositions were used in the Roman period for Roman bronze mirrors, one in the British Museum having 69.5% copper, 22.8% tin and 6.8% lead (GR 1814, 7-4, 1063, Meeks 1993: 66) and a Ptolemaic ring, also in the British Museum, was cast in an alloy of: 69.2% copper, 24.6% tin and 4.5% lead (GR 1930, 7-15.3, Meeks 1993: 66). Looking at the ternary alloy phase diagram, we can see that increasing the lead content and decreasing the amount of tin, raises the liquidus of the resulting alloy, and with increasing lead concentrations, not only is there a danger that pools of the immiscible lead would form in the cast bronze alloy rather than being finely dispersed, but the resulting colour tends towards a silver-grey hue rather than a silver-white.

On the other hand, increasing the amount of tin to around 40% and reducing the percentage of copper while leaving the lead content at around 10%, produces lower melting point alloys only at great expense in terms of tin consumption, increases the brittleness of the alloy still further, and creates no improvement in the colour of the resulting alloy. Reducing the tin content to lower levels, such as 12%, produces an alloy of golden yellow colour, and the whiteness is lost. We can therefore explain why ancient metalsmiths arrived at the compositional parameters for their high-quality ternary alloys: it was the one perfectly adapted for making a silver-coloured bronze mirror.

In terms of lead, tin and copper contents, the overall analytical spread of the Cotsen mirrors is shown on the ternary Cu-Sn-Pb diagram in Figure 10.9. The cluster of green dots towards the 25% tin area on the right hand side of the diagram, is in agreement with previous analytical studies.

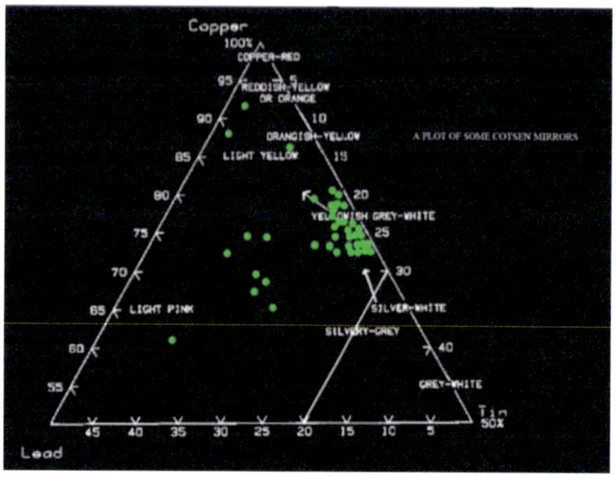

Figure 10.9. Plot of ternary compositions of the Cotsen mirrors. Most of the mirrors conform to the classic compositional field where the majority of green dots cluster. The spread of other green dots across the diagram are generally from mirrors of later date, very early date, or unusual composition.

After the Song Dynasty, the impression method was sometimes used, in which a mould was made by pressing a finished mirror into soft clay before firing the clay. This method is said to have been widely used during the Song Dynasty (960-1279 AD), the Jin Dynasty (1115-1234 AD), to make usable reproductions of classical mirrors and, in Japan, to produce copies of Chinese mirrors. A possible example of a later copy of a Han original is the Cotsen mirror, 0-0233, which contains a little zinc, but which is otherwise perfectly acceptable for being old.

Figure 10.10 Tang Dynasty mirror from the Cotsen Collection, 0-0234, 7th-8th century AD, with a perfect example of the "quicksilver white" patina. Photograph by courtesy of the Lloyd Cotsen Corporation

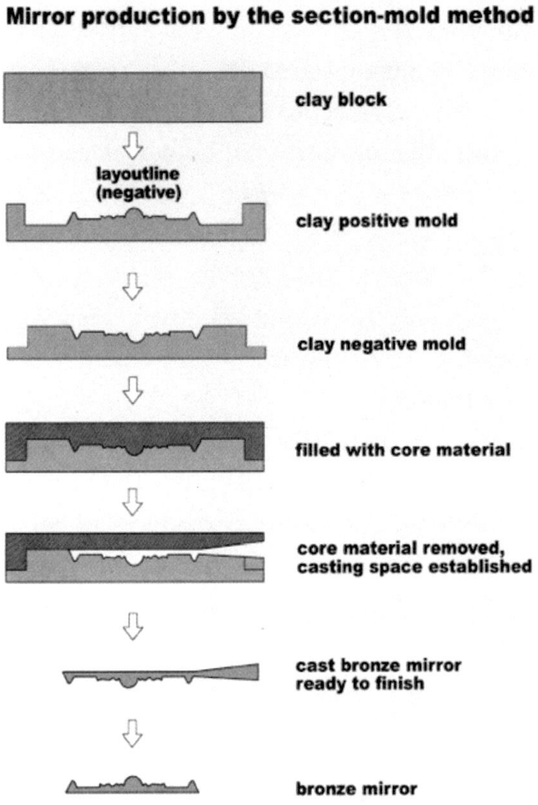

Figure 10.11 Ceramic piece-mold technique of casting Chinese bronze mirrors (After Murakami et al 2004).

According to the *Duo Neng Bi Shi* written by Liu Ji in the Ming period, this *Xuan Xi* consists of 1 part of tin, 1 part mercury, 6 parts alum and 1 part of deerhorn ash (Zhu and He 1993: 57). The original colour of *Xuan Xi* is a grayish-white powder which gradually changes to black. Experiments by Chen et al (1987) showed that after treatment with this mixture, a thin silvery surface is formed on the bronze. Exactly how this works has been clarified by investigations by Meeks (1988a,1988b; 1993 a,1993b). When mercury and tin are mixed together, a very sluggish reaction occurs at room temperature, and the mixture must be heated to form a mercury-tin amalgam; heating at 200 – 250 °C produces a pasty γ-phase $HgSn_6$. The γ phase decomposes at 214 °C, and mercury can be driven off by heating above 357 °C with tin diffusion into the substrate. Further details of the binary mercury-tin equilibrium diagram are explained in Meeks (1993b).

A thin layer of ν intermetallic compound develops at the outer surface over a thick layer of ε intermetallic compound; if the surface of the object is heated above 520 °C, a layer of $\alpha + \delta$ eutectoid phase develops under this layer and is contiguous with the copper substrate. Very little evidence exists to support the use of tinned surfaces made by applying a mercury-tin amalgam to the mirrors, and this applies equally to the mirrors of the Cotsen collection. Meeks (1993a) discovered, however, that a cold, mechanical application of a mercury-tin amalgam to the polished surface of a 25% tin bronze did produce a subtle surface effect. The treatment involved rubbing the polished bronze mirror with a small crushed pellet of the amalgam. After two minutes, the treated surface appeared more noticeably silver coloured. Optical microscopy revealed that a silver-coloured film obscured all but a faint trace of the morphology of the underlying eutectoid microstructure. Examination of the surface with a scanning electron microscope confirmed that this layer is very thin, perhaps 10 – 100 nm, and appears to consist of tin with some mercury. Meeks thinks that this extremely thin coating may result from localized heating of the amalgam when it is rubbed onto the surface of the mirror; the heating breaks down the gamma compound into its constituent parts, allowing the tin to bond to the surface. If that is the case, it is possible that mercury could have been slowly lost by volatilization over the centuries, although usually in metallurgical processes we find evidence for mercury within surface layers. Thus, *Xuan Xi* could have been used to enhance the silvery finish of the Chinese mirrors as the Marquis claimed over 2000 years ago. Chemical treatments could also be used to improve the silver-colour: vinegar was used, for example, and this would selectively dissolve a little copper from the outer surface of the mirror, increasing the silvery colour as the percentage of the δ phase would be increased at the surface.

It is the black-surfaced Chinese mirrors which have created the most controversy as to the origin of the black colour, and this argument has been going on for the past one hundred and forty years at least, as it is difficult to adduce sufficient evidence to decide whether it was a deliberate patina created when the mirror was made, or a type of corrosion known to afflict high-tin bronze alloys in which the tin-oxide enriched surface assumes a black colour from the effects of burial.

Attention to these black-surfaced mirrors was first drawn by Morin in 1874. Many excavated mirrors retain a silvery-coloured surface which is simply the colour of the polished metal. During corrosion, the alpha-phase solid solution and the alpha component of the eutectoid present the common mode of attack, frequently engulfing the lead globules in the metal, to a depth that varies from 25 to 200 um. The delta phase remains uncorroded and retains or imparts a silver-coloured surface appearance.

The black-surfaced mirrors show the same kind of corroded surface with the important difference that the delta phase of the eutectoid is also corroded, often to a depth of only a few micrometers, but this is sufficient to produce a black colouration. Soil minerals are absorbed into the corroded surface layer, which, in addition to some cassiterite or stannous oxides, may include silicon, aluminium, iron and sulphur.

Collins (1934) found a Han dynasty mirror that was only partially buried in the typical Chinese loess soil. The buried part had a blue-black patina while the exposed part remained silver-coloured. Analysis of the typical black patina by Taube et al (1997) found it to be composed of $Sn_{1-x}(Cu,Fe,Pb,Si)_xO_2$, while recent studies by Wang et al (1995) suggest a nanocrystalline SnO_2. Meeks (1988a & b; 1993a & b) and Zhu and He (1993) are of the opinion that all occurrences of the black patinated bronzes can be accounted for by natural phenomena during burial. On the other hand, Chase (1977, 1983), Chase and

Franklin (1979), Chase and Wang (1997), and Wang et al (1989, 1995) think that many of these patinas are man-made at the time of manufacture to produce a deliberate black-surfaced mirror. It may be possible that both suggestions are correct, as we know that burial can produce a black surface, but perhaps this could also have been artificially created as well, but since the end results are so similar, it will be hard to resolve the question.

Corrosion and Patina.

The corrosion of tin bronzes usually results in an enrichment in tin and loss of copper from the surface, creating a corroded surface of a variety of colours, ranging from the red of cuprite, the green of malachite, the deep blue of azurite, to the black of the largely amorphous tin oxides which make up the remarkable black surfaced mirrors discussed above. The cast leaded bronze alloy consists of a multi-phase microstructure consisting of alpha phase islands in a matrix of the $\alpha+\delta$ eutectoid phase with small lead globules interspersed in the matrix. In the process of corrosion, the α phase tends to be removed and converted during corrosion to copper salts which may be partially leached away. The small globules of lead may similarly be corroded to basic lead carbonates or oxides. The tin content becomes converted to one of the tin oxides. The dendritic structure of this mirror is seen in the surface, suggesting that this mirror is not made in the classic high-tin alloy and is yellow in colour, a fact confirmed by XRF analysis which found only 18% of tin, and by visual observation as the colour of the exposed metal is yellow rather than white. The lead content of this alloy is also very low.

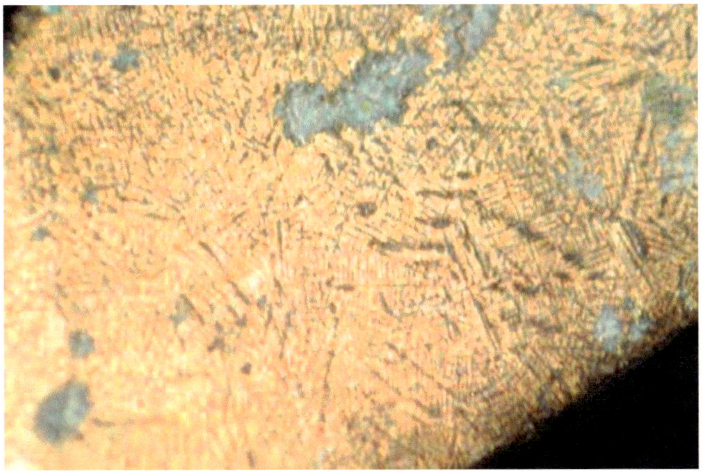

Figure 10.12 Photomicrograph of the surface of Cotsen mirror 0-0648 showing dendritic morphology of the cast structure, indicative of a lower tin content alloy, confirmed by XRF analysis.

Figure 10.13 Mirror 0-0720. Warring States with etched surface areas of the central design. Photograph by courtesy of the Lloyd Cotsen Corporation.

Figure 10.14. Surface photomicrograph of the unusual patina on mirror 0-0421 which is flaking away from the surface revealing a grey-purple underlayer. Magnification x 55.

Taken in conjunction with the observation of the surface bubbles, the unsatisfactory patina of these mirrors makes the argument conclusive. As part of a scientific evaluation of the evidence from a metallurgical perspective, a cross-section was cut from the edge of mirror 0-0421 and examined using the metallurgical microscope and the electron microprobe. The resulting optical photomicrograph is shown in Figure 10.14, in which the extent of penetration of the patina into the remaining sound metallic grains can be seen to be very superficial.

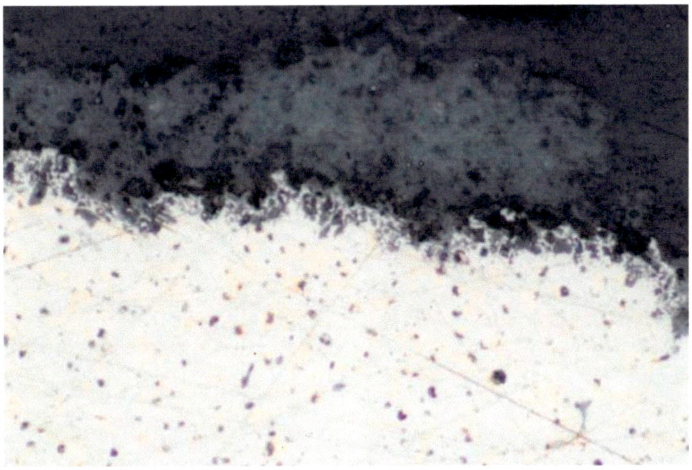

Figure 10.15 Photomicrograph of a cross-section through mirror 0-0421, revealing very superficial penetration of the leaded bronze alloy of only a few microns in depth. The small spots in the unetched section are of lead, and the alpha + delta eutectoid has very limited corrosion indeed. There is no cuprite present and the rather fluffy-looking corrosion crust which can be seen has no penetration into the metal at all and lacks the appearance of a genuine corrosion or patina. Photomicrograph x 400, unetched.

The conclusion that the patina was not consistent with the presumed age of the bronze was confirmed by an examination of the polished section under polarized light, shown in Figure 10.14 in which a thin, yellowish copper oxide corrosion crust can just be seen under the light green corrosion crust.

Figure 10.16. Photomicrograph under crossed polars of the cross-section of 0-0421, showing that the patina is of recent origin. Magnification x 400. Unetched.

These thin, yellowish cuprite crusts adjacent to the metallic matrix are an indication that the cuprite is not especially crystalline and has grown as a result of superficial corrosion which has been chemically induced.

Figure 10.17 Tang Dynasty Mirror 0-0753.Photgraph by courtesy of the Lloyd Cotsen Corporation.

The sample taken from one edge preserves part of the original surface of the mirror in the corroded matrix as shown in Figure 10.16. The thin blue-grey line separating the corroded interior and exterior corrosion products can clearly be seen. The same view under crossed polar illumination reveals the green, outer corrosion products overlying the original surface and the red cuprite with the darker grey-black of the metallic matrix underneath. The lead inclusions and the alpha + delta eutectoid phase have been preferentially attacked here, leaving isolated islands of the alpha phase, showing that this mirror contains less tin content than the two other Tang examples examined here. During the process of corrosion, the outer region of the microstructure cracked, allowing septae of cuprite to then invade the already corroded matrix. The retention of the alpha phase and corrosion of the delta phase is less commonly seen in these Chinese mirrors, but that is what has occurred here, and it is a consequence of the different environmental conditions to which the mirror is exposed during burial. The texture of this mirror was further investigated with the electron microprobe, a backscattered electron view being shown in Figure 10.15. The bright white spots are lead globules, the jagged light grey regions are the alpha-phase and the darker, granular background is the eutectoid mixture.

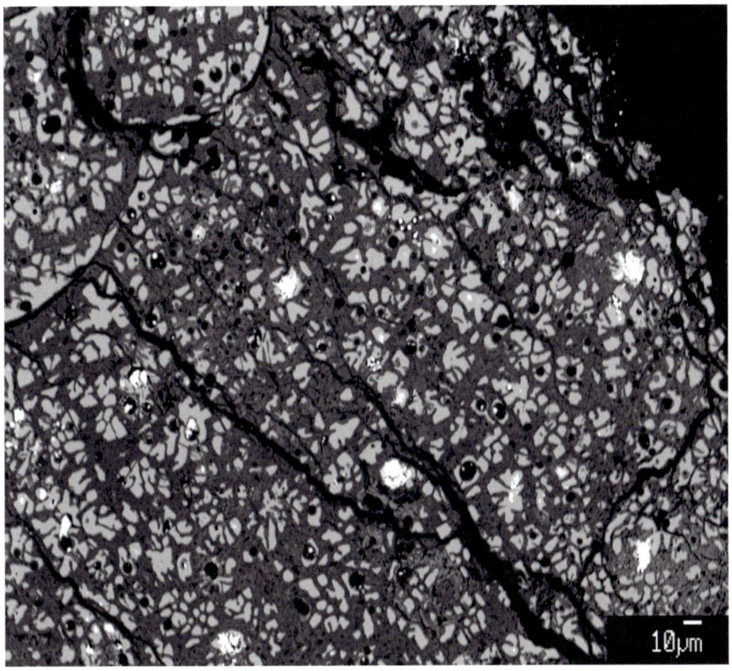

Figure 10.18 Mirror 0-0753 7th-10th century AD, Tang, made in a ternary copper-tin-lead alloy shows an unusual amount of internal corrosion which penetrates 300 microns into the metal, corroding the alpha + delta eutectoid completely and leaving unusual islands of alpha phase behind. The bright white spots are larger lead globules. Electron microprobe backscattered electron image, small bar represents 10 microns.

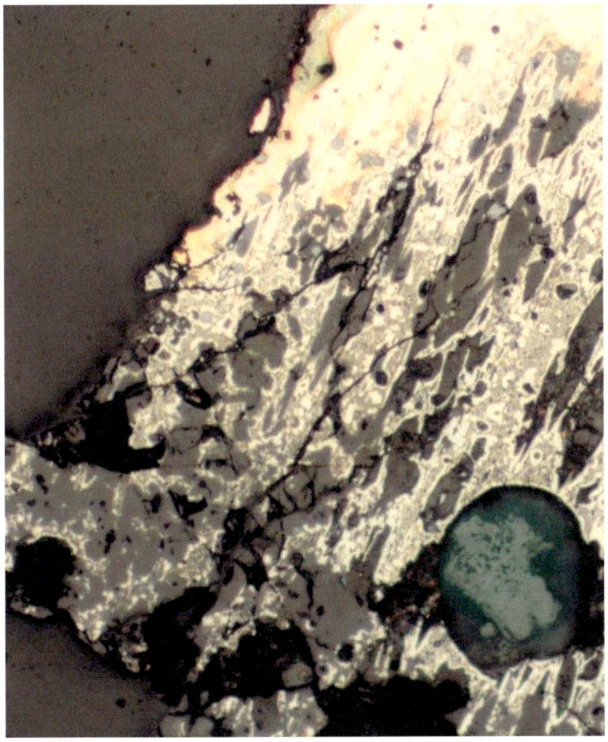

Figure 10.19. Photomicrograph of a black-patinated mirror from the Eastern Han, 0-0233, revealing extensive corrosion and loss of the alpha phase. Lead globules are large and also corroded away. Unetched, x200.

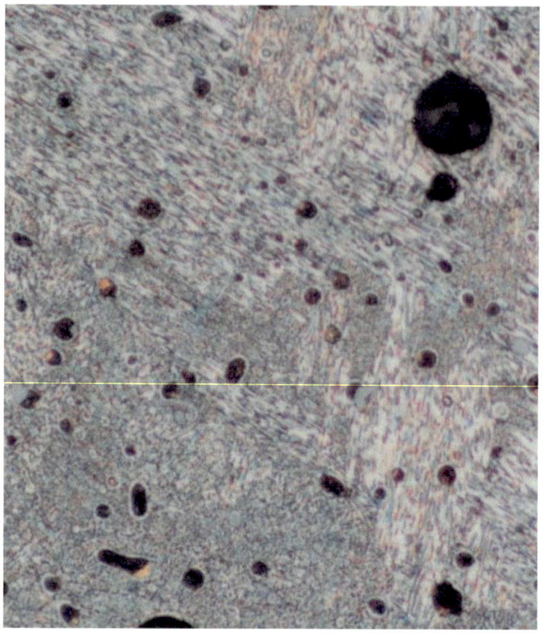

Figure 10.20 A typical Widmanstätten cast leaded bronze alloy with 22% tin, 4% lead, and 74% of copper. Etched in alcholic ferric chloride x470.

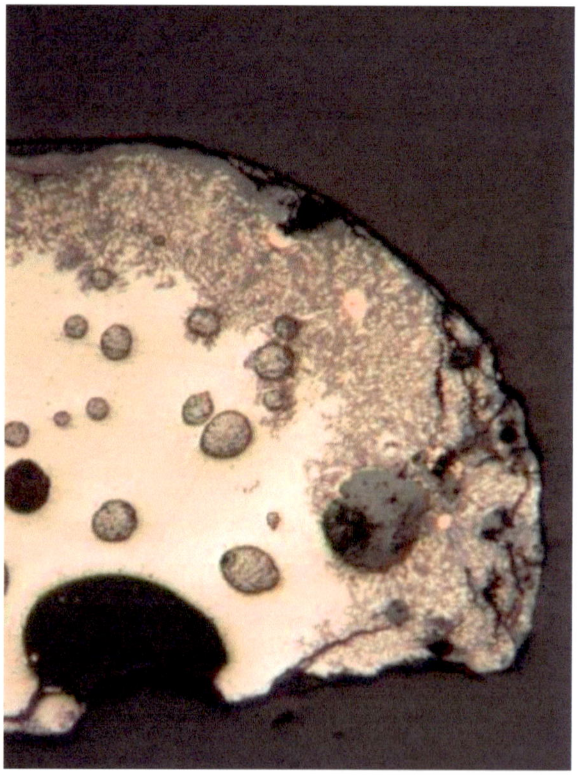

Figure 10.21 A Roman cast copper-tin-lead alloy with 7% lead, 22% tin and 71% copper, cast in the horizontal mode shown by this photomicrograph so that the lead content has had time to agglomerate and sink to some extent to the bottom of the mold. Direct metallographic evidence of the casting technology of this mirror. Unetched, Magnification x 210.

Chapter 11: Alloys of Copper and Zinc: Brass

Chapter 11. Frontispiece. Standard Metallurgical Alloy B3, Set 4, which is a 55Cu 45Zn alloy chill cast.

Brass is an alloy which has been used for thousands of years and it has been a difficult story to unravel all of its occurrences in different cultures at different times. This is partially due to the fact that zinc production as a metallic element in its own right is difficult to achieve by normal smelting activities, as zinc metal boils at a temperature below that needed for its extraction. Alloys with small amounts of zinc are often problematic in early cultural settings and demand careful analysis and further archaeological investigation.

Development of brass

Brass, the alloy of copper and zinc, was restricted in early usage, because to smelt zinc, the temperature used must exceed the boiling point of the metal. By co-smelting zinc and copper ores, brass did become important during the Roman period, and the a brass coinage was being produced from 45 BC. The early history of developments in metal mining and production is the subject of detailed reviews by Craddock (1995), and Craddock and Eckstein ; smelting and alloying are discussed by Tylecote (1976).

Figure 11.1 Casting a brass Shiva in Southern India.

Figure 11.2. Indian craftsman shaping a vessel.

Bidri vessels have been manufactured in India from at least the 17th century until the present day, and are known to be made of a zinc alloy, decorated with inlays of silver, brass or, occasionally, gold. The inlays contrast strikingly with the matt black patina of the vessel. It is this patina, easily damaged by standard cleaning techniques used on the silver inlays, which prompted an investigation by LaNiece and Martin (1987). A number of vessels were analyzed, and all were found to be alloys of zinc with between 2% and 10% copper (i.e. falling within the epsilon + eta phase field of the binary copper-zinc equilibrium diagram). Some have lead, tin, or iron also present in the alloy. The compound or compounds forming the patina have not been identified. However, x-ray fluorescence (XRF) analysis and replication experiments have proved that the copper content of the alloy is critical for the patina formation. The black patina forms instantaneously on immersion of the alloy in a warm solution of potassium nitrate (one part) and ammonium chloride (four parts). Standard brasses, as used for the inlay, are unaffected by the solution which induces only a whitish-gray patina on pure zinc. The ammonium chloride preferentially dissolves the zinc from the Bidri alloy. The rest of the process is unproven but it is suggested that the resulting copper-enriched surface is oxidized by potassium nitrate.

Early brass

Extraordinary early brass objects have been found at Taxila in Pakistan. Brass coinage began to appear in Bithynia and Phrygia in southwest Asia from the first century BC. The colour of brass containing 10–20% zinc is very much like gold, and the presence of zinc confers hardness and strength much as arsenic and tin provide to bronze. Brass was especially popular in India, where it has been in continuous use for two thousand years for temple roofs, furniture, and cooking and storage vessels (Lambert 1997). By the Renaissance, many objects identified as bronze were, in fact, cast from leaded brass alloys, which continues to be used.

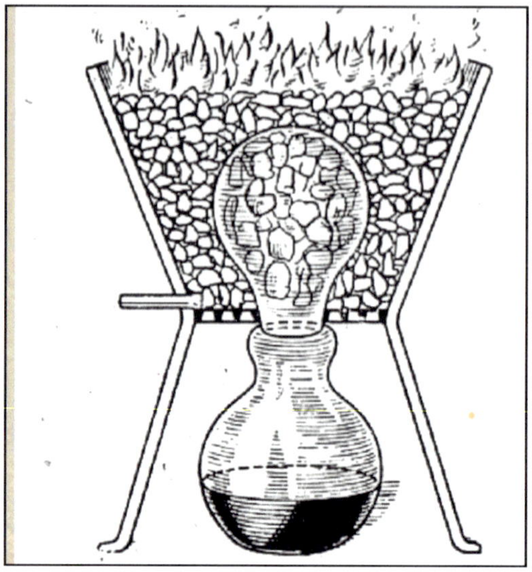

Figure 11.3 The melting point of zinc is 420 °C but more critical is the boiling point at 950 °C, below the temperature needed for its extraction. Copper-zinc alloys can be made by smelting zinc ores and copper ores mixed together. The figure here shows one version of the Hindu metallurgical process for making metallic zinc. The co-smelting of zinc ores and copper ores can be used to make brasses with up to about 28% zinc content, although they are usually lower than that. The drawing shows how Indian metalsmiths may have made metallic zinc by downward distillation of zinc vapour. Confirmed as correct by the research carried out by Craddock (1998).

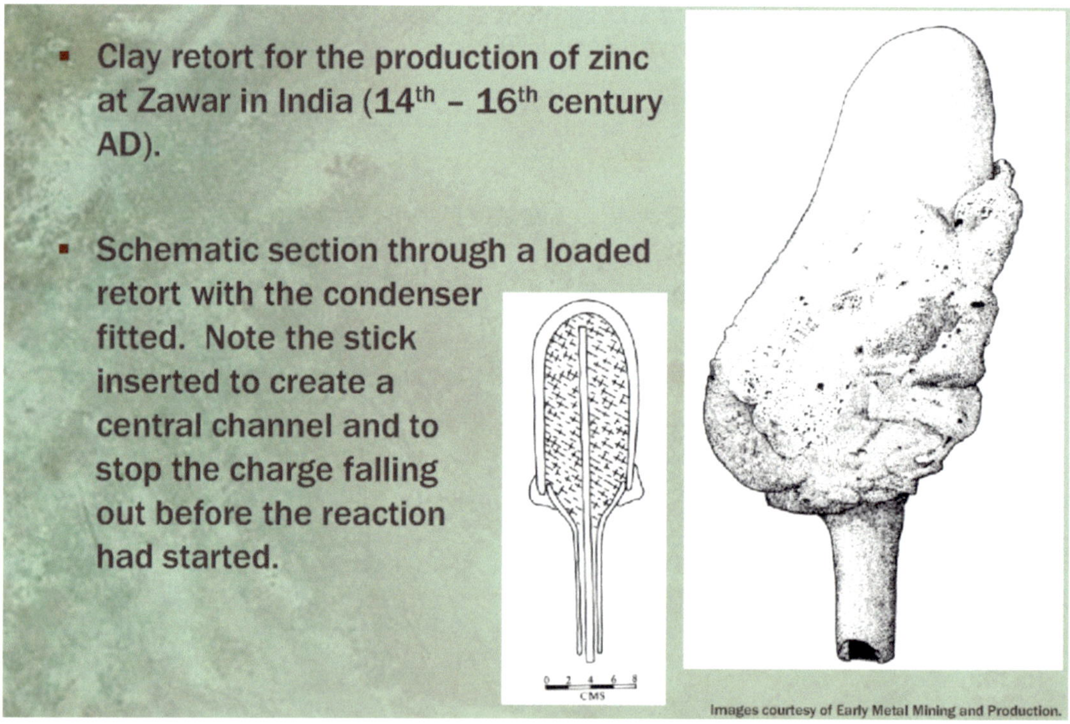

Figure 11.4 The Indian crucibles used at Zawar for the production of metallic zinc. After Craddock (1998)

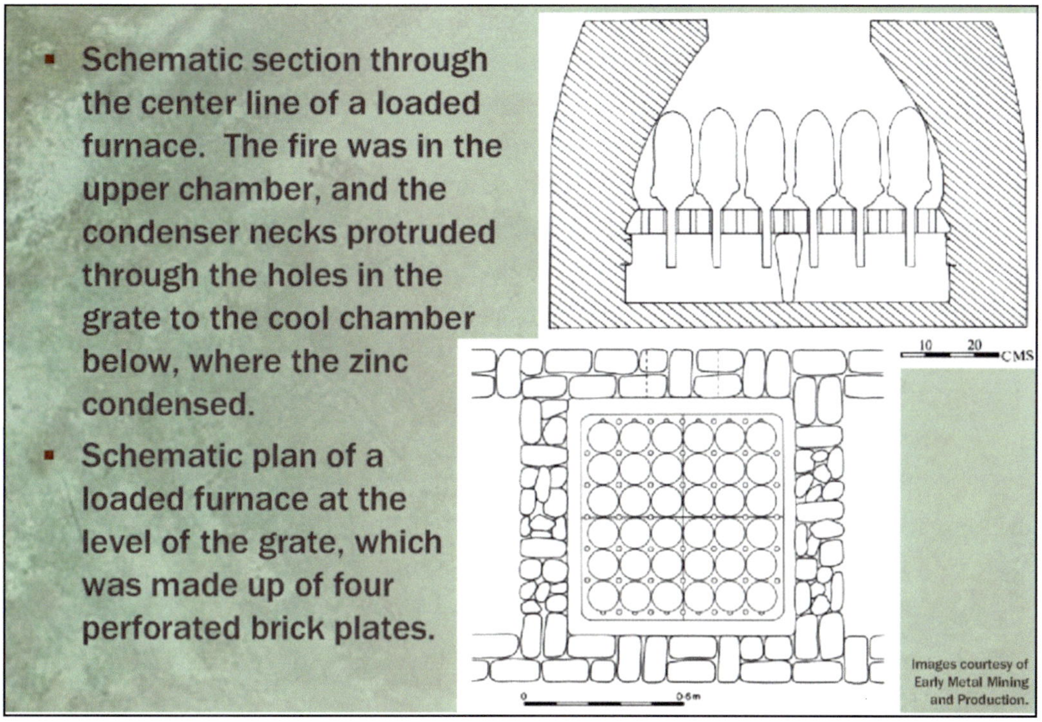

Figure 11.5 The multiple furnaces at Zawar were capable of producing 36 small crucibles of zinc in the protected chamber at the bottom at one firing. A remarkable feat of Indian metallurgy. After Craddock (1998)

The alloys of copper and zinc were highly valued for their golden colour and could be mistaken for gold. Those alloys that most counterfeited gold were given particular names, such as ormolu, prince's metal, Mannheim gold, or pinchbeck. Gold-coloured brass alloys could be beaten into thin foil or leaf, which is often called by the misnomer "bronze powder." If carefully lacquered and untarnished, such items could only be distinguished from gold by chemical analysis.

Early brass manufacture

Early metallurgical practice may have used the co-smelting of copper and zinc ores in which some of the zinc is absorbed into the copper, producing a brass by the *cementation process*. The approximate upper limit of zinc which can be absorbed in this process is 28%, still within the alpha solid solution region of the copper-zinc system.
Some of the most ancient finds of brass are from Nuzi in northern Iraq around 1350 BC. One had 14.4% zinc, 4.7% lead, 0.5% tin, 80.4% copper. Similar compositions with 12% zinc and 3% tin are known from Ugarit in Syria from 1400 BC (Schaeffer-Forrer et al 1982).

The Romans spread the use of brass through the European area, an alloy they called *aurichalum* or *orichalcum* (Caley 1964: 31). The earliest dated artefacts are a series of coins issued by Mithradates VI, the king of Pontus from about 75 - 65 BC. This brass was made either by reacting zinc ores with copper metal in a closed crucible (Cowell et al 2000), or by reacting zinc oxide with molten copper (Craddock and Eckstein 2003). The Roman emperor, Augustus began production of brass coinage at Rome around 23 BC and by the first century AD brass was in use for brooches, medical tools and military equipment (Bayley 1990; Bayley and Butcher 1995; Ponting 2002). Istenic and Smit (2008) carried out some analyses of Roman brooches from Slovenia, which showed that brass was already in use in the region by 60 BC, with contents of zinc ranging from 18% - 22%, a substantial zinc content. Strangely the amount of zinc in some the later Roman coins begins to diminish: during the 1st century AD, the content fell from about 27% in 45 BC to 15.9% in 79 AD and 7.8% in 161 AD while the contents of tin and lead increased (Caley 1964; Moesta 1993; Weisgerber 2008).

In China, the use of brass did not begin until the Late Tang period, around 900 AD, according to the analyses of mirrors in the Carter collection carried out by Christman (1995:358), although zinc does not become common until sometime around 1100-1200 AD.

Ming Dynasty (1368 -1644) examples from the Carter collection were found to contain over 10% of zinc (Chou 1995: 370), while several from the Song and Jin (1115-1234 AD) dynasties contained over 20% of lead (Chou 1995:381). The increase in the percentage of cheap lead would have cut the cost of production while impairing the quality of the polished surface of the mirror. But the presence of substantial amounts of zinc during the Tang dynasty does not accord with the chronology of zinc use proposed by Zhou Weirong (2007), and Tang-style mirrors with high zinc contents may well be later reproductions.

Whenever analysis detects small amounts of zinc in Chinese alloys of objects alleged to date much earlier than the Tang Period, this indicates that the objects in question must be either later copies or modern reproductions. It is true that very early finds of brass artifacts are reported from Late Neolithic China, which have no later parallels. A brass flake and brass tube were found from the Yangshao 仰韶 culture site at Jiangzhai 姜寨 in Lintong 臨潼 (Shaanxi), dated to 4100-3600 BC , and two fragments of brass cones from the Longshan culture site at Sanlihe 三里河, Jiao Xian 膠縣 (Shandong), dated to 2300-1800 BC (Zhou Weirong 2007: 182), but these anomalous developments should be seen, not as proof of an exceedingly early brass-working tradition in China, but rather as evidence for the lack of continuity between early experimental metallurgy and the culturally embedded concept of alloying which followed in later periods, during which the technology of fabrication of copper-zinc alloys, by all indications, was completely lost. In later antiquity up to the Tang dynasty there is no evidence whatever for the use of zinc in China. According to Zhou (2007: 182), the earliest indigenous brass began in the 10th century CE, during the transition from the Five Dynasties (CE 907-960) to the Northern Song. From the 11th century CE onwards, brass became better known. By the Jiajing 嘉靖 period (1521-1567) of the Ming dynasty, the zinc content of brass was from 10 to 20%, and by the Wanli 萬▪ period (1573-1620) it had risen to over 30%, often with no tin content

Indian metallic zinc

The work carried out by Craddock (1998) at the important Indian zinc smelting site of Zawar is now well-known. One of the ways in which metallic zinc could be produced, rather than relying on the cementation process to produce in situ copper-zinc alloys, was to invent a process which collected zinc vapour by downward distillation. The Indian process relied on being able to hold the reaction vessel upside down without the ore and charge falling out! Sticky balls of reactants were made up and a very clever internal tube ran through the centre of the flask as shown in Figure 11.4. This would allow the zinc vapour to escape and the fumes to be condensed in the cooler part of the furnace shown in Figure 11.5. The multiple collection of small inverted furnaces were placed above a perforated ceramic plate into which the nozzles rested, while collection cups could be placed in the lower area to collect the zinc. How this kind of alchemical process evolved is not known, but it represents one of the triumphs of ancient Indian metallurgy, at a time when zinc metal was unknown in the western world. Later on, after the 12th century AD, the Chinese began to produce zinc too and knowledge of the new metal gradually spread so that by the 15th century AD brass was becoming quite common. Our problems are with the early use of brass in different cultures, which is now an interesting area of research as the complexities of the early use of brass are increasingly recognized due to careful excavation and reliable analysis of early finds. The copper-zinc equilibrium diagram, shown in Figure 11.6, possesses a large region in which the α phase exists, where copper and zinc are mutually soluble in each other. This is the region of the alpha brasses, which accounts for most of the known occurrences in antiquity. The limit of solid solubility is about 28% of zinc at room temperature, although the diagram shows that the curve of the solid solubility is greater at around 400 ^{o}C than it is at higher or lower temperatures. Beyond 28% of zinc content the alloy can become a two-phase system, with the β phase precipitating out in conjunction with the alpha phase, although the standard limit of the alpha solid solution is up to 35% of zinc.

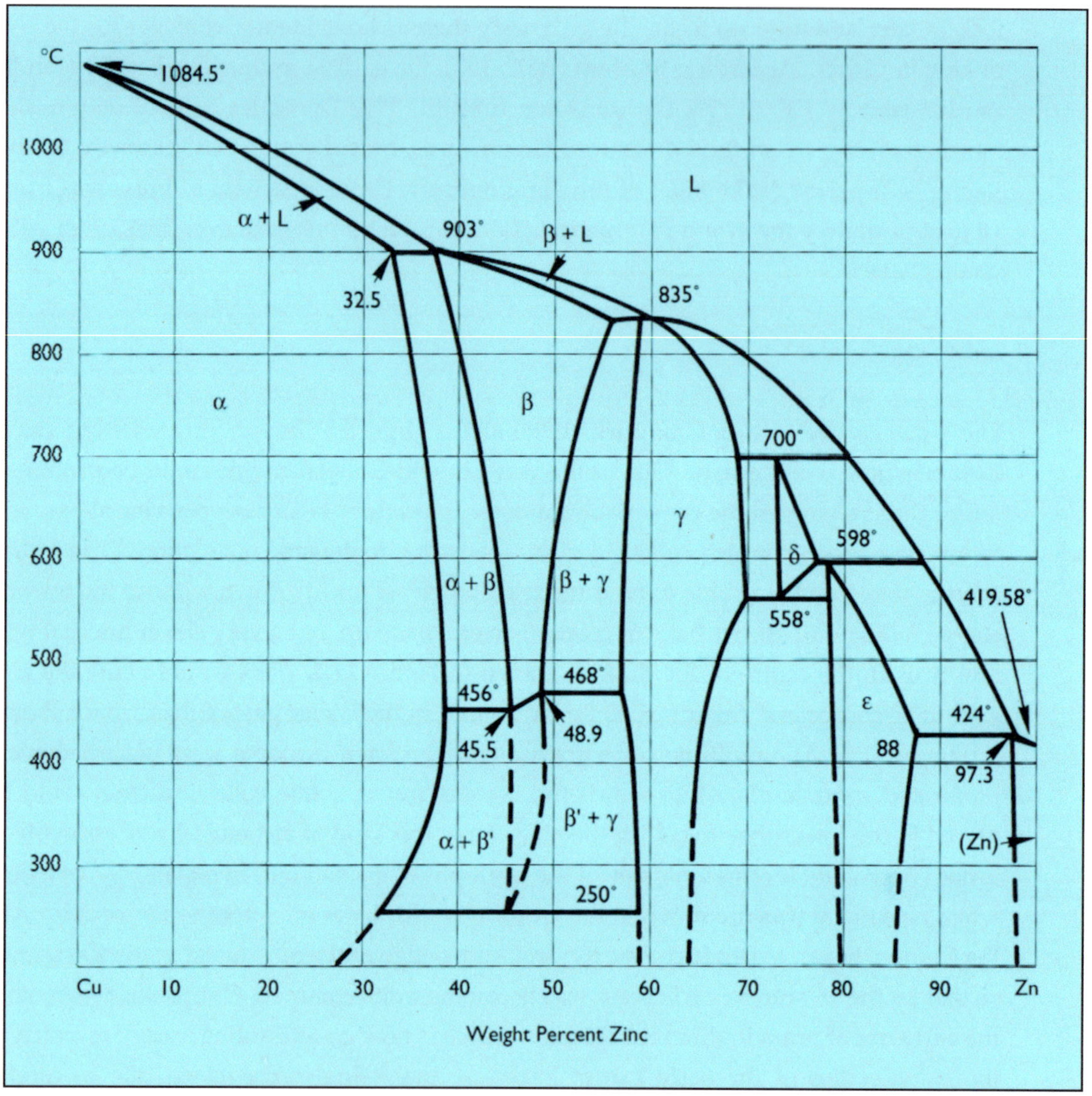

Figure 11.6. The copper-zinc system. After Scott (1991) Most ancient alloys are less than 28% zinc. Indian bidri alloys may be eta phased as they have over 90% zinc content. Historical period alloys may be of alpha + beta or beta composition. The beta + gamma alloys are rarely used as too brittle.

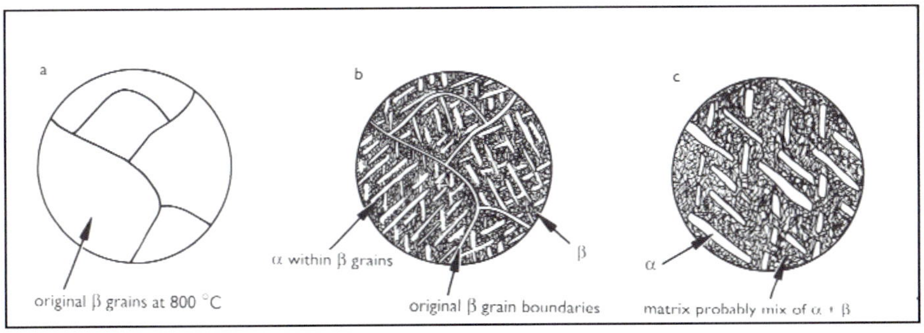

Figure 11.7 Microstructure of the precipitation of alpha and beta phases in brass alloys with about 45% zinc content. After Scott (1991). The phases are precipitated at grain boundaries or in orientated precipitation within the grains

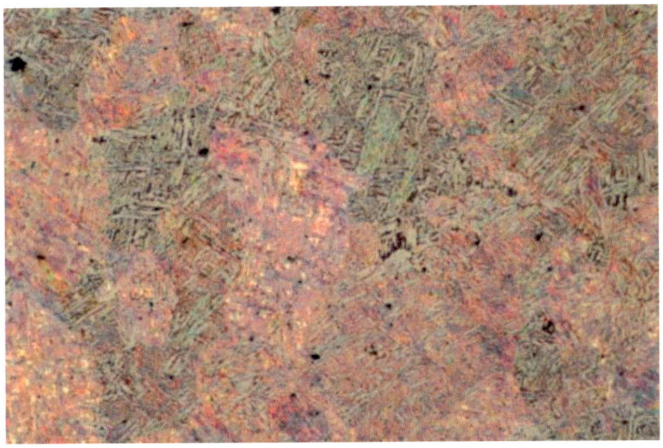

Figure 11.8 Laboratory Alloy. Annotated Metallographic Specimens Sample BU13, a 40/60 brass, that is 40% zinc, 60% copper, cast ingot, showing well-developed Widmanstätten precipitation of alpha and beta phases, from the decomposition of the original beta grains. Etched in ammonium persulphate pre-etch for 10 seconds followed by 180 second etch in Klemm's number 1 reagent, viewed under partially polarized illumination,

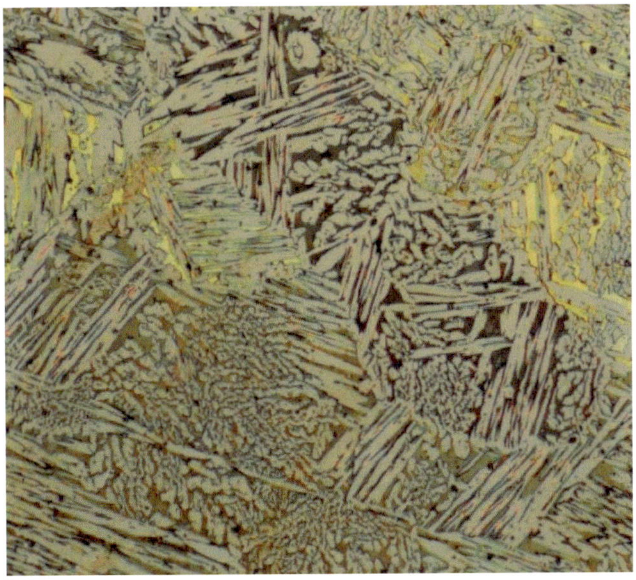

Figure 11.9 Laboratory Alloy. Annotated Metallographic Specimens Sample BU13, a 40/60 brass, that is 40% zinc, 60% copper, cast ingot, showing well-developed Widmanstätten precipitation of alpha and beta phases. Etched in ammonium persulphate pre-etch for 10 seconds followed by 180 second etch in Klemm's number 1 reagent, viewed under partially polarized illumination, x200.

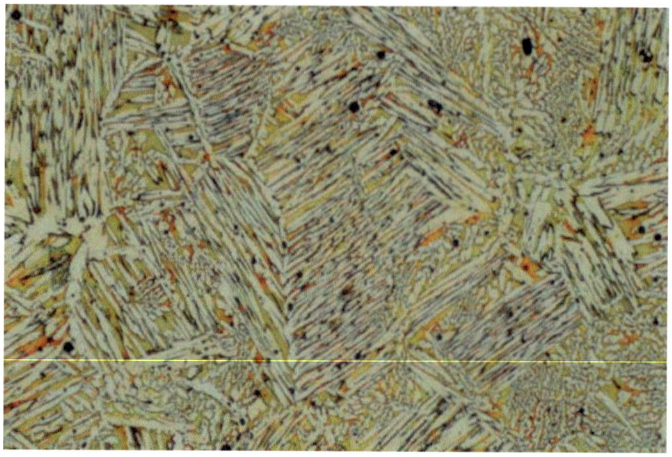

Figure 11.10 Laboratory Alloy. Annotated Metallographic Specimens Sample BU13, a 40/60 brass, that is 40% zinc, 60% copper, cast ingot, showing well-developed Widmanstätten precipitation of α and β phases. Etched in ammonium persulphate pre-etch for 10 seconds followed by 180 second etch in Klemm's number 1 reagent, viewed under partially polarized illumination, x400.

The α-β brasses, as they are called, are common in historic brass alloys as they have different colours and hardening properties than the α brasses; typical compositions range from 30 - 45% of zinc. Beyond 50% of zinc, the γ phase appears which severely impacts the ductility and workability of the higher-zinc brasses.

Alpha-phase alloys can be cold-worked and then annealed followed by further working as necessary. This phase has a face-centered cubic lattice, but the β alloys are body-centered cubic which is a less ductile lattice and considerably harder. α-β brass with 35%-45% zinc can be hot-worked only. An alloy of 58% copper 42% zinc is completely within the β field at high temperatures and becomes a mixture of $\alpha + \beta$ on cooling. This kind of alloy can also be worked hot and when cooled under different conditions and reheated to certain temperatures, a variety of microstructural textures are created. Some of these are shown below Figures 11.11, 11.12 and 11.13 and in Figures 11.14 and 11.15.

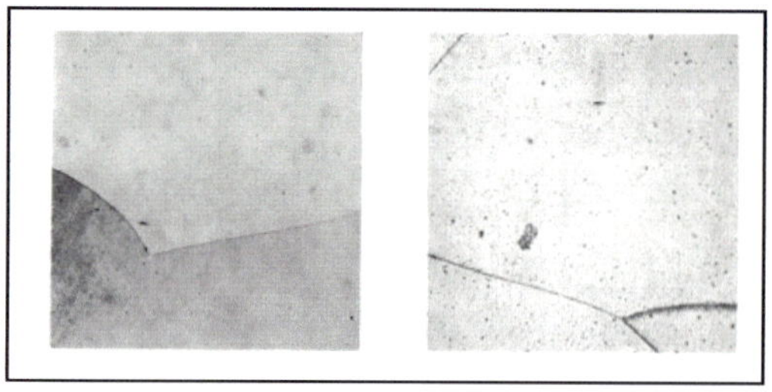

Figure 11.11. 58% Cu 42% Zn Left: as quenched from 800 °C. The structure is completely β solid solution. Right: quenched and reheated for 30 min at 200 °C. Very small particles of darker etched α start to form. x100, etched in ammonia/hydrogen peroxide. After Samans (1963:23

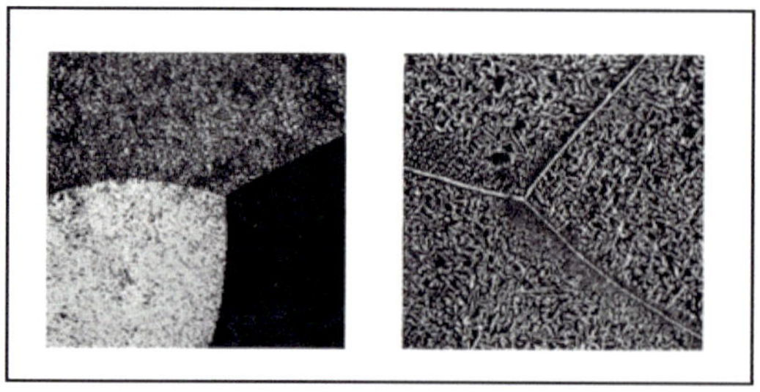

Figure 11.12. 58%Cu42%Zn Left: quenched and reheated for 30 minutes at 300 °C. Right: Quenched and reheated thirty minutes at 400 °C. Etched in ammonia/hydrogen peroxide. x100. After Samans (1963:238).

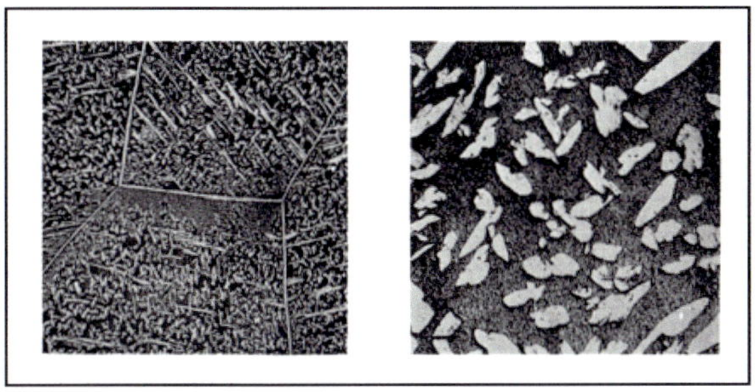

Figure 11.13. Left: 58%Cu48%Zn quenched and reheated for thirty minutes at 500 °C. Right: Quenched and reheated for 30 minutes at 600 °C. Etched in Grard's number 2 etchant. x100. (After Samans 1963: 238. Grard's etchant is essentially the same as Klemm's and uses saturated sodium thiosulphate with one gram potassium metabisulphite.)

The morphology of the precipitated β phase in Figure 11.13, left, is that of a typical Widmanstätten precipitation.

Figure 11.14 shows part of the phase diagram relevant to the above microstructures, with the compositional line of the 58% copper 42% zinc alloy marked. A selection of mechanical properties of the alloys up to this level of zinc are also illustrated.

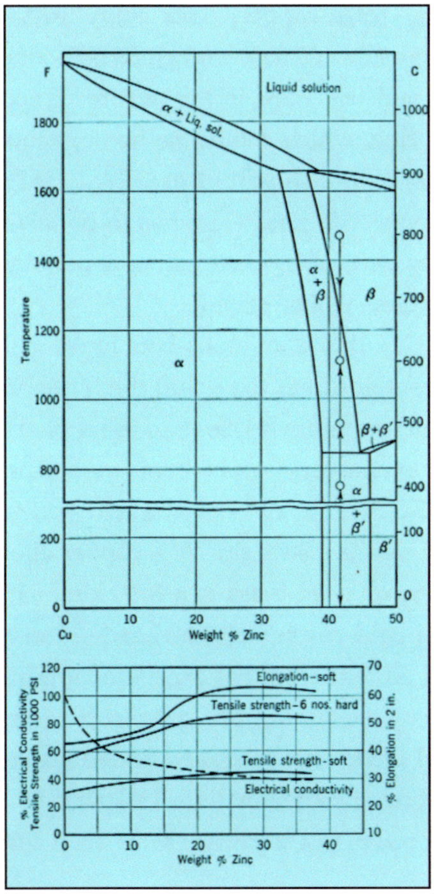

Figure 11.14 After Samans (1963) showing tensile strength in PSI for brass in different conditions. In the hardened condition the tensile strength can reach just over 80,000 PSI.

All of the α brasses have a similar microstructure, which tends towards rather large grains sometimes with dendritic ghosting within them.

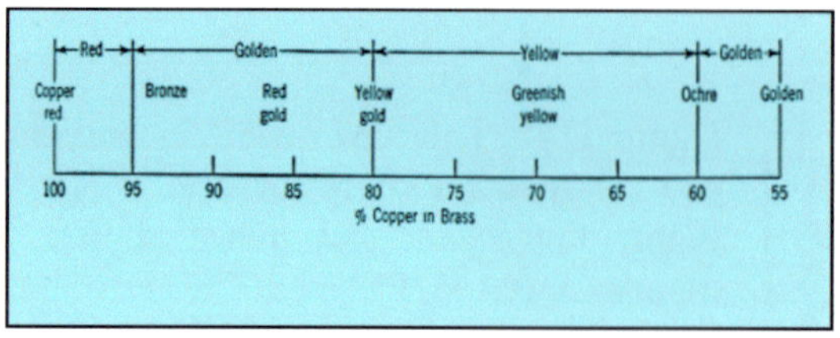

Figure 11.15. Some colour variations of binary brasses (After Samans 1963)

The colour of the brass alloys changes as more zinc is added as shown in Figure 11.15. At 5-10% of zinc, a "bronze" colour develops, which merges into a red-gold at 15% zinc and a yellow-gold at 20% zinc. From 25%-38% zinc the alloys are greenish-yellow and from 38%-45% zinc they are again reddish, passing through ochre into another series of golden hues.

Different brass alloys have been used for so many applications in the past, especially from the 17th-20th century, that a number of descriptive terms are employed in the literature, a selection of some are briefly discussed below.

Admiralty brass: An alloy of 30% zinc, 1% tin, 69% copper. This alloy resists dezincification in most environments. Aich's alloy: contains 60.66% copper, 36.58% zinc, 1.02% tin and 1.74% iron, has a golden appearance and was used in the past to protect ships. Alpha brasses: alloys with less than 35% of zinc, which should be homogenous and single-phased. Prince Rupert metal: named after Prince Rupert of the Rhine (1619-1682 AD), is a good yellow colour and has 75% copper 25% zinc. Care has to be taken with assuming that these compositions are accurately defined by their name: depending on which sources are used, slightly different compositions will be stated.

Muntz Metal, named after Frederick Muntz (1843-1897): this is an alpha-beta brass with 35%-45% zinc which can be hot-worked. The alloy contains both the α and the β' phase. The β' phase is body-centered cubic and much harder and more brittle than the α phase. We will examine more of Muntz metal later in this chapter. Alpha-beta brasses are usually worked hot. Beta brasses: have between 45-50% zinc, can only be hot worked and are harder and suitable for castings. Cartridge brass is a 30% zinc 70% copper alloy with good cold-working properties. Common brass, or rivet brass is a 37% zinc 63% copper alloy, and often used for cheap castings and cold-worked sheet. Leaded brass: this is an alpha-beta brass with a few percent of lead added which has excellent working and shaping properties. Low brass: a copper-zinc alloy containing 20% zinc with a light golden colour. Used for flexible metal hoses. Naval brass: this is similar to Admiralty brass, and is 40% zinc, 59% copper, 1% tin. Tombac brass: is an alloy with 15% zinc, 85% copper used for small decorative items. White brass: refers to alloys with more than 50% zinc. These are very brittle.

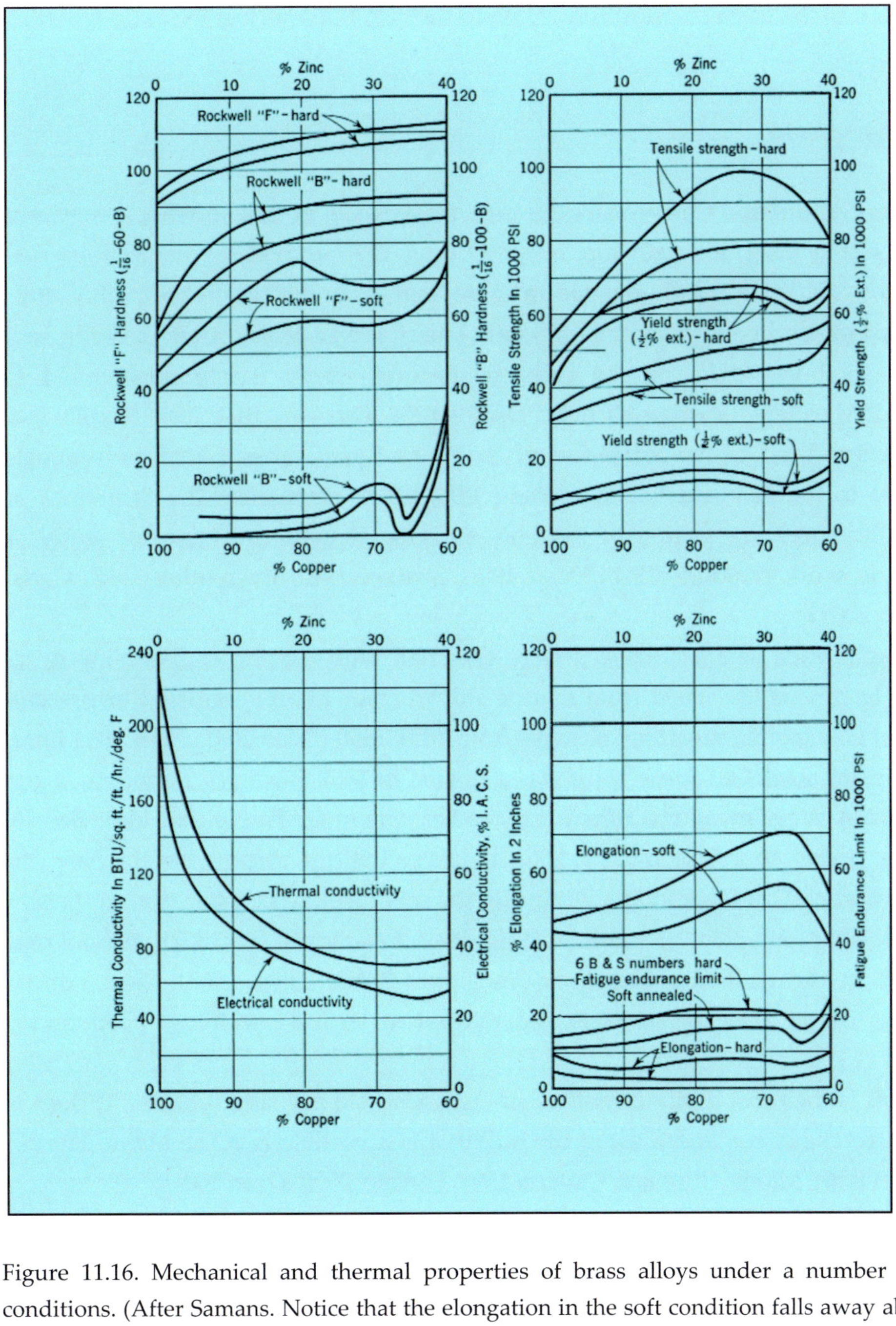

Figure 11.16. Mechanical and thermal properties of brass alloys under a number of different conditions. (After Samans. Notice that the elongation in the soft condition falls away above 30% of zinc).

Case Study Number 13. The Muntz Metal from the Shipwreck of the SnowSquall. An investigation of the brass compositions, close to the beta boundary of the brass system.

Historical Background

The brothers Alford and Cornelius Butler built *Snow Squall* in 1851 in what is now South Portland, Maine, USA. A small ship, only 742 tons, 157 feet register length, it carried a crew of 19 to 22. *Snow Squall* operated in trade, mainly to the West Coast, USA, and to China, until 1864, when it grounded off South America in Le Maire Strait. Leaking badly, the ship sailed to Stanley in the Falkland Islands, where it was condemned and abandoned. Maritime historian Nicholas Dean's 1979 discovery that *Snow Squall*'s lower hull still existed led to the formation of the *Snow Squall* Project. Five archaeological expeditions to the Falkland Islands during 1982-1987 documented the shipwreck and removed the surviving section of the bow to bring it back to Maine for study and conservation work (Yalouris 1988, Dean 1990, Switzer 1992, Bayreuther 1993, Carlson 1994, Dean 2001).

Wooden hulls such as *Snow Squall's* were sheathed with copper, copper-alloy or zinc sheathing to protect the wood from marine shipworms. Dean speculated or presented evidence of six possible sheathing changes from 1851-1860 (Dean 2001). The 1864 Bureau Veritas register notes that *Snow Squall* was sheathed in 1862, however, no month is given (*Registre* 1864). A review of the *Directory of the Shipping in the Port of New York, Brooklyn, etc.*, for the period after September 5, 1862, indicates that the ship was in the New York Sectional Floating Dry Dock at Pier 49 during the first week of October, and probably its last re-sheathing took place at that time (*Shipping & Commercial List* 1862). A final repair job is noted in May of 1863, in Singapore (Singapore Straits Times, 1863). Over a thirteen year sailing career *Snow Squall* likely experienced seven full sheathings (Carlson et al 2010).

Snow Squall is the most easily examined of American clipper ship sections. It does not survive intact, however, and most of the hull that is accessible is at the Maine Maritime Museum in Bath, Maine. (Rice and Carlson 1997, Switzer 2000, Dean 2001).

The condition of the bottom sheathing varies from intact plates to fragments retained only by lapping plates or fastenings. Many different coloured corrosion products and iron stains present on the sheathing and exposed underlying padding were analyzed by X-ray diffraction to inform conservation cleaning reported elsewhere. In 1858, I.R. Butts, author of *The Merchant's and Mechanic's Assistant* published tables that record Muntz metal sheathing plate thickness was gauged by weight; thus, twenty-four gauge was a plate weighing twenty-four ounces per square foot (7.32 kg/m^2) (Butts 1858). This was sufficient to last three years (Ronnberg 1980).

The sheathing plates were fastened to the hull with flat-head, cast copper alloy nails driven so their heads were flush with the surfaces of the plates. Beneath the sheathing there is padding made of loose fibers with thick black residue on the side that once adhered it to the outer planking (Carlson et al 2010).

Other ships with Muntz Metal examined

The wreck of the Tasmanian-built 195 ton bark *Petrel* (1847-1853) resurfaced on Hope Beach, South Arm, Tasmania in July of 2006 (Knott, 2008). *Petrel*'s bow sheathing bears double oval 'Muntz's 26 Patent' stamps. Muntz patent stamps refer to Birmingham, England's George Frederick Muntz Sr., the creator of the alloy called Muntz metal that was patented in 1832 (Flick 1975). Other known Muntz family patent stamp designs include double circles with the gauge in the center ring and the outer ring words 'Muntz's' and 'Patent' separated by '40' or '41' (McCarthy 2005, Stanbury 2009). The '40' and '41' numbers remain a mystery, possibly some internal Muntz company code (Callcut 2010). A piece of ship sheathing recovered from the Eighty Mile Beach in Western Australia bears a double oval stamp with three crowns and the words 'PH Muntz 1872' (Stanbury 2009). George Frederick Muntz's original 1832 patent No. 6325 for sheathing allowed for a mix of copper and zinc 'in any proportions between fifty percent of copper to fifty percent of zinc, and sixty-three percent copper to thirty-seven percent of zinc', but 'I prefer the allow to consist of about sixty percent of copper to forty percent of zinc.' (Bruzelius 2006).

At 40 percent zinc content, the alloy would be two-phased with grains of alpha phase and an infill of beta phase. Muntz rigorously defended his 60:40 brass alloy composition from piracy in a landmark 1844 court case and until his 1832 patent expired in 1846 (Flick 1975). X-ray fluorescence analysis was carried out on an unstamped degraded sheathing sample 'Port PO2' with associated nail sample 'Nail PO2', and a sheathing sample in good condition, 'Port 80E' with associated nail sample 'Nail 80E', to contrast their condition and to determine their composition. Some XRF data was obtained from corroded metallic unpolished face surfaces. Other results are from mounted, ground and polished cross section surfaces so that no gross corrosion effects of the XRF data occurred (Scott, in Carlson et al 2010).

Electron microprobe analysis (EPMA) of two sheathing samples each of 'Starboard 32C' and 'Port 80C', that directly bore the 'Muntz's 24 Patent' stamps were examined using a Joel Superprobe Electron Microprobe. *Snow Squall*'s sheathing padding was comprised of vegetable bast fibres, further characterized as jute, *Corchorus* sp., species undetermined. Shellac and a degraded tree resin that best matched common yellow resin or *Pinus sylvestris* were detected by FTIR. Sheathing paper, canvas and felt attached to ship planking by tar or pitch have been documented beneath sheathing (Staniforth 1985). Jute padding sized with shellac has not been previously reported. XRF results are reported in Table 11.1. The degraded 'Port PO2' sheathing surface samples may have copper and zinc loss with the result that some zinc is present in the corrosion products enhancing the detected zinc percentage. The good condition 'Port 80E' polished cross section subsurface sheathing sample 2.3 presented reliable composition data of 67.9% copper, 30.1% zinc, 0.8% lead and 0.39% iron. This is close in composition to a 70:30 copper:zinc alloy called cartridge brass and also suggests nergandin. Nergandin is a (Philip Henry) PH Muntz & Co., Ltd. Term for the alloy comprised of 70% copper, 28% zinc with 2% lead. It is listed in Woldmans Engineering Alloys, edited by John P. Frick, American Society of Metals, 8th edition (1994) 786, as being used for condensing tubes and being resistant to seawater corrosion. No patent or time period for this alloy has been located to date.

X-ray Fluorescence analysis ofSnow Squall's unstamped sheathing and nails									
Material tested	Elements Tested % (no detection for Bi, Ag, Sb, Mn)								Comments
	Iron	Nickel	Copper	Zinc	Arsenic	Lead	Tin	Total	
Unstamped Sheathing									
Good Condition 'Port 80E' sample									
2.1 face section	4.50	0.40	67.50	16.50		2.00		90.90	This is a surface sample. Ingress of iron corrosion products is present on the surface of the sample. Low zinc levels reflect corrosion and dezincification.
2.2 face section	3.20	0.20	68.50	18.00		1.00		90.90	This is a duplicate surface sample.
2.3 cross section	0.39	trace	67.90	30.10		0.80		99.20	This is a polished sub-surface sample unaffected by corrosion products and reflects the true composition of sample 'Port 80E'.
2.4 cross section	0.40		66.90	30.00		0.80		98.10	This is a duplicate polished sub-surface sample.
Degraded 'Port PO2' sample									
2.5 face section	0.40	1.20	59.80	37.00		0.80		99.20	This is a surface sample that is a very degraded alloy. Both copper and zinc loss may have occurred here, with the result that some zinc is present in the corrosion products, influencing the apparent zinc percentage.
2.6 face section	0.40	1.00	60.20	36.70		0.70		99.00	This is a duplicate surface sample.
Nails									
Nail PO2' sample									
2.7 head of nail	1.15		38.50	54.90		1.90		96.45	Heavily degraded, unreliable composition figures.
Nail 80E' sample									
2.8 head of nail	1.60	2.80	63.70	23.20	0.60	6.00	0.68	98.58	Alloy is closer to Admiralty brass due to the presence of arsenic and tin.

Table 11.1. XRF Results for the SnowSquall Muntz metals. After Carlson et al (2010).

Figure 11.17 Snow Squall, Sheathing 80E. Homogeneous alpha brass microstructure but with large grain size which would have seriously weakened the sheet. Etched in sodium thiosulphate and potassium metabisulphite under partially polarized illumination showing large twinned crystals and no second phase. X130

Two distinct compositions of nails are in evidence. 'Nail-PO2' contained copper, zinc, iron and lead. 'Nail-80E' contained the same elements plus nickel, arsenic and tin that suggests it might have been an Admiralty brass. Metallographic examination of the unstamped sheathing and associated nails revealed that both of the nail tips' microstructure were a cored dendritic cast morphology that was α phase, with no β phase precipitate present. Two samples of degraded 'Port PO2' sheathing and two samples of good condition 'Port 80E' sheathing were all α phase, confirming the impression from the analytical work that these are α phase brasses with a zinc content of about 30%. Binary copper-zinc alloys of 30% zinc content can absorb all of the zinc into the α solid solution, and therefore there is no secondary β phase.

Figure 11.18 Snow Squall 32 starboard, showing large grain-size alloy with straight twin lines and intergranular corrosion. In this case, corrosion may have removed small traces of the beta grains, but otherwise it is an alpha brass, not a typical Muntz microstructure of a 60-40 brass. Etched in alcoholic ferric chloride x 130.

Metallographic examination of the stamped sheathing samples 'Starboard 32C', 'Port 80C' and an additional 'Muntz's 24 Patent' stamped sample, 'Port 70C', revealed identical microstructures. Their structure is essentially an α phase brass but just at the boundary where the β crystals are beginning to form. The β phase is electrochemically anodic to the α phase in seawater, which results in the selective corrosion of the β phase, the weakening of the structure, and often the re-deposition of copper where the β phase once was. The weakness of a two-phase brass composition is revealed in the microstructure of stamped 'Port 70C' seen in Figure 11.18 and 11.19.

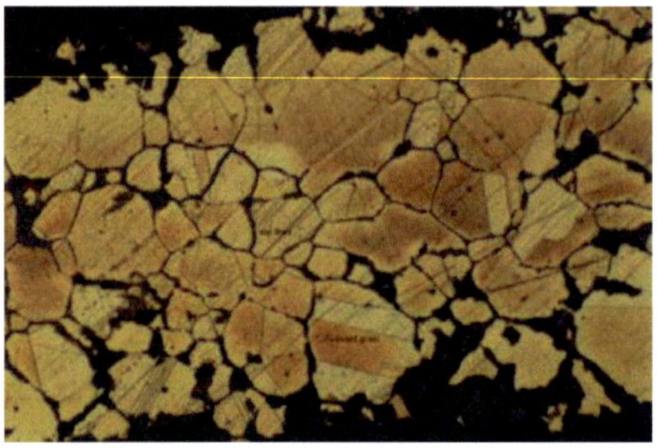

Figure 11.19. *Snow Squall*'s degraded 'Port PO2' α phase sheathing. Twin grains are visible with many slip lines in some areas showing cold-working of the sheet after final annealing of the brass. This could reflect George Frederick Muntz Jr.'s 1858 patent No. 1134. 80x.

Scanning Electron Microprobe analysis of Muntz-stamped sheathing from *Snow Squall* and *Petrel*							
Snow Squall	'Muntz's 24 Patent' stamp		Elements Tested %				
1851-1864	'Strb. 32 C' sample	Copper	Iron	Sulfur	Zinc	Lead	Total
The last sheathing was	pt. 1	64.20	0.05	0	37.60		101.85
in 1862, New York,	pt. 2	64.30	0.01	0	37.40		101.71
with repairs in 1863	pt. 3	64.50	0.06	0	37.50		102.06
in Singapore.	Average	64.33	0.04	0	37.50	0.10	
American built.	Standard deviation	0.15	0.03	0	0.10		
	'Port 80C' sample						
	pt. 1	64.03	0.10	0	37.68		101.91
	pt. 2	64.53	0.08	0	37.25		101.87
	pt. 3	64.46	0.05	0	37.43		101.94
	pt. 4	65.03	0.06	0	37.13		101.22
	Average	64.54	0.07	0	37.37	0.10	
	Standard deviation	0.37	0.02	0	0.24		
Petrel	'Muntz's 26 Patent' stamp						
1847-1853	'M2' sample						
Tasmanian built.	pt.1	63.90	0.045	<0.02	36.70	<0.06	100.70
	pt.2	63.70	0.042	<0.02	36.90	<0.06	100.60
	pt.3	63.40	0.042	<0.02	37.00	<0.06	100.60
	pt.4	63.60	0.046	<0.02	37.30	<0.06	101.00
	pt.5	63.30	0.042	<0.02	37.10	<0.06	100.50
	pt.6	63.10	0.047	<0.02	37.30	<0.06	100.50
	pt.7	64.70	0.034	<0.02	36.90	0.08	101.70
	pt.8	63.50	0.049	<0.02	37.10	0.09	100.70
	pt.9	63.90	0.038	<0.02	36.70	<0.06	100.70
	pt.10	63.50	0.037	<0.02	36.80	0.10	100.40
	Average	63.60	0.042	*	37.00	*	
	Standard deviation	0.40	0.005	*	0.20	*	
	3s precision +/-	0.40	0.01	0.02	0.30	0.05	
	* statistical calculation not possible as some or all individual values below detection limit.						

Table 11.2 Scanning electron and microprobe data for the Muntz metal from some ships. After Carlson et al (2010).

SEM data from stamped samples is reported in Table 11.2. The average composition of sample 'Starboard 32C' from three analyzed matrix points was 64.3 ± 0.15% copper, 37.5 ± 0.10% zinc, 0.04 ± 0.03% iron and 0.1% lead. The average composition of sample 'Port 80C' from four analyzed matrix points was 64.5 ± 0.37% copper, 37.4 ± 0.24% zinc, 0.07 ± 0.02% iron and 0.1% lead.

Snow Squall's 'Muntz's 24 Patent' stamped sheathing is essentially a Muntz metal, but care may have been taken to try to prevent excessive precipitation of the β phase, as the two-phase alloys are more susceptible to general corrosion.

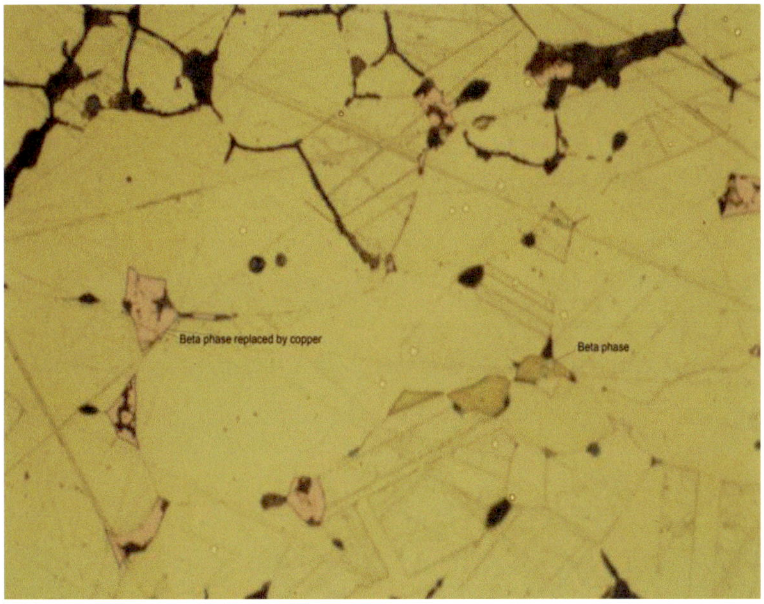

Figure 11.20. *Snow Squall* sample 'Port 70C' showing the small islands of the beta phase within the alpha grains. Note that some of the beta phase is preferentially corroded and replaced by re-deposited copper. 300x.

These findings are based upon a small sample set and cautiously raise awareness of suggested changes in Muntz ship sheathing alloy metallurgy. Two sheathing nails from the Snow Squall were cast and have two distinct compositions. Technical analysis of *Snow Squall*'s unstamped sheathing suggests an alpha phase binary copper and zinc metal with a trace of lead and iron. This composition is close to cartridge brass or the PH Muntz & Co., Ltd. alloy nergandin.

Vessel	Time period	Last sheathing	Stamp	Composition	Microstructure
Petrel	1847-1853	unknown	Double oval 'Muntz's 26 Patent'	63.6 ± 0.4% copper, 37.0 ± 0.3% zinc with a trace of iron and lead	Alpha phase brass with small islands of beta phase present.
			unstamped	63.5 ± 0.3% copper, 36.9 ± 0.2% zinc with a trace of iron and lead	Alpha phase brass with small islands of beta phase present.
Eglinton	1848-1852	1849	Oval 'P. Grenfell & Sons 26 oz'	59.9 ± 1.1% Cu and 40.8 ± 0.9% Zn	not recorded
Snow Squall	1851-1864	1862 with repairs in 1863	Double oval 'Muntz's 24 Patent'	'Starboard 32C' 64.3 ± 0.15% copper, 37.5 ± 0.10% zinc, 0.04 ± 0.03% iron and 0.1% lead.	Alpha phase brass at the boundary where beta phase begins to form.
			Double oval 'Muntz's 24 Patent'	'Port 80C' 64.5 ± 0.37% copper, 37.4 ± 0.24% zinc, 0.07 ± 0.02% iron and 0.1% lead	Alpha phase brass at the boundary where beta phase begins to form.
			unstamped	'Port 80E' sample 2.3: 67.9% copper, 30.1% zinc, 0.8% lead and 0.39% iron.	Alpha phase brass

Table 11.3 Analytical data for brass alloys compared with other ships.

Snow Squall's 'Muntz's 24 Patent' sheathing, 64.3-64.5% copper, 37.5-37.4% zinc with traces of lead and iron, is an α phase alloy right at the edge of β phase precipitation. This similarity of sheathing composition and microstructure in two vessels approximately ten years apart suggests a change in the Muntz sheathing metallurgy towards a superior, more corrosion resistant product for ship sheathing from the 60% copper, 40% zinc Muntz metal formula with a corresponding alpha-beta microstructure.

Appendix 1: Glossary of Etchants.

NOTE: This appendix includes all of the etchants which will be used in the next few volumes as well as this volume. The intention here is to bring together all of the formula for the various etchants for different alloys in one place.

Before etching, ensure that the sample is as scratch-free as possible and that the surface is free of oil, water, or ethanol staining, as these will all interfere with the correct etching of the sample. Some samples are liable to be stained on etching due to the presence of different phases, holes, or interdendritic porosity. In such cases the sample should be carefully washed in ethanol after etching and dried as thoroughly as possible to prevent excessive staining. Samples can be repolished on one micron diamond paste, cleaned in ethanol and then dried and re-etched. This is often advisable if the finish is less than desired on the final polishing and first etching.

For etching the sample can be held in a pair of tongs, of which there are many different types available. Make sure that there is enough room in the etching tray, often a crystallizing dish, for the sample diameter and the tongs to be immersed into the etching solution. Use gentle agitation if possible while etching to avoid stagnant conditions and uneven etching. Samples after etching can be stored in an efficient desiccator or simply repolished and dried if exceptionally valuable. Samples with etched surfaces will continue to corrode, and in the case of iron artefacts, this can be severe with filiform corrosion of exposed surfaces being a common phenomenon. Sound metallographic principle dictates that all specimens should be examined in the unetched state before proceeding to etching. This is partly because etching may dissolve out any non-metallic inclusions or adversely affect the examination of patina and corrosion which can be essential in authenticity determinations.

Health and Safety

Observe all health and safety regulations pertaining to the country in which you are working. Many etchants, such as nital, are only partially stable and tend to alter over time in their bottle. Make up small quantities of etchants if just for your own use, as this not only ensures that the etchants are fresh, but also saves money on chemical supplies. Many of the preparations described here are poisonous or have to be used in a fume hood, depending on local regulations.

Etchants for Copper and Bronze Alloys

Aqueous ferric chloride: 240 ml distilled water, 60 ml hydrochloric acid and 20 g ferric chloride. Produces grain contrast, very useful for all copper alloys, etching time from a few seconds to 60 seconds with 3-5 seconds being common. The author prefers to use alcoholic ferric chloride as faster drying.

Alcoholic ferric chloride: 240 ml ethanol, 60 ml hydrochloric acid, 20 g ferric chloride. This is the most common etchant used in this book.

Aqueous ammonium persulphate: 200 ml distilled water, 20 g ammonium persulphate. Produces grain contrast. Must be use fresh; will not keep.

Ammonia/hydrogen peroxide: 50 ml distilled water, 50 ml ammonium hydroxide, 10-50 ml hydrogen peroxide. Make up and use fresh only. Adding larger amounts of hydrogen peroxide creates better grain contrast, adding less hydrogen peroxide creates better grain boundary etching. An interesting etchant and can be used with benefit on a number of different alloy types.

Potassium ferricyanide: 5g potassium ferrocyanide, 100 ml distilled water. This etchant can be used to darken some precipitates such as Cu_3P which may co-exist with the alpha + delta eutectoid. Cu_3P can be seen as a purple shade against the very pale blue of the delta phase in bronzes. After etching in alcoholic ferric chloride, this distinction is lost. Some difficulties may be overcome by using a relatively high amount of aqueous ammonia in the final polish, such as 5% ammonia.

Etchants for iron, steel and cast iron

Nital: 200 ml ethanol, 4 ml nitric acid. The most useful general etchant for most iron alloys.

Picral: 200 ml ethanol, 4 g picric acid. This is a very useful etchant for iron and carbon steels. Often used for iron and heat-treated steels, pearlite and martensite. Cementite is stained a light yellow. This can be very useful in distinguishing ferrite from cementite. Nital and picral can be used mixed in a 1:1 proportion.

Oberhoffer's reagent: 100 ml distilled water, 100 ml ethanol, 8 ml hydrochloric acid, 6 g ferric chloride and 0.1 g stannous chloride. After etching the section should be rinsed in a 4:1 mixture of ethanol and hydrochloric acid. Useful for steels in segregation studies such as arsenic segregation.

Heyn's reagent: 40 ml distilled water, 40 g copper (II) ammonium chloride. Copper precipitates must be wiped from the surface with distilled water or washed off with distilled water from a wash bottle. Useful for phosphorus segregation in steels.

Klemm's reagent: 50 ml saturated aqueous solution of sodium thiosulphate and 1 g potassium metabisulphite. Useful for some colour etching and for phosphorus segregation in cast steels and cast iron.

Baumann's print solution: 100 ml distilled water and 5 ml sulphuric acid. Silver bromide paper is saturated with the solution and firmly
pressed against the specimen surface. After one to five minutes, rinse and fix with a solution of 6 g sodium thiosulphate in 100 ml distilled water. Wash and dry. Useful for verification, arrangement and distribution of iron and manganese sulphide inclusions.

Beraha's reagent: 100 ml distilled water, 24 g sodium thiosulphate, 3 g citric acid, 2 g cadmium chloride. Use for 20-40 seconds. The chemicals must be dissolved in the order given. Each component must dissolve before adding the next. Store in the dark under 20 ^{o}C. Filter before use. Solution remains active for only four hours. Used as a tint etchant for iron and carbon steels. Ferrite is stained brown or violet. Carbides, phosphides, and nitrides may be only lightly stained. With Beraha's reagent, ferrite is coloured but cementite is not, so that both proeutectoid cementite and cementite in pearlite are in strong contrast to the ferrite, which can make even thin films of cementite easily visible.

Alkaline sodium picrate: 2 g picric acid, 25 g sodium hydroxide and 100 ml distilled water. This solution is useful for distinguishing between iron carbide and ferrite in steels. It can be used as a boiling solution for ten minutes or longer if required. The cementite is darkened by this reagent, while ferrite is unaffected. Etching in this solution may provide a good indication of pearlite lamellae spacing.

Beaujard's reagent: 20 g sodium bisulphate, 100 ml distilled water. To ensure that the specimen is evenly wetted, it is often a good idea to etch in nital for two to four seconds beforehand. Beaujard's reagent can then be used for 10-25 seconds: the surface should be very carefully washed and dried, otherwise the deposited surface film will be disturbed.

The reagent produces good contrast between ferrite grains and between lightly tempered martensite and ferrite, as well as delineating cementite networks. The reagent works by depositing a complex oxide-sulphide-sulphate film on the metal surface, in various shades of brown.

Dimethlyglyoxime nickel reagent: some ancient steels contain nickel as an important impurity. A simple test is to take a nickel print by pressing a blotting paper soaked in dimethylglyoxime against the polished section when the nickel-rich areas are brown-staining on the blotting paper.

Villella's reagent: 1 g picric acid, 100 ml ethanol, 5 ml hydrochloric acid. A reagent that can be used for 5-40 seconds and reveals clearly the needles of plate martensite. It is useful for exposing the austenitic grain size of quenched and tempered steels if this feature is at all discernible.

Whiteley's method: 5 g silver nitrate and 100 ml distilled water. A technique for revealing sulphide inclusions in steel and other alloys. Soak the polishing cloth in the freshly prepared solution. The cloth is then washed to remove all excess solution. The polished surface of the sample is then rubbed carefully over the prepared cloth. Any sulphide inclusions that are present should be stained a dark brown.

Etchants for gold alloys.

Aqua regia: 40 ml nitric acid, 60 ml hydrochloric acid. Use for a few seconds or up to one minute, although longer times may result in over-etching. Use fresh. Aqua regia is a strong oxidizing agent and highly corrosive. In most alloys it produces grain contrast.

Hydrogen peroxide/ferric chloride: 100 ml distilled water, 100 ml hydrogen peroxide and 32 g ferric chloride. Sometimes useful for variable carat gold jewelry and ternary gold-copper-silver alloys.

Ammonium persulphate/potassium cyanide: 100 ml distilled water, 10 g ammonium persulphate. 100 ml distilled water, 10 g potassium cyanide. These two solutions must be mixed and used fresh. Use in proportion of 1:1 usually. Very useful with many gold alloys, although etch pits are sometimes hard to prevent. Never mix with acid solutions as HCN could be produced. The solution of potassium cyanide is poisonous. Make up the persulphate solution just before use. After mixing the solution is essentially converted to potassium cyanate which can be disposed of down the sink with plenty of water.

Etchants for silver alloys

Acidified potassium dichromate: 10 ml sulphuric acid, 100 ml potassium dichromate, saturated in water, 2ml of a saturated solution of sodium chloride in water, this solution can be diluted 1:9 with distilled water before use. Can also be used without the sulphuric acid addition. Can be used for silver-copper alloys as well as silver.

Ammonium persulphate/potassium cyanide: the same solution as described above for gold alloys can be used on silver.

Ammonia/hydrogen peroxide: 50 ml ammonium hydroxide, 50 ml hydrogen peroxide. Must be used fresh. Sometimes successful where the dichromate etch, which is the more common choice, is ineffectual.

Acidified thiourea: 10% solution of thiourea in distilled water, 5-10 drops of either nitric acid or hydrochloric acid. An uncommon etchant, especially as the use of thiourea has been increasingly not used in conservation.

Etchants for tin

Ammonium polysulphide: a saturated aqueous solution of ammonium polysulphide. Use for 20-30 minutes. Wipe off with cotton wool after etching. Can be used for all types of tin alloys. Nital or picral are sometimes useful as well.

Etchants for zinc

Zinc alloys are difficult to prepare in usual metallographic methods. Fake or altered microstructures are common.

Palmerton's reagent: 100 ml distilled water, 20 g chromic acid, 1.5 g anhydrous sodium sulphate or 3.5 g of hydrous sodium sulphate. Can be used for seconds or minutes. Seconds are more advisable.

50% Hydrochloric acid: 50 ml hydrochloric acid, 50 ml distilled water. This solution can be swabbed over the surface or the sample attacked by immersion.

Etchants for lead alloys.

Glycerol etchant: 84 ml glycerol, 8 ml glacial acetic acid, 8 ml nitric acid. Use fresh only. Gives grain boundary contrast. Nital can sometimes be used as a replacement. If kept, the solution starts to smell of various aldehydes.

Glacial acetic acid: 15 ml glacial acetic acid, 20 ml nitric acid. Useful for lead solders and lead-tin alloys. If difficulty is experienced in the preparation of lead alloys, a good technique is to use etch/polish methods (Scott 1994).

Etchants for oxide layers on iron

Solution One: 10 ml distilled water, 5 ml 1% nitric acid, 5 ml 5% citric acid, 5 ml 5% thioglycollic acid. Swab for 15-60 seconds. Etches haematite and maghemite, magnetite is not attacked.

Solution Two: 15ml distilled water, 5 ml formic acid. 15ml of distilled water and 5mlo fluoboric acid. Swab firstly with the formic acid solution followed by the fluoboric solution for two seconds. Etches magnetite only.

Interference film etchants for colour effects.

It is sometimes useful to investigate the application of colour metallography to ancient metallic specimens. One of the potential advantages of tint etching is that coring and segregation may be revealed when conventional techniques fail. Colour metallography may also be used to enhance the visual appreciation of microstructural features. Further details can be found in Yakowitz (1970), Petzow and Exner (1975), and Phillips (1971). Some useful recipes are set out here.

Interference etchants for copper alloys

Acidified selenic acid. 2ml 35% solution of hydrochloric acid, 0.5 ml selenic acid, 300 ml ethanol. The solution can be used after first etching the polished sample for a few seconds in a 10% solution of ammonium persulphate. Dry after etching in persulphate before using the selenic acid etch.

Acidified thiosulphate/acetate. 12 g sodium thiosulphate, 1.2 g lead acetate, 1.5 g citric acid, 50 ml distilled water. Use after first etching with ammonium persulphate. A very useful etchant for copper alloys, but must be made up fresh as the solution will not keep. Dissolve the ingredients in the order given. Make sure that the thiosulphate is dissolved first. Best made by placing the prepared solution in a refrigerator for twenty-four hours when a clear lead thiosulphate solution should form with a sulphur precipitate. If the solution is kept for a long time, lead thiosulphate precipitates onto the glass wall of the container and is hard to clean.
Pre-etching should be carried out in 10% ammonium persulphate. It can be difficult to follow the change in colour of small immersed specimens. The sequence of colours should be from yellow, orange, red, violet blue and finally bright silver. Etching should not be allowed to continue beyond the blue or violet stage. The thiosulphate etchant, for example, when used on an alpha-beta brass results in the alpha grains turning light green, while the beta grains become orange-yellow. With a 5% cast tin bronze, the dendritic alpha regions are coloured blue, yellow or brown depending on orientation; the alpha + delta eutectoid can also be coloured brown in the etchant.

Beraha's tint etch for aluminium alloys: 20 g chromium trioxide, 2 g sodium sulphate, 1.7 ml hydrochloric acid and 100 ml distilled water. Pre-etch with 10% aqueous sodium hydroxide followed by 50% aqueous nitric acid. Rinse in water, dip immediately into the tint etchant for 1-5 seconds. Rinse and dry. Colours the matrix grains, outlines the second phase particles.

Lienard and Pacque etch: 1g ammonium molybdate, 6 g ammonium chloride, 200 ml distilled water. This is the tint etch used to colour CuAl2. Immerse for about 2 minutes.

Beraha's tint etch for Fe, Ni or Co alloys: (a) stock solution of 1:2. 1:1 or 1: 0.5 of hydrochloric acid and distilled water. (b) 100 ml stock solution with addition of 0.6 – 1.0 g potassium metabisulphite. (c) optional additions of 1-3 g ferric chloride or 1 g copper (II) chloride or 2-5 g ammonium bifluoride. This tint etch is good for heat resistant alloys. Colours the matrix, carbides and nitrides are unaffected. Immerse the sample in solution at room temperature for 60-150 seconds. Move sample during etching. Start with the lowest HCl concentration, if colouration does not occur, increase the HCl or etch longer.

Klemm's Number One: 50 ml saturated aqueous sodium thiosulphate, 1 g potassium metabisulphite. This is a good multi-purpose tint. Immerse for 3 minutes for beta brass, alpha-beta brass and bronzes. Use for 10-60 minutes for alpha brass. Use 40-100 seconds for colouring ferrite in steels, reveals phosphorus segregation and overheating. Longer times produce line etching of ferrite. Etch for 30 seconds for zinc alloys.

Klemm's Number Two: 50 ml saturated aqueous sodium thiosulphate, 5 g potassium metabisulphite. Immerse for 6 minutes or more for alpha brass. Immerse 30 – 90 seconds for steels. Reveals phosphorus segregation. Good for austenitic manganese alloys. Immerse 60 -90 seconds for tin and its alloys.

Klemm's Number Three: 5ml saturated aqueous sodium thiosulphate, 45 ml water, and 20 g potassium metabisulphite. Immerse for 3-5 minutes for bronze, immerse 6-8 minutes for monels.
Beaujard and Tordeux's etch: 21-28% aqueous sodium bisulphate. A tint etch for steels. Immerse 10 – 25 seconds. Reveals grain boundaries and ferrite orientations. Darkens as-quenched martensite.

Metabisulphite etch. 8-15 g sodium metabisulphite 100 ml water. Darkens as-quenched martensite. Immerse about 20 seconds.

Beraha's tint etch for steels: 1 g sodium molybdate 100 ml distilled water. Add nitric acid to pH 2.5-4.0 (about 0.4 ml). immerse 20-30 seconds for cast iron. Fe3P and Fe3C, yellow-orange, ferrite, white. For low carbon steel add 0.1 g ammonium bifluoride and immerse for 45 -60 seconds. For medium carbon steel add 0.2 g ammonium bifluoride. For high carbon steel add 0.3 – 0.4 g ammonium bifluoride. Carbides, yellow-orange to violet, ferrite, white or yellow.

Beraha's tint for steels: 3 g potassium metabisulphite and 10 g anhydrous sodium thiosulphate in 100 ml distilled water. Good tint etch for iron and steel. Immerse 1-15 minutes. Colours ferrite, martensite, pearlite and bainite; sulphides are brightened.

Beraha's tint for tool steels: 0.5 – 1.0 ml hydrochloric acid, 100 ml water and 1 g potassium metabisulphite. A good tint for irons, steels and tool steels. Agitate strongly during etching, then hold motionless until surface is coloured; 10-60 seconds total time. Colours ferrite, martensite, pearlite and bainite. Reveals grain boundaries.

Beraha's tint for stainless steels: stock solution 20 ml hydrochloric acid and 100 ml water with 2.4 g ammonium bifluoride. Before use add 0.6-0.8 g potassium metabisulphite (0.1-0.2 g for martensitic grades). After mixing, reagent is good for use for 2 hours. Use plastic tongs and beaker. Immerse 20-90 seconds, shake gently during etching. Colour matrix phases.

Hasson's tint etch: 40-60 ml ferric chloride and 1000ml water with 25 ml hydrochloric acid with 75 ml ethanol. Tint etch for molybdenum. Immerse without agitation for 40-50 seconds. Ferric chloride can be dissolved in ethanol but etch is slower. Etch for 2-3 minutes.
Oak Ridge etch: 20 ml hydrogen peroxide and 10 ml sulphuric acid. Tint etch for molybdenum alloys. Immerse for two minutes, wash and dry. Swab removes colours, produces grain boundary attack.

Acidified sulphate/ chromic oxide: 50 g chromium trioxide, 5 g sodium sulphate, 4.25 ml 35% hydrochloric acid, 250 ml distilled water. I have never used this solution so am unable to comment on its effectiveness.

Interference etchants for iron, steel and cast iron

Metabisulphite etch: 6 g potassium metabisulphite, 100 ml distilled water. This etchant is useful for both carbon and alloy steels. *Molybdate etch*: 2 g sodium molybdate, 200 ml distilled water. Acidify to a pH of 2.5 – 3 using dilute nitric acid. Useful for cast irons after first etching with nital for a few seconds.

Thiosulphate etch: 60 g sodium thiosulphate, 7.5 g citric acid, 5 g cadmium chloride, 250 ml distilled water. Useful for cast irons and steels without pre-etching. Must be made up fresh and will not keep longer than a week.

Saturated thiosulphate solution: 100 ml saturated sodium thiosulphate, 2 g potassium metabisulphite. Can be made up fresh and used for cast irons and steel alloys.

Selenic acid solution: 10 ml 35% hydrochloric acid, 4 ml selenic acid, 200 ml ethanol. Can be used for both ferrite and martensitic steels.

Interference etchants for aluminium alloys

Molybdate/bifluoride etch: 2g sodium molybdate, 5ml 35% hydrochloric acid, 1 g ammonium bifluoride, 100 ml distilled water. This unusual combination can be used for etching both aluminium and titanium alloys.

Author

David A. Scott, BSc in Chemistry from the University of Reading; BA in Conservation from the Institute of Archaeology, London; Ph.D. from University College London in 1982. Awarded Fellow, Royal Society of Chemistry (FRSC) in 1991 and Fellow, International Institute for Conservation (FIIC) in 1994. From 1981-87, Lecturer in Conservation at University College, Institute of Archaeology, Department of Conservation and Materials Science. From 1987-2003, Head of the GCI Museum Research Laboratory, J. Paul Getty Museum, Malibu.
Professor in Art History, 2003-present, and Director of the UCLA/Getty Programme Archaeological and Ethnographic Conservation, 2003-2011. His principal interests are the analysis of Museum objects, the characterization of pigments, ancient metals and microstructure, the teaching of conservation, and the archaeometallurgy of preHispanic Colombia and Ecuador. Dr. Scott is the author of five previous books and has published over one hundred papers in the peer-reviewed literature. His book "Copper and Bronze in Art: Corrosion, Colorants, Conservation" won the prize in 2002 for the best scholarly/art book published that year from the American Association of Publishers, as well as several other awards, and has just gone into its second printing.

Bibliography for Volume I

Anheuser, Kilian and France, Philip. 2002
Silver plating technology of the late 3rd century Roman coinage.
Historical Metallurgy 36, 17-23

An Zhimin 2000
On early copper and bronze objects in ancient China. In: (Ed. Linduff, K. M., Han Rubin, and Sun, Sh.) *The Beginnings of Metallurgy in China.* 195-199.
Edwin Mellen Press: New York.

Archbutt, S. L. and Prytherch, W. E. 1937
Effect of impurities in copper: antimony.
Research Monograph No. 1.
British Non-Ferrous Metals Research Association: London

Barnard, Noel 1961
Bronze casting and bronze alloys in ancient China.
Monumenta Serica Monograph XIV
Australian National University and Monumenta Serica: Canberra.

Bassett, Jane 2002
An opportunity taken: the project design for a technical study of 26 bronzes in the exhibition *Adriaen deVries: Imperial Sculptor* at the J. Paul Getty Museum. *Objects Specialty Group postprints (American Institute for Conservation of Historic and Artistic Works. Objects Specialty Group)* 9 68-78.

Bayley, Justin 1990
The production of brass in antiquity with particular reference to Roman Britain.
In: (Ed. P. T. Craddock) *2000 years of Zinc and Brass.* British Museum Occasional Paper No 50. British Museum Press: London.

Bayley, Justin and Butcher, Susan 1995
The composition of Roman brooches found in Britain. In: (Ed S. T. A. M. Mols, A. M. Gerhartl-Witteveen and H. Kar) *Ancient Bronzes,* 113-119.
Provinciaal Museum G. M. Kam: Nijmegen.

Bayreuther, W. 1993
The clipper ship Snow Squall, translating a dream into reality', *Cultural Resources Management,* National Park Service, 16, 24-26.

Becker, Lawrence; Pilosi, Lisa; and Schorsch, Deborah. 1994
An Egyptian silver statuette of the Saite period--a technical study.
Metropolitan Museum Journal 29 37-56.

Beraha, E. 1965
Metallographic reagent for the identification of cementite in iron and steel.
Journal of the Iron and Steel Institute of Japan 203 454-456

Beraha, E. 1970
Staining metallographic agent for copper-base alloys.
Practical Metallography 7 131-134

Bingham, Hiram 1930
Machu Picchu, A Citadel of the Incas.
Yale University Press: New Haven

Borić , D. 2009
Absolute dating of metallurgical innovations in the Vinča culture of the Balkans. In: (Ed. T. L. Kienlin and B. W. Roberts) *Metals and Societies: papers in honour of Barbara S. Ottaway* 191-245. Universitätsforschungen zur prahistorischen Archäologie, Habelt: Bonn.

Bourgarit, David 2007
Chalcolithic copper smelting.
In: (Ed S, LaNiece, D. R. Hook and P. T. Craddock) *Metals and Mines: Studies in Archaeometallurgy* 3-14. British Museum Press: London.

Breeks, James Wilkinson 1873
An account of the Primitive Tribes and Monuments of the Nilgiris.
Seeley & Co: London.

Bruzelius, L. 2006
The Maritime History Virtual Archives. Shipbuilding and naval architecture. George Frederick Muntz patents.
http://www.bruzelius.info/Nautica/Nautica.html

Buchwald, V. F. and Leisner, P. 1990
A metallurgical study of twelve prehistoric bronze objects from Denmark.
Journal of Danish Archaeology 9 64-102.

Budd, Paul and Ottaway, Barbara 1989
The properties of arsenical copper alloys: implications for the development of Eneolithic metallurgy.
Paper presented at Archaeological Sciences Meeting, Bradford 1991.

Butts, Alison and Coxe, Charles D. 1967
Silver: Economics, Metallurgy, Use.
D. Van Nostrand: Princeton, NJ.

Butts, I.R. 1858
The Merchant and Mechanics Assistant. 3rd. ed. 82-83.
Published by I.R. Butts: Boston.

Caley, Earle R. 1964
Orichalcum and related ancient alloys.
American Numismatic Society: New York.

Cammann, Schuyler V. R. 1958
The Bactrian nickel theory.
American Journal of Archaeology 62 409-414.

Caredo Muro P., and Shimada, I. 1985
Behind the golden mask: the Sican gold artifacts from Batán Grande, Peru. In: *The Art of Pre-Columbian Gold: the Jan Mitchell Collection.* (Ed. J. Jones and H. King) 60-75. The Metropolitan Museum of Art: New York.

Carlson, Molly O'Guiness 1994
The clipper ship *Snow Squall*. The first three years of conservation in a fledgling museum' in *Proceedings of the 5th ICOM Group on Wet Organic Archaeological Materials Conference Portland, Maine, 16-20 August, 1993*, ed. P. Hoffmann, T. Daley and T. Grant. International Council of Museums (ICOM), Committee for Conservation Working Group on Wet

Organic Archaeological Materials, Bremerhaven, (1994), 485 – 488.

Carlson, Molly O'Guiness, Lipfert, Nathan, Ronnberg, Erik A. R. Jr., and Scott, David A. 2010
Technical analysis of Muntz metal sheathing from the American Clipper Ship Snow Squall (1851-1864).
In (Ed. David Hallam) *ICOM Metals Conference,* Virginia. 235-241.
ICOM: Printed by Lulu: North Carolina.

Carroll, Diane L. 1987
Tools of the Renaissance Jeweller: a goldsmith's workshop in 1576.
Archaeomaterials 1 153-172.

Case, Henry and Coghlan, Herbert H. 1957
Early metallurgy of copper in Ireland and Britain.
Proceedings of the Prehistoric Society 23 91-123.

Chadwick, R. 1939
The effect of composition and constitution on the working and on some physical properties of the tin bronzes.
Journal of the Institute of Metals 64 331-346.

Chase, W. Thomas III 1968
The examination of two Sasanian silver plates.
Ars Orientalis 7 75-93.

Chase, W. Thomas III 1977
What is the smooth lustrous black surface on ancient bronze mirrors?
In: *Corrosion and Metal Artefacts, a dialogue between conservators, archaeologists and corrosion scientists.* 191-203.
National Bureau of Standards, Special Publication No. 479.
U. S. Department of the Interior: Washington, D.C.

Chase, W. Thomas III 1983
Bronze casting in China: a short technical history.
In: *The great Bronze Age of China: A symposium.* (Ed. George Kuwayamei)
Seattle and London
Chase, W. Thomas III and Franklin, Ursula M. 1979
Early Chinese black mirrors and pattern-etched weapons.
Ars Orientalis 11 215-258.

Chase, W. Thomas III., Barnes, I. L. and Joel, E. C. 1995
Appendix 4: Lead Isotope Ratios.
In: *Eastern Zhou Ritual Bronzes from the Arthur M. Sackler Collections* Vol.III.
Arthur M. Sackler Foundation. Harry Abrams: New York.

Chase, W. Thomas III and Wang, Quanyu. 1997
Metallography and corrosion product studies on archaeological bronze fragments from the Qu Cun site.
In: *Materials Issues in Art and Archaeology* V (Eds. P. B. Vandiver, J. R. Druzik, J. P. Merkel, and J. Stewart) 73-79.
Materials Research Society, Symposium Proceedings, Volume 462.
Materials Research Society: Pittsburgh.

Chen, Kunlong, Rehren, Thilo, Mei, Jianjun, and Zhai, Congcang 2009
Special alloys from remote frontiers of the Shang Kingdom: scientific study of the

Hanzhong bronzes from southwest Shaanxi, China.
Journal of Archaeological Science 36 2108-2118.

Cheng, C. F. and Schwitter, C. M. 1957
Nickel in ancient bronze.
American Journal of Archaeology 61 352-367.

Chernykh, E. N. 1992
Ancient Metallurgy in the USSR: The Early Metal Age.
Cambridge University Press: Cambridge

Childe, V. Gordon 1930
The Bronze Age.
Biblio and Tannen: New York.

Chou, Ju-his 2000
Circles of Reflection: The Carter Collection of Chinese Bronze Mirrors. Cleveland: The Cleveland Museum of Art.

Christman, Bruce 2000 Technical Appendix. In: (Ed. Chou, Ju-his) *Circles of Reflection: The Carter Collection of Chinese Bronze Mirrors.* 97-128. Cleveland: The Cleveland Museum of Art.

Chubb, J. P., Billingham, J., Hancock, P., Dimbylow, C., and Newcome, G. 1980
The effect of alloy and residual elements on the strength and hot ductility of cast cupro-nickel.
Philosophical Transactions of the Royal Society of London, Series A 295 no. 1413 123-129.

Cieza de Leon, P. de 1564
The travels of Pedro de Cieza de Leon AD 1532-50 contained in the first part of his chronicle of Peru. (Trans. And Ed. C. R. Markham) Hakluyt Society, Series I, 33 (1864). Hakluyt Society: London.

Coghlan, Herbert H. 1968
A metallurgical examination of some prehistoric flat and flanged axes.
Archaeologia austriaca 44 61-82.

Coghlan, Herbert H. 1969
A metallurgical examination of eleven palstaves.
Bulletin of the Historical Metallurgy Group 1, no. 9 7-21.

Coghlan, Herbert .H. 1970
British and Irish Bronze Age implements in the Borough of Newbury Museum.
Borough of Newbury Museum: Berkshire.

Coghlan, Herbert H. 1971
A metallurgical examination of a bronze sword found in the Thames at Battersea.
Bulletin of the Historical Metallurgy Group 5 71-73.

Coghlan, Herbert H., 1975
Notes on the prehistoric Metallurgy of Copper and Bronze in the Old World.
Pitt Rivers Occasional Papers on Technology No. 4.
Pitt Rivers Museum: Oxford.

Coghlan, Herbert H. 1979
Analyses of Bronze Age artifacts from Irish Museums.
Historical Metallurgy 13 98-106.

Coghlan, Herbert H. 1980
Examination of some continental Bronze Age decorative objects.
Historical Metallurgy 14 80-93.

Coghlan, Herbert H. and Macphail, I. 1967
An examination of two Bronze Age flat axes from Ireland.
Archaeologia Austriaca 41 48-65.

Coghlan, Herbert H. and Parker, G. 1970
A report on the board of Bronze Age tools and weapons from Yattendon, near Newbury, Berks.
Newbury Museum: Berkshire.

Coghlan, Herbert H. and Tylecote, Ronald F. 1978
Medieval iron artifacts from the Newbury area of Berkshire: metallurgical examination.
Historical Metallurgy 12 12-18

Collins, William F. 1934
The mirror-black and 'quicksilver' patinas of certain Chinese bronzes.
Journal of the Royal Anthropological Institute LXIV 69-79.

Cooke, S. R.B. and Aschenbrenner, S. 1975
The occurrence of metallic iron in ancient copper.
Journal of Field Archaeology 2 251-266.

Corporate Author 1985
Metallography and Microstructures.
American Society for Metals. 9th Edition.
ASM: Metals Park: Ohio

Cowell, M. R., Craddock, P. T., Pike, A. M. and Burnett, A. M. 2000
An analytical survey of Roman provincial copper-alloy coins and the continuity of brass manufacture in Asia Minor In: (Ed. B. Kluge and B. Weisser). *XII Internationaler Numismatischer Kongress Berlin 1997, Akten, Proceedings-Actes I. 670-677.*
Staatliche Museen Preussischer Kulturbesitz, Münzkabinett: Berlin.

Craddock, Paul T. 1976
Copper alloys used by the Greek, Etruscan and Roman civilizations: 1.
Journal of Archaeological Science 3 93-113.

Craddock, Paul T. 1977
Copper alloys used by the Greek, Etruscan and Roman civilizations: 1.
Journal of Archaeological Science 4 103-123.

Craddock, Paul T. 1978
The composition of the copper alloys used by the Greeks, Etruscan and Roman civilizations: the origins and early use of brass.
Journal of Archaeological Science 5 1-16.

Craddock, Paul T. 1995
Early Metal Mining and Production.
Edinburgh University Press: Edinburgh.

Craddock, Paul T. (Ed.) 1998
2000 years of Zinc and Brass. 2nd Ed.
British Museum Occasional Paper No 50.
British Museum Press: London.

Craddock. Paul T. 1999
Paradigms of metallurgical innovation in prehistoric Europe. In (Ed. A. Hauptmann, E. Pernicka, T. Rehren, and U. Yalcin) *The Beginnings of Metallurgy* 175-192.
Der Anschnitt, Beiheft 9. Bergbau-Museum: Bochum.

Craddock, Paul T. and Meeks, Nigel D. 1987
Iron in ancient copper.
Archaeometry 29 187-204.

Craddock, Paul T, and Eckstein, K. 2003
Production of brass in antiquity by direct reduction. In (Ed. Paul T. Craddock and Janet Lang) *Mining and Metal Production through the Ages.* 216-230.
British Museum Press: London.
Craddock, Paul T., Cowell, Michael and Stead, Ian 2004
Britain's first brass.
Antiquaries Journal 84 339-346.

Cutler, Anthony 1990
The Disputa Plate in the J. Paul Getty Museum and its Cinquecento Context.
The J. Paul Getty Museum Journal 18 5-32.

Dandridge, Peter 2000
A study of the gilding of silver in Byzantium.
In: *Gilded Metals: History, Technology and Conservation.* (ed. Terry Drayman-Weisser) 123-143. Archetype Publications: London.

Dean, N. 1990
Last of her breed; the Yankee clipper *Snow Squall', Seaways,* 1 13-9.

Dean, N. 2001 *Snow Squall: The Last American Clipper Ship with expedition accounts by David Switzer.* Tilbury House Publishers and Maine Maritime Museum: Maine.

Demoule, J. P., and Perles, C. 1993
The Greek Neolithic : A new review.
Journal of World Prehistory 7 355-416.

Elam, S. F. 1932
Some bronze specimens from the Royal Graves at Ur.
Journal of the Institute of Metals 48 97-99.

Etchant's Database 2010
http://www.kaker.com/etch/demo/list
[accessed March 2010].

Fasnacht, W. and Northover, Peter J. 2002
Ancient raw copper from primary smelting sites in Cyprus.
In: (Ed. Pamela B. Vandiver, Marta Goodway and Jennifer L. Mass) *Materials Issues in Art and Archaeology VI,* 119-205.
Materials Research Society Volume 712.
Materials Research Society: Warrendale.

Fitzgerald-Huber, Louisa G. 1995
Qijia and Erlitou: the question of contacts with distant cultures.
Early China 20 17-67.

Flick, C. 1975
Muntz metal and ships' bottoms; the industrial career of G. F. Muntz', *Birmingham & Warwickshire Archaeological Society Transactions,* 87 70-88.

Flight, Walter 1868
On the chemical composition of a Bactrian coin.
Numismatic Chronicle (new series) 8 305-308.

Flight, Walter 1882
Contributions to our knowledge of the composition of alloys and metalwork.
Journal of the Chemical Society 14 135-139.

Foltz, E. 1975
Zur Herstellungstechnik der byzantine Silberschaler aus dem Schatzfund vor Lambousa.
Jahrbuch des Roemisch-Germanischen Zentralmuseums Mainz 22 221-245.

France-Lanord, Albert 1949
The conservation of the Merovingian helmet of Baldenbeim.
Cahiers d'archéologie et d'histoire d'Alsace 130 277-280. [Authors note: the first account of iron preserved in a case filled with nitrogen].

France-Lanord, Albert 1963
Les lingots de fer proto-historiques.
Revue d'histoire de la sidérurgie 4 167-178.

France-Lanord, Albert 1964
La fabrication des épées de fer gauloises.
Revue d'histoire de la metallurgie 5, no 4, 315-327.

Gale, N. H., Stos-Gale, Z., Raduncheva, A., Panayotov, I., Ivanov, I., Lilov, P., and Todorov, T. 2003
In: (Ed. P.T. Craddock and J. Laing) *Mining and Metal Production Through the Ages.*
Butterworths: London.

Gänsicke, Susanne and Newman, Richard. 2000
Gilded silver from ancient Nubia.
In: *Gilded Metals: History, Technology and Conservation.* (Ed. Terry Drayman-Weisser) 73-96. Archetype Publications: London.

Gettens, Rutherford J. 1959
Examining tables in use at the Freer Gallery of Art.
Studies in Conservation 4 23-27

Getty Museum 2002
Handbook of the Collections.
J. Paul Getty Museum: Malibu.
Getty Trust Publications: Brentwood.

Ghandour, Naila 1981
Elemental and metallographic analysis of objects from Tell Ajjul, Palestine. MA Dissertation (unpublished).
Institute of Archaeology, University College London: London.

Gibbons, D. F., Ruhl, K. C., and Shepard, D. G. 1979
Techniques of silversmithing in the Hormizd II plate.
Ars Orientalis 11 163-176

Giovannelli, G.; Natali, S.; and Bozzini, B. 2006
Silver coated lead coins: an appraisal of ancient technology.
Journal of Applied Electrochemistry 36 951-956

Gold of the Andes 1963
Gold of the Andes: Treasure from Peru. A special exhibition. Brooklyn Museum and M. H. de Young Memorial Museum: New York.

Goodway, Martha and Conklin, H. 1987 .
Quenched high-tin bronzes from the Philippines.
Archaeomaterials 2 21-27.

Gordon, R., and Knopf, R. 2006
Metallurgy of bronze used in tools from Machu Picchu, Peru.
Archaeometry 48 57-76.

Gordon, R., and Knopf, R. 2007
Late horizon silver, copper, and tin from Machu Picchu, Peru.
Journal of Archaeological Science 34 38-47.

Gullapalli, Praveena 2009
Early metal in South India: copper and iron in Megalithic contexts.
Journal of World Prehistory 22 439-459.

Han Rubin, Ma Zhaozeng, Wang Zengjuan and Ke Jun 1983
Studies related to the black passive oxide film on bronze arrowheads unearthed with the terracotta warriors near the tomb of the first emperor of Qin (210 BC).
Studies in the History of Natural Sciences 4 295-302.

Hanks, Bryan and Doonan, Roger 2009
From scale to practice: a new agenda for the study of early metallurgy on the Eurasian steppe.
Journal of World Prehistory 22 329-356.

Hansen, M., and Anderko, K. 1958
The copper-iron system.
The Constitution of Binary Alloys.
McGraw-Hill: New York.

Hanson, D. and Marryat, C. B. 1927
Effects of arsenic on copper.
Journal of the Institute of Metals 37 121-143

Hanson, D. and Pell-Walpole, W. T. 1951
The Chill-Cast Tin Bronzes.
Edward Arnold: London.

Hardness comparison charts 2010
http://www.gordonengland.co.uk [accessed March 8th 2010].

Harper, Prudence O. and Meyers, Pieter 1981
Silver vessels of the Sasanian Period, Volume 1: Royal Imagery.
Metropolitan Museum of Art: New York.

Hauptmann, Andreas, Pernicka, E., Rehren, T., and Yalcin, U) 1999.
The Beginnings of Metallurgy.
Der Anschnitt, Beiheft 9. Deutsches Bergbau-Museum: Bochum.

Hauptmann, Andreas 2007
The Archaeometallurgy of Copper. Evidence from Faynan, Jordan.
Springer: Berlin,

Heubeck, A. 1986
Die Würtzburger Alphabettafel.
Würtzburger Jahrbücher für die Altertumswissenscaft Neue Folge 12 7-20.

Heycock, C. T. and Neville, F. H. 1901a
On the constitution of copper-tin alloys.
Proceedings of the Royal Society of London 69 320-329.

Heycock, C. T. and Neville, F. H. 1901b
On the results of chilling copper-tin alloys.
Proceedings of the Royal Society of London 68 171-178.

Heycock, C.T. and Neville, F. H. 1904
Bakerian lecture: on the constitution of the copper-tin series of alloys.
Philosophical Transactions of the Royal Society of London, Series A. 202 1-70.

Higgins, G. 1967
Minoan and Mycenaean Art.
Thames and Hudson: London.

Hillard, J. E. 1964
Estimating grain size by the intercept method.
Metal Progress 85 99-102.

Höppner, B., Bartelheim, M., Husijmans, M., Krauss, R., Martinek, K., and Pernicka, E. 2005
Prehistoric copper production in the Inn Valley, Austria, and the earliest copper production in central Europe.
Archaeometry 47 293-315.

Howe, E. G., and Peterson, U. 1994
Silver and lead in late prehistory of the Mantaro Valley, Peru. In: (Ed. D. A. Scott and P. Meyers) *PreColombian Archaeometry. Papers presented at the 1992 Archaeometry Conference*, UCLA March 1992, 214-230.
The Getty Conservation Institute: Los Angeles.

Ingo, G.M.; Balbi, S.; de Caro, T.; Fragalà, I.; Riccucci, C.; and Bultrini, G. 2006
Microchemical investigation of Greek and Roman silver and gold plated coins: coating techniques and corrosion mechanisms.
Applied Physics. A, Materials Science & Processing 83 623-629.

Istenic, J., and Smit, Z. 2008
The beginning of the use of brass in Europe with particular reference to the southeastern Alpine region.
In: In: (Ed. S. LaNiece, D. Hook and P. T. Craddock) *Metals and Mines: Studies in Archaeometallurgy* 140-147.
Archetype Press & British Museum Press: London.

Jett, Paul 1993
A study of the gilding of Chinese Buddhist bronzes.
In: (Ed. S. La Niece and P. Craddock) *Metal Plating and Patination*, 193-210.
Butterworth-Heinemann: Oxford.

Jones, T. N. 2004
The Mica shipwreck: deepwater nautical archaeology in the Gulf of Mexico, Master of Arts Thesis, Texas A&M University: Texas.

Karmazin, Lubomir; Brezina, Josef; and Jelinkova, Dagmar. 1996
Metallography and electron probe microanalysis of the material of silver temple rings found in the Slavic cemetery at Musov, Czech Republic.
Archaeologia Austriaca 80 277-290.

Kent, W. L. 1926
The brittle ranges of bronzes.
Journal of the Institute of Metals 35 45-53.

Kerr, Rose 1990
Later Chinese Bronzes.
Victoria and Albert Museum Far Eastern Series.
Oxford University Press & Bamboo Publishing.
Victoria & Albert Museum: London.

Khandekar, Narayan 2003
Preparation of cross-sections from easel paintings.
Reviews in Conservation 4 52-64.

Killick, David 2009
Cairo to Cape: the spread of metallurgy through eastern and southern Africa.
Journal of World Prehistory 22 399-414

Klemm, H. 1952
Farbätzung des Feingefüges der Metalle mit Natriumthiosulfat.
Metallkundliche Berichte 45 123-125.

Klemm, H. 1968
Uses of sodium thiosulphate (Klemm's reagent) as an etchant.
Practical Metallography 5 163-165.

Knott, P. 2008
Ship Parts
Newsletter of the Australasian Institute for Maritime Archaeology (AIMA) 27 21-22.

Kruse, Susan E.; Robert, D.; and Starling, Katharine. 1988
Experimental casting of silver ingots.
Historical Metallurgy 22 87-92.

Lambert, Joseph B. 1997
Traces of the past: unraveling the secrets of archaeology through chemistry.
Addison-Wesley: New York.

Lang, Janet and Hughes, Michael J. 1984
Soldering Roman silver plate.
Oxford Journal of Archaeology 3 77-108.

LaNiece, Susan, and Martin, Graham 1987
The technical examination of Bidri ware.
Studies in Conservation 32 97-101

Lechtman, Heather N. 1971
Ancient methods of gilding silver: examples from the Old and the New Worlds. In: (Ed. Robert H. Brill). *Science and archaeology: papers, Symposium on Archaeological Chemistry (Atlantic City, NJ)].* 2-29.
MIT Press: Massachusetts.

Lechtman, Heather. 1979
A pre-Columbian technique for electrochemical replacement of plating of gold and silver on copper objects.
Journal of Metals 3 154-160

Lechtman, Heather N. 1981
Copper-arsenic bronzes from the north coast of Peru.
In: (Ed. Ann-Marie Cantwell, James B. Griffin and Nan A. Rothschild) *The research potential of anthropological museum collections*. 77-121. Annals of the New York Academy of Sciences, v. 376.
New York Academy of Sciences: New York

Lechtman, Heather N., Erlij, Antonia, and Barry, B. J. Jr. 1982
New perspectives on Moche metallurgy: techniques of gilding copper at Loma Negra, northern Peru.
American Antiquity 47 3-30.

Lechtman, Heather N. 1984
Andean value systems and the development of prehistoric metallurgy.
Technology and Culture 25 1-36.

Lechtman, Heather N. 1988
Traditions and styles in Central Andean Metallurgy. In: (Ed. R. Maddin) *The Beginning of the use of Metals and Alloys*. 344-378. MIT Press: Cambridge.

Lechtman, Heather and Klein, Sabine 1999
The production of copper-arsenic alloys (arsenic bronze) by cosmelting: modern experiment, ancient practice.
Journal of Archaeological Science 26 497-526.

Lechtman, Heather 2003
Middle Horizon bronze: centers and outliers.
In: (Ed. Lambertus van Zelst) *Patterns and process: a Festschrift in honor of Dr. Edward V. Sayre*. 246-268.
Smithsonian Center for Materials Research and Education: Washington DC.

Lloyd-Morgan, Georgina 1981
Description of the collections in the Rijksmuseum, G. M. Kam at Nijmegen, IX The Mirrors.
Ministry of Culture, Recreation and Social Welfare: Amsterdam.

Lord, John O. 1949
Alloy Systems.
Pitman: New York.

Lutz, J. and Schwab, R. 2008
Kupfer für die Kelten.
Archaologie in Deutschland 2/08, Stuttgart.

MacLeod, I.D. and Schindelholz, E. 2004
Surface analysis of corroded silver coins from the wreck of the San Pedro de Alcantara (1786).
In: *Metal 04: proceedings of the International Conference on Metals Conservation.* Canberra, Australia, 4-8 October 2004. (Ed. John Ashton and David Hallam) 114-125. National Museum of Australia: Sydney.

McCall, James L. 1974
Metallographic specimen preparation: optical and electron microscopy.
Proceedings of the 19th Metallographic Symposium, Beverly Hills, California. September 23rd – 24th 1973. Plenum Publishing: New York.

McCall, J. L. and Buchheit, R. D. 1971
Metallographic studies of archaeological; artifacts from Ecuador.
Metallography 4 209-222.

McCarthy, M., and Stanbury, M. 2003
The structure of the *Eglinton*, its associated fastenings and fittings, in *The barque Eglinton wrecked Western Australia 1852. The history of its loss, archaeological excavation, artefact catalogue and interpretation,* (ed. M. Stanbury). Australian National Centre of Excellence for Maritime Archaeology, Special Publication No. 6. The Australasian Institute for Maritime Archaeology, Special Publication, No. 13: 47–70.

McCarthy, M. 2005
Ship's fastenings from sewn boat to steamship.
Texas A&M University Press, College Station: Texas.

McGeehan-Liritzis V., and Gale, Noel H. 1988
Chemical and lead isotope analyses of Greek Late Neolithic and Early Bronze Age Metals.
Archaeometry 30 199-225.

McKerrell, Hugh and Tylecote, Ronald F. 1972
The working of copper-arsenic alloys in the Early Bronze Age and the effect on the determination of provenience.
Proceedings of the Prehistoric Society 38 209-219.

McNeil, Ian 2002
Encyclopedia of the History of Technology
Routledge: London.

Ma, Qing-lin and Scott, David A. 2004
Gold and silver gilding techniques in the Western Han dynasty of China.
Wen wu bao hu yu kao gu ke xue (In Chinese) 16 21-26

Marshall, Sir John 1951
Taxila. 3 vols.
Cambridge University Press: Cambridge

Maryon, Herbert 1948
The Mildenhall Treasure: Some Technical Problems.
Man 48 25-27; 38-41.

Maryon, Herbert 1963
The making of a Chinese bronze mirror, part 2.
Archives of the Chinese Art Society of America 17 23-25.

Mathewson, C. H. 1915
A metallographic description of some ancient Peruvian bronzes from Machu Picchu.
American Journal of Science, New Series, 240: 525-616.

Matsuda, T. 1928
Of the quenching and tempering of brass, bronze and aluminium bronze.
Journal of the Institute of Metals 39 67-107.

Matttusch, Carol C. 1995
Two bronze Herms: questions of mass production in antiquity.
Art Journal 54 53-59.

Meeks, Nigel D. 1988a
A technical study of Roman bronze mirrors.
In: *Aspects of ancient mining and metallurgy:* ACTA of the British School at Athens, Centenary Conference, Bangor 1986 (Ed. J. Ellis-Jones) 66-79
University College of North Wales: Wales.

Meeks, Nigel D. 1988b
Surface studies of Roman bronze mirrors, comparative high-tin bronze Dark Age material and black Chinese mirrors.
In: *Proceedings of the 26th International Archaeometry Symposium* (Eds R. M. Farquhar, R. G. V Hancock, and L. A. Pavlish) University of Toronto, Canada, May 16th – 20th 1988, 124-127.
The Archaeometry Laboratory: Toronto.

Meeks, Nigel D. 1993a
Patination phenomena on Roman and Chinese high-tin bronze mirrors and other artefacts.
In: *Metal Plating and Patination* (Ed. S. LaNiece and P. T. Craddock)
Butterworth-Heinemann: Oxford 63-84.

Meeks, Nigel D. 1993b
Surface characterization of tinned bronze, high-tinned bronze, tinned iron and arsenical bronze.
In: *Metal Plating and Patination* (Ed. S. LaNiece and P. T. Craddock)
Butterworth-Heinemann: Oxford 247-275.

Meggers, Betty J., Evans, Clifford, and Estrada, Emilo nd
Field Notes: Excavation at La Compañia, Los Rios Province, Ecuador 1961. On file at the Museo Antropologico del Banco Central del Ecuador, Guayaquil.

Mei, Jianjun 2000
Copper and Bronze Metallurgy in Late Prehistoric Xinjiang: its cultural context and relationship with neighbouring regions.
BAR International Series 865.
Archaeopress: Oxford.

Mei, Jianjun; Kunlong, Chen and Wei Cao 2009
Scientific examination of Shang dynasty bronzes from Hanzhong, Shaanxi Province, China.

Journal of Archaeological Science 36 1881-1891.

Melikian-Chirvani, A. S. 1974
The white bronzes of early Islamic Iran.
Metropolitan Museum Journal 9 123-151.

Merkel, John 1990
Experimental reconstruction of Bronze Age copper smelting based on archaeological evidence from Timna. In: (Ed. Beno Rothenberg) *Researches in the Arabah 1959-1984. Vol. II The Ancient Metallurgy of Copper* 78-122.
Institute of Archaeometallurgical Studies. Institute of archaeology: London.

Meyers, Pieter 2003
Production of silver in antiquity: ore types identified based upon elemental compositions of ancient silver artifacts.
In: *Patterns and process: a Festschrift in honor of Dr. Edward V. Sayre*. (Ed. Lambertus Van Zelst). 270-288.
Smithsonian Center for Materials Research and Education: Washington, D.C.

Meyers, Pieter 1981
Technical study of Sasanian silver.
MASCA Journal 1, no 3 242-245

Moestra, H. 1993
Einige mikkoskopische Beobachtungen am Römischen Münzmessing aus Kleinasien.
Blesa 1 461-464.

Morin, H. 1874
Über chinesische und japanische Bronzen mit dunkler Patina.
Comptes Rendus Hebdomadaires des Séances de l'Académie des Sciences 78 811-814.

Morse, Birgit Faber and Gordon, Robert B. 1986
Metallographic examination of pre-Columbian Mexican copper and silver artifacts from Mitla, Oaxaca (Mexico).
Archeomaterials 1 57-67.

Muhly, J. D. 1986
The beginnings of metallurgy in the Old World. In: *The Beginning of the Use of Metals and Alloys: Papers from the Second International Conference on the Beginning of the Use of Metals and Alloys, Zhengzhou, China*, 21-26 October 1986. (Ed. R. Maddin) 2-20.
MIT Press: Cambridge, Massachusetts.

Muhly, James D. 1999
Copper and bronze in Cyprus and the Eastern Mediterranean .
MASCA research papers in science and archaeology 16, 15-25.

Nakano, Toru (Editor) 1994
Bronze Mirrors from ancient China. Donald H. Graham Jr. Collection. Preface, Essays and Catalogue by Toru Nakano. Essays by Tseng Yuho Ecke and Suzanne Cahill. (Trans. Tetsuro Kono and Ace Translation Service.
Donald H. Graham Jr. : Honolulu.

Needham, Joseph ;Wang, Ling; Gwei-Djen Lu; Tsuen-hsuin, Tsein; Kuhn, Dieter and Golas, Peter J. 1974
Science and Civilization in China, Vol. 5 Part II, 237-250.
Cambridge University Press: Cambridge.

North, Neil and MacLeod, Ian D. 1987
Corrosion of Metals

In: (Ed. Colin Pearson) *Conservation of Marine Archaeological Objects* 94-95. Butterworths: London.

Northover, Peter 1992
Materials issues in the Celtic coinage.
In: (Ed. M. Mays) Celtic Coinage: Britain and beyond. *British Archaeological Reports* 222, 235-300. BAR: Oxford.

Northover, J. Peter 1996
Metal analysis and metallography of early metal objects from Denmark. In (Ed. H. Vandkilde) *From Stone to Bronze: the metalwork of the late Neolithic and earliest bronze age in Denmark.* Aarhus: Jutland Archaeological Society 32 321-358.

Northover, Peter 1998a
Exotic alloys in antiquity. In: (Ed. Th. Rehren, A. Hauptmann, and J. D. Muhly) *Metallurgica Antiqua* 113-121. Deutsches Bergbau-Museum: Bochum.

Northover, Peter 1998b
Analysis of copper alloy metalwork from Arbedo T1.
In: (Ed. M. P. Schindler) Der Depotfund von Arbedo TI und Bronzedepotfunde des Alpenraums vom 6, bis zum Beginn des 4 Jh v Chr. *Antiqua* 30 289-315.

Northover, Peter 2004
Interdisziplinare Untersuchgen zu den Metallfunden. In: (Ed I. Bauer, B. Ruckstuhl, J. Speck) *Kantonales Museum fur Urgeschichte Zug Zug* 102-143.

Oddy, W. Andrew 1988
The gilding of Roman silver plate.
In: *Argenterie Romaine et Byzantine.* (Ed. François Baratte). 9-25. De Boccard: Paris.

Oddy, W. Andrew 1991
Gilding: an outline of the technological history of the plating of gold on to silver or copper in the Old World.
Endeavour. New Series 15 29-33.

Ottaway, Barbara S. 2003
Experimental archaeometallurgy. In: (Ed. Th. Stollner, G. Korlin, G. Steffens and J. Cierny) Man and Mining- Mensch und Bergbau. Studies in honour of Gerd Weisgerber on the Occasion of his 65th Birthday. *Der Anschnitt* 16 341-348.

Ottaway, Barbara S. and Seibel, S. 1997
Dust in the wind: experimental casting of bronze in sand moulds. In: *Instrumentum Monographies 5. Paléométallurgie des Cuivres.* Actes du colloque de Bourg-en-Bresse et Beaune, 17th-18th October 1997.

Ottaway, Barbara S. and Wang, Quanyu 2004
Casting experiments and microstructure of archaeologically relevant bronzes. BAR International Series 1331.
Archaeopress: Oxford.

Panseri, Carlo and Leoni, Massimo 1959
Analysis of 2100 year old Gallic swordblades.
La metallurgia italiana 51 5-12.
Panseri, Carlo and Leoni, Massimo 1957
The manufacturing technique of Etruscan mirrors.
Studies in Conservation 3 49-63.

Panseri, Carlo and Leoni, Massimo 1959
Analysis of 2100 year old Gallic swordblades.
La metallurgia italiana 51 5-12.

Panseri, Carlo and Leoni, Massimo 1965
Research on an iron spearhead from the Etruscan sanctuary of Fanun Votumnae 4th -3rd C. BC.
In: (Ed. M. Levey) *Archaeological Chemistry* Vol III, 205-229.
American Chemical Society: Washington.

Pappa, Maria and Besios, Manthos 1999
The Neolithic settlement at Makriyalos, Northern Greece. Preliminary report on the 1993-1995 excavation.
Journal of Field Archaeology 26 177-195.

Park, Jang Sik, Park, Cheol Woo, and Lee, Keun June 2009
Implications of peritectic composition in historical high-tin bronze metallurgy.
Materials Characterization 60 1268-1275

Partington, J. R. 1935
Origins and Development of Applied Chemistry.
Longmans, Green and Co. London.

Pernicka, E., Begemann, F., Schmitt-Strecker, S., Todorova, H., and Kuleff, I. 1997
Prehistoric copper in Bulgaria: its composition and provenance.
Eurasia Antiqua 3 41-180.

Pernicka, E., Schwab, R., and Willer, F. 2007
Die Analyse eines Silberlechfragments an Helm Kat. Nr 1, in: (Ed. R. Meijers, F. Willer), *Achter het zilveren masker. Hinter der silbernen Maske, Ausstellungskatalog* 51-56.
Museum Het Valkhof Nijmegen. Reinisches Landesmuseum, Bonn : Nijmegen and Bonn.

Philip, Graham and Rehren, Thilo 1996.
Fourth millennium BC silver from Tell Esh-shuna, Jordan: archaeometallurgical investigations and some thoughts of ceramic skeuomorphs.
Oxford Journal of Archaeology 15 129-150.

Piaskowski, Jerzy 1958
The determination of the technique of production and the chronology of metal objects.
Z. otchlani wieków 25 283-286.

Piaskowski, Jerzy 1992
Distinguishing between directly and indirectly smelted iron and steel.
Archeomaterials 6 no 2 169-173.

Piaskowski, Jerzy 1993
Metallographic examination of a Japanese sword.
Historical Metallurgy 27 no 2 110-117.

Piaskowski, Jerzy 2002
Technology of gonjos – separate piece of early Indonesian daggers (keris).
Kinzoku Hakubutsukan kiyō 35 48-63.

Pistofidis, N.; Vourlias, G.; Pavlidou, E.; Dilo, T.; Civici, N.; Stamati, F.; Gjongecaj, S.; Prifti, I.; Bilani, O.; Stergioudis, G.; and Polychroniadis, E.K. 2006
On the comparative study of three silver coins of the 3rd century BC minted in Korkyra, Dyrrachion and by the Illyrian king Monounios.
Applied Physics. A, Materials Science & Processing 83 637-642.

Pleiner, Radomir 1963
Rediscovering the techniques of early European blacksmiths.

Archaeology 16 no 4 234-242.

Pleiner, Radomir 1969
Iron working in ancient Greece.
Národni Technická Muzeum.

Pleiner, Radomir 1979
The technology of three Assyrian iron artifacts from Khorsabad.
Journal of Near Eastern Studies 39 no 3 83-91.

Pleiner, Radomir 1998
A Celtic bronze-coated iron sword blade from Manching.
In: (Ed. T. Rehren, A. Hauptmann and J. D. Muhly) *Metallurgica Antiqua*. 163-170.
Montanhistorische Zeitschrift Der Anschnitt Beiheft 8.
Bergbau Museum: Bochum

Podany, Jerry and Scott, David A. 1997
Looking through both sides of the lens: why scientists and conservators should know each other's business.
In: (Ed. Susan Bradley) *The Interface between Science and Conservation*. 211-220. British Museum Occasional Publication No. 116.
British Museum Press: London.

Ponting, Matthew J. 2002
Roman military copper-alloy artefacts from Israel: questions of organization and ethnicity.
Archaeometry 44 555-571.

Radivojević, M. 2007
Evidence for early copper smelting in Belovode, a Vinča culture site in eastern Serbia. M.Sc. Thesis (unpublished) University College London: London.

Rajpitak, Waranghkana and Seeley, Nigel J. 1979
The bronze bowls from Ban Don Ta Phet, Thailand: an enigma of prehistoric metallurgy. *World Archaeology* 11 26-31.

Rapp, George, Jr. 1986
On the origins of copper and bronze alloying. In: *The Beginning of the Use of Metals and Alloys: Papers from the Second International Conference on the Beginning of the Use of Metals and Alloys, Zhengzhou, China*, 21-26 October 1986. (Ed. R. Maddin) 21-27.
MIT Press: Cambridge, Massachusetts.

Ravich, Irina G, 1991
Study of the composition of Scythian and Sarmatian bronze mirrors and technologies of their manufacture.
Bulletin of the Metals Museum 16 20-31.

Ravich, Irina G. 1993
Annealing of brittle archaeological silver: microstructural and technological study. *ICOM Committee for Conservation 10th triennial meeting: Washington, DC, 22-27 August 1993*: preprints. 792-795. (Ed. Janet Bridgland) ICOM Committee for Conservation: Paris.

Ravich, Irina G. and Ryndina, N. V. 1995
Early copper-arsenic alloys and the problems of their use in the Bronze Age of the North Caucasus.
Bulletin of the Metals Museum 23 1-18.

Reeve, M. R., Bowden, J. S. and Cuthbertson, J. W. 1953a
Tin Bronzes: A report on the microstructure of those containing 15-28 per cent tin.
Metal Industry 143 23-25 and 49-52

Reeve, M. R., Bowden, J. S. and Cuthbertson, J. W. 1953b
A report on the microstructure of tin bronzes containing 15-28% tin.
Part Two.
Metal Industry 78 49-52

Rehren, Thilo 1999
Small size, large scale: Roman brass production in Germania Inferior.
Journal of Archaeological Science 26 1083-1087.

Rice, R.W., Carlson, M. 1997
Stabilization of the *Snow Squall* using a lumber drying kiln. In *Proceedings of the 6th ICOM Group on Wet Organic Archaeological Materials Conference York, England, 9-13 September, 1996*, (ed. P. Hoffmann, T. Daley and T. Grant, J. Spriggs.) International Council of Museums (ICOM), Committee for Conservation Working Group on Wet Organic Archaeological Materials, Bremerhaven, (1997), 47-55.

Registre International de Classification de Navire 1864
Bureau Veritas: Paris.

Renfrew, Colin 1967
Before civilization.
Penguin Books: London.

Renfrew, Colin 1969
The autonomy of the south-east European Copper Age.
Proceedings of the Prehistoric Society 35 12-47.
Rhines, F. N. 1956
Phase Diagrams in Metallurgy.
McGraw-Hill Book Co: New York.

Roberts, Benjamin W. 2009
Production networks and consumer choice in the earliest metal of western Europe.
Journal of World Prehistory 22 461-481

Roberts, Benjamin, Thornton, Christopher and Piggot, Vincent 2009
Development of metallurgy in Eurasia.
Antiquity 83 123-136

Ronnberg, E. A. R. Jr. 1974
Copper sheathing of whaleships.
Nautical Research Journal, 20 183-185.

Ronnberg, E. A. R. Jr. 1980
The coppering of 19th century American merchant sailing ships; some historical background with notes for modelmakers.
Nautical Research Journal, 26 125-148.

Ryndina, Natalia 2009
The potential of metallography in investigations of early objects of copper and copper-based alloys.
Historical Metallurgy 43 no 1 1-18.

Samans, Carl 1963
Metallic Materials in Engineering
Macmillan: New York.

Samuels, L. E. 1980
Optical Microscopy of Carbon Steels.
American Society for Metals: Metals Park Ohio.

Schaeffer-Forrer, C. F. A., Zwicker, U., and Nigge, K. 1982
Untersuchungen an metallischen Werkstoffen und Schlacken aus dem Bericht von Ugarit (Ras Shamra, Syrien)
Mikrochimica Acta 1 35-61.

Scholten, Fritz 2000
Adriaen de Vries 1556-1626.
Rijksmuseum, Amsterdam.
Waanders Publishers: Zwolle.

Schwab, Roland 2010
Personal communications to the author, December 5th 2010.

Schwab, Roland 2011
Kupferlegierungen und Kupferverarbeitung im Oppidum auf dem Martberg.
Berichte zur Archäologie an Mittelrhein und Mosel 15, Koblenz (in press).

Schwab, Roland; Eggert, Gerhard; Pernicka, Ernst and Willer, Frank. 2008
Zu den Bronzefunden aus dem Schiffswrack von Mahdia.
Bonner Jahrbucher des LVR-Landesmuseums Bonn. Band 208.
Verlag Philipp von Zabern: Mainz am Rhein.

Schweizer, François and Meyers, Pieter. 1978a
Structural changes in ancient silver alloys: the discontinuous precipitation of copper.
ICOM Committee for Conservation 5th triennial meeting: Zagreb, 1-8 October :preprints. 16-18. International Council of Museums: Paris.

Schweizer, François and Meyers, P. 1978b
Authenticity of ancient silver objects: a new approach.
MASCA Journal 1 9-10.

Scott, David A. 1986
Gold and silver alloy coatings over copper: an examination of some artefacts from Ecuador and Colombia.
Archaeometry 28 33-50.

Scott, David A. 1990
A technical and analytical study of two silver plates in the collection of the J. Paul Getty Museum.
The J. Paul Getty Museum Journal 18 33-52.

Scott, David A. 1991
Metallography and Microstructure of Ancient and Historic Metals.
J. Paul Getty Press: Los Angeles.

Scott, David A. 1996
Technical study of a ceremonial Sican tumi figurine.
Archaeometry 38 305-311.

Scott, David A. 1998
Technical examination of ancient South American metals: some examples from Colombia, Peru and Ecuador.
Boletín, Museo del Oro, Bogotá 15 24-35.

Scott, David A. 2002
Copper and Bronze in Art: Corrosion, Colorants, Conservation.
J. Paul Getty Press: Los Angeles.

Scott, David A. 2008
Laboratory notes: homemade aids for metallographic sample preparation.
The Conservator 31 87-91.

Scott, David A. 2011
The La Tolita-Tumaco Culture: Master Metalsmiths in Gold and Platinum from the Ecuadorian Littoral.

Latin American Antiquity 22 65-95.

Scott, David A. and Bray, Warwick 1980
Ancient platinum technology in South America.
Platinum Metals Review 24 147-157.

Scott, David A. and Podany, Jerry 1991
Some metallurgical and technological studies of Greek and Roman bronzes. In (Ed. Marion True) *The J. Paul Getty Museum Bronze Symposium* (Getty Museum April 1989).
Getty Museum Press: Malibu.

Scott, David A. and Woodard, Roger D. 2004
Technical examination of two ancient Greek copper plaques.
In: *Metal 2001. Proceedings of the International Conference on Metals Conservation.* (Ed. Ian D. MacLeod, Johanna M. Theile and Christian Degrigny), Santiago, Chile 2nd-6th April 2001. 138-145.
Western Australian Museum: Perth.

Scott, David A., Woodard, R. D., McCarter, P. K/. Zuckerman, B., and Lundberg, M. 2005
Greek Alphabet (MS 108) In: (Ed. R. Pintaudi) Papyri Graecae Schoyen 149-160.
Schoyen: Florence.

Scott, George S. 1912
Copper and its alloys in early times.
Journal of the Institute of Metals 7 23-49.

Shalev, Shariel and Northover, Peter 1993
The metallurgy of the Nahal Mishmar hoard reconsidered.
Archaeometry 35 35-47.

Shipping & Commercial List 1862: New York, Oct. 1, 4, 1862.

Shoukang, Z 1986
Ancient metallurgy of non-ferrous metals in China.
Bulletin of the Metals Museum, 2, 1-13.

Shoukang, Zhu and He, Tangkun 1993
Studies of ancient Chinese mirrors and other bronze artefacts.
In: *Metal Plating and Patination* 50-62. (Ed. Susan LaNiece and Paul T. Craddock)
Butterworth-Heinemann: Oxford.

Singapore Straits Times 1863: May 2, 5, 16, 19, 1863

Smith, Charles O. 1969
The Science of Engineering Materials.
Prentice-Hall: Englewood Cliffs, NJ.

Smith, Cyril Stanley 1960
A History of Metallography
University of Chicago Press: Chicago.

Smith, Cyril Stanley 1965
The Interpretation of Microstructures of Metallic Artefacts.
In: (Ed. W. J. Young) *Application of Science in Examination of Works of Art: Proceedings of the Seminar*, 7-16 September 1965. 20-52.
Museum of Fine Arts: Boston

Smith, Cyril Stanley 1971
The techniques of the Luristan smith.
Science and Archaeology 32-52.
Museum of Fine Arts: Boston.

Smith, Cyril Stanley 1973
Bronze technology in the east: : a metallurgical study of early Thai bronzes. In: (Ed. M. Teich

and R. Young,) Changing Perspectives in the History of Science: Essays in honour of Joseph Needham.21-32. Cambridge University Press: London

Smith, Cyril Stanley 1981
A Search for Structure: Selected Essays on Science Art and History.
MIT Press: Massachusetts.

Smith, Cyril Stanley 1992
A history of martensite: early ideas on the structure of steel
In: (Ed. G. B. Olson and W. S. Owen) *Martensite: A Tribute to Morris Cohen.*
ASM International: Ohio.

Sotheby's 1993
Catalogue of pre-Columbian Art. Sale 6500 no. 36
Sotheby's: New York.

Srinivasan, Sharada 1998
High-tin bronze working in Kerala.
In: (Ed. V. Tripathi) *Archaeometallurgy in India. Proceedings of the National Seminar held in 1991.* 241-250.
Sharda Publishing House: New Delhi

Srinivasan, Sharada and Glover ,Ian 1998
The use of tin and bronze in prehistoric southern Indian metallurgy.
Journal of Metals 50 44-50.

Stanbury, M. 2009
Department of Maritime Archaeology, Western Australian Museum, personal communication to Molly Carlson (September 2009).

Staniforth, M. 1985
The introduction and use of copper sheathing – a history.
Newsletter of the Australasian Institute for Maritime Archaeology (AIMA), 9 21-48.

Stollner, Th. and Schwab, R. 2009
Hart oder weich? Worauf es ankommt. Pickel aus dem prahistorischen Bergbau in den Ostalpen, Festschrift Fritz Eckart Barth.
Mitteilungen der Anthropologischen Gesellschaft in Wein 159 149-166.

Stos-Gale, Z. A. and Gale, N. H. 1982
The sources of Mycenaean silver and lead.
Journal of Field Archaeology 9 467-485.

Strabo 1960
The Geography (Ed. H. L. James)
Loeb Classical Library: London.

Switzer, D. 1992
The clipper ship Snow Squall: the last of her breed.
Seaways 3(6) 16-19.

Switzer, D. 2000
The rescue and preservation of the only existing American clipper ship.
In: *Down the River to the Sea, Proceedings of the Eighth International Symposium on Boat and Ship Archaeology, Gdańsk (1997),* (Ed. J. Litwin), 197-200. Polish Maritime Museum: Gdańsk.

Thornton, Christopher P. 2009
Of brass and bronze in prehistoric Southwest Asia.
Journal of World Prehistory 22 123-135.

Thornton, Christopher P. and Roberts, Benjamin W. 2009
Introduction: the beginnings of metallurgy in global perspective.
Journal of World Prehistory 22 181-184.

Teisser, Beatrice 1984
Cylinder seal studies.
Personal communication to the author, London March 1984.

Timberlake, Simon 2003
Excavations on Copa Hill, Cwmystwyth (1986-1999): An early Bronze Age copper mine within the uplands of central Wales.
British Archaeological Reports (British Series) 348 .
Archaeopress: Oxford.

Turner, Evan 1993
Silver plating in the 18th century.
In: *Metal Plating and Patination.* (Ed. Susan La Niece, Susan and Paul Craddock) 211-222. Butterworth-Heinemann Publishers: London.

Tylecote, Ronald F. 1962
The Prehistory of Metallurgy in the British Isles.
Edward Arnold: London.

Tylecote, Ronald F. 1968
Metallographic examination of a spin auger from Letchworth, Hertfordshire.
Bulletin of the Historical Metallurgy Group 2 68-69.

Tylecote, Ronald F. 1973
Casting copper and bronze into stone molds.
Bulletin of the Historical Metallurgy Group 7 1-5.

Tylecote, Ronald F. 1976
A History of Metallurgy.
The Institute of Metals: London

Tylecote, Ronald F. 1987
The Early History of Metallurgy in Europe.
Longman: London.

Tylecote, Ronald F. 1990
Oxidation enrichment bands in wrought iron.
Historical Metallurgy 24 33-38.

Tylecote, Ronald F.; Balmuth, Miriam S.; and Massoli-Novelli, R. 1976
Copper and bronze metallurgy in Sardinia.
Historical metallurgy 17 63-78.

Tylecote, Ronald F. and Boydell, D. 1978
Experiments on copper smelting based upon early copper furnaces found at Timna. In : (Ed Beno Rothenberg) *Archaeometallurgy: Chalcolithic copper smelting.* 27-49.
Institute of Archaeometallurgical Studies 1.
Institute of Archaeology: London.

Tylecote, Ronald F., Balmuth, Miriam S., Massoli-Novelli, R. 1983a
Copper and bronze metallurgy in Sardinia.
Journal of the Historical Metallurgy Society 17 63-78.

Tylecote, Ronald F., Balmuth, Miriam S., Massoli-Novelli, R. 1983b
Copper and bronze metallurgy in Sardinia.
In: (Ed. M. S. Balmuth and R. J. Rowland Jr.) *Studies in Sardinian Archaeology* 115-162.
University of Michigan: Michigan.

Tylecote, Ronald E. and Gilmour, Brian J. J. 1986
The metallography of early ferrous edge tools and edged weapons.
BAR British Series Number 155. British Archaeological Reports: Oxford.

Vander Voort, George F. 1999
Metallography: Principles and Practice.
ASM International: Materials Park, Ohio.

Vander Voort, George F. 2004
Colour Metallography In: (Ed. G. F. Vander Voort)
ASM Handbook Volume 9: *Metallography and Microstructures.*493-579.
The Materials Information Society: Materials Park: Ohio

Vilella, J. R. and Kindle, W. F. 1959
Sodium bisulphite as an etchant for steel.
Metals Progress 76 99-103

Voce, E. 1951
Examination of specimens from the Pitt Rivers Museum. (In: Coghlan H. H.) *Notes on the Prehistoric Metallurgy of Copper and Bronze in the Old World.*
Pitt Rivers Museum Occasional Papers on Technology 4 110-111.

Wang, C., Lu, B., Zuo, J., Zhang, S., Tau, M., Suzuki, M., and Chase, W. T. 1995
Structural and elemental analysis on the nanocrystalline SnO2 in the surface of ancient Chinese black mirrors.
Nanostructured Materials 5 489-496.

Wanhill, Richard J. H. 2003
Brittle archaeological silver: a fracture mechanisms and mechanics assessment.
Archaeometry 45 625-636.

Wayman, Michael and Duke, John 1999
The effects of melting on native copper. In: (In (Ed. A. Hauptmann, E. Pernicka, T. Rehren, and U. Yalcin) *The Beginnings of Metallurgy* 55-63
Der Anschnitt, Beiheft 9. Bergbau-Museum: Bochum.

Weisberger, Gerd 2008
Roman brass and lead ingots from the western Mediterranean.
In: (Ed. S. LaNiece, D. Hook and P. T. Craddock) *Metals and Mines: Studies in Archaeometallurgy* 148-158.
Archetype Press & British Museum Press: London.

White, Joyce C. and Hamilton, Elizabeth G. 2009
The transmission of early bronze technology to Thailand: new perspectives.
Journal of World Prehistory 22 357-397

Woodard, Roger D. 1997
Greek Writing from Knossos to Homer.
Oxford University Press: Oxford

Yalouris, E. 1988
The Falkland Islands expeditions; resurrecting the last Yankee clipper ship (1982-1987), *Symbols,* Harvard University, 14 14-23.

Zhou Weirong. 2007.
The Origin and Invention of Zinc-Smelting in China. In: (Ed. S. LaNiece, D. Hook and P. T. Craddock) *Metals and Mines: Studies in Archaeometallurgy,* 179-186. London: Archetype Publications.

Zhu, Shoukang and He, Tangkun 1993
Studies of ancient Chinese mirrors and other bronze artefacts. In: (Ed. S. La Niece and P. Craddock) *Metal Plating and Patination*, 50-62.
Butterworth-Heinemann: Oxford.

Zwicker, Ulrich, Oddy, Andrew, and La Niece, Susan 1993
Roman techniques of manufacturing silver-plated coins.
In: *Metal Plating and Patination.* (Ed. Susan La Niece, Susan and Paul Craddock) 223-246. Butterworth-Heinemann Publishers: London.

==

Index for Volume I

Made in the USA
Las Vegas, NV
03 November 2026

57421600R00200